D1011299

ELECTRICITY & ELECTRICAL APPLIANCES HANDBOOK

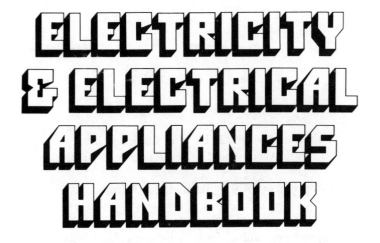

ELECTRICITY & ELECTRICAL APPLIANCES HANDBOOK

Jeannette T. Adams

ARCO PUBLISHING COMPANY INC.

219 Park Avenue South, New York, N.Y. 10003

Published by Arco Publishing Company, Inc.
219 Park Avenue South, New York, N.Y. 10003

Library of Congress Cataloging in Publication Data

Adams, Jeannette T
  Electricity and electrical appliances handbook.

  Bibliography: p.
  Includes index.
  1. Electric engineering—Amateurs' manuals. 2. Electric lighting—Amateurs' manuals. 3. Household appliances, electric—Amateurs' manuals. I. Title.

TK9901.A3   621.319′24   75-22746
ISBN 0-668-03855-1 (Library Edition)

Acknowledgments

THE AUTHOR desires to acknowledge with thanks the assistance of the following national organizations, colleges, and branches of the government that have cooperated in the production of this book:

Alabama Polytechnic Institute; American Blower Division American Radiator and Standard Sanitary Corp.; American Kitchens; AVCO Manufacturing Corp.; Bendix Home Appliances; College of Agriculture of the University of Wisconsin; Crosley Home Appliances; Edison Electric Institute; Electric Energy Association; Electric Furnace-Man, Inc.; Electric Storage Battery Co.; Emerson Electric Manufacturing Co.; Franklin Institute; General Electric Co.; Hotpoint, Inc.; Illuminating Engineering Society; Industry Committee on Interior Wiring Design; Kansas State College; Kester Solder Co.; Landers, Frary & Clark (Universal); Minneapolis-Honeywell Regulator Co.; Munsell System of Color Notation; National Adequate Wiring Bureau; National Board of Fire Underwriters; National Carbon Co.; National Electrical Manufacturers Association; Norge Division Borg-Warner Corp.; Philco Corp.; Proctor Electric Co., Toastmaster Products Division McGraw Electric Co.; University of Illinois; U.S. Department of Agriculture; U.S. Department of Commerce; U.S. Naval Bureau; U.S. War Department; Westinghouse Electric Corp.

Table of Contents

ELECTRICITY & ELECTRICAL APPLIANCES HANDBOOK

Preface

—————————•————————————————
•

THIS BOOK is designed to close the gap between the person
with little or no knowledge of electricity who watches
equipment and machinery perform without understanding
the basic principles of its operation, and the experienced
technician who understands what is actually happening.

Today there is scarcely a person who does not come in
contact with electrical equipment.

Most electrical items, that only a few years ago were
considered luxuries, are now regarded as essential to our
normal way of life. Equipment such as the electric light,
telephone, radio, television, electric refrigerator, electric
stove, and other appliances are accepted as a matter of
course. Most people turn on a light switch without giving a
thought to the vast electrical system to which it is connected or to what happens when the light comes on. The
same attitude prevails when turning the switch on a radio,
a television, and other items. As long as the equipment
functions properly the results are accepted as a matter of
course with little concern about what is actually taking
place.

Today electricity projects itself into our life at every
turn.

The ELECTRICITY AND ELECTRIC APPLIANCES
HANDBOOK has been produced with the conviction that
it will be a welcome and valued aid to help American families understand the growing dependence of every home on
its electrical wiring system, so they may plan wisely for the

comfort, convenience, economy, and efficiency which can be obtained from electric service.

All the procedures and projects described in this book are well within the scope of any person who desires to use tools intelligently and has the incentive to enhance the beauty and value of the home by maintaining it at all times in top-notch condition.

J.T.A

CHAPTER 1

Fundamentals of Electricity

WHAT IS ELECTRICITY?

The word *electric* is derived from the Greek word meaning *amber*. The ancient Greeks used the word to describe the strange forces of attraction and repulsion that were exhibited by amber after it had been rubbed with a cloth. The question, *what is electricity*, has been baffling the world's greatest scientists for many years. Although it is not known exactly what electricity is, knowing what it does has made it possible to develop productive theories; and the laws by which electricity operates are becoming more widely known and better understood. Today, all *matter* is considered to be essentially electrical in nature.

The Molecule. The objects that make up the world around us are said to be made of *matter* (Fig. 1). Matter is the physical substance of common experience. The conception of matter may be summarized as follows.

The familiar forms of matter are of a particle nature in their last analysis. *For example,* if a crystal of common table salt were divided into very small particles, and then if one of these particles were divided again and again, finally a particle would be reached that was so small that no further division could be made which would leave the material still in the form of salt. This ultimate particle of salt is called a *molecule.*

Now, it is known that salt is composed of two kinds of material—sodium and chlorine. The salt molecule is then the smallest possible physical form of this compound, or chemical union, of the two constituent elements. The mole-

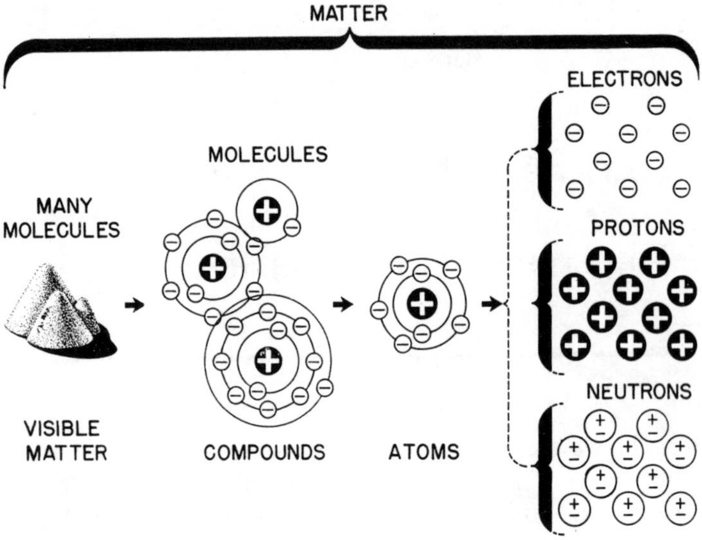

Fig. 1. Electricity and matter.

cule is the particle that is involved in most of the *chemical* changes: the baking of bread, the explosion of dynamite, the changes involved in converting food into a component of blood. These are a few of the actions in which molecules are created and destroyed.

The Atom. The molecule is far from being the ultimate particle into which matter may be subdivided. The salt molecule may be decomposed into radically different substances—sodium and chlorine. These particles that make up molecules called *atoms*, can be isolated and studied separately.

The atom is the smallest particle that makes up that type of material called an *element*. An element retains its characteristics when subdivided into atoms. More than 100 elements have been identified. They can be arranged into a table of increasing weight and can be grouped into families of materials having similar properties. This arrangement is called the *periodic table of the elements*.

The idea that all matter is composed of atoms dates back more than 2,000 years, to the Greeks. Many centuries passed before the study of matter proved that the basic idea of atomic structure was correct. Physicists have explored the interior of the atom and discovered many subdivisions in it. The core of the atom is called the *nucleus*. Most of the mass of the atom is concentrated in the nucleus. It is comparable to the sun in the solar system, around which the planets revolve. The nucleus contains *protons* (positively charged particles) and *neutrons* which are electrically neutral.

Most of the weight of the atom is in the protons and neutrons of the nucleus. Whirling around the nucleus are one or more smaller particles of negative electric charge. *These are the electrons.* Normally there is one proton for each electron in the entire atom so that the net positive charge of the nucleus is balanced by the net negative charge of the electrons whirling around the nucleus. Therefore, *the atom is electrically neutral.*

The electrons do not fall into the nucleus, even though they are strongly attracted to it. Their motion prevents it, as the planets are prevented from falling into the sun by their centrifugal force of revolution.

The number of protons, which is usually the same as the number of electrons, determines the kind of element in question. Figure 2 illustrates several atoms of different materials, based on the conception of planetary electrons describing orbits about the nucleus. For example, hydrogen has a nucleus consisting of one proton, around which rotates one electron. The helium atom has a nucleus containing two protons and two neutrons with two electrons encircling the nucleus. Near the other extreme of the elements is curium (not shown in the illustration), an element discovered in the 1940s, which has 96 protons and 96 electrons in each atom.

The *periodic table of the elements* is an orderly arrangement of the elements in ascending atomic number (number of planetary electrons) and also in atomic weight (number of protons and neutrons in the nucleus). The various kinds of atoms have distinct masses or weights with respect to each other. The element most closely approaching unity (meaning one) is hydrogen, whose atomic weight is 1.008, as compared with oxygen, whose atomic weight is 16. Helium has an atomic weight of approximately 4, lithium 7, fluorine 19, and neon 20, as shown in the figure.

The electrons in the outer orbits of certain elements are easily separated from the positive nucleii of their parent atoms and caused to flow in metals, in vacuums, or in tubes containing gas. Electrons have many important characteristics. The weight of an electron is very small compared with

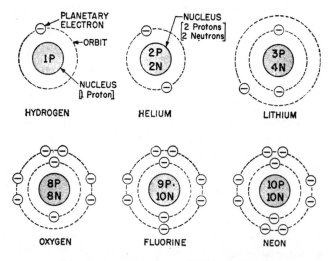

Fig. 2. Atomic structure of different elements.

that of a proton or neutron (about 1/1845 of the weight of the proton of the lightest atom—that of hydrogen). The electron has a weight of 9×10^{-28} gram and a negative charge of 1.6×10^{-19} coulomb. This combination makes the electron an extremely active particle with many possibilities for practical use.

Ionization. Ordinarily an atom is most likely to be in that state in which the internal energy is at a minimum, called the *normal* state. If the internal energy of the atom is raised above that of the normal state, it is said to be *excited.* Excitation may be produced in a number of ways, such as collision of the atom with high-speed positive or negative particles which may give up all or part of their energy to the atom during the collision. The excess energy absorbed by an atom may become sufficient to cause loosely bound outer electrons to leave the atom against the force that acts to hold them within. An atom that has lost or gained one or more electrons is said to be *ionized*. If the atom loses electrons it becomes positively charged and is referred to as a *positive ion*. Conversely, if the atom gains electrons, it becomes negatively charged and is referred to as a *negative ion*. Actually then, an ion is a small particle of matter or group of such particles having a net positive or negative charge.

Free Electrons. When an orbital electron is removed from an atom it is called a *free electron*. Some of the electrons of certain metallic atoms are so loosely bound to the nucleus that they are comparatively free to move from atom to atom. Therefore, a very small force or amount of energy will cause such electrons to be removed from the atom and become free electrons. It is these free electrons that constitute the flow of an electric current in electrical conductors.

CONDUCTORS, SEMICONDUCTORS, AND INSULATORS

Substances that permit the free motion of a large number of electrons are called *conductors*. Copper wire is considered a good conductor because it has many free electrons. Electrical energy is transferred through conductors by means of the movement of free electrons that migrate from atom to atom inside the conductor. Each electron moves a very short distance to the neighboring atom where it replaces one or more electrons by forcing them out of their orbits. The replaced electrons repeat the process in other nearby atoms until the movement is transmitted throughout the entire length of the conductor. The greater the number of electrons that can be made to move in a material under the application of a given force the better are the conductive qualities of that material. A good conductor is said to have a low opposition or low resistance to the current (electron) flow.

In contrast to good conductors, some substances such as rubber, glass, and dry wood have very few free electrons. In these materials large amounts of energy must be expended in order to break the electrons loose from the influence of the nucleus. Substances containing very few free electrons are called *poor conductors, nonconductors,* or *insulators.* Actually, there is no sharp dividing line between conductors and insulators, since electron motion is known to exist to some extent in all matter. Electricians simply use the best conductors as wires to carry current and the poorest conductors as insulators to prevent the current from being diverted from the wires.

In the following list, some of the best conductors and insulators are arranged in order of their respective abilities to conduct or resist the flow of electrons:

Conductors	*Insulators*
Silver	Dry Air
Copper	Glass
Aluminum	Mica
Brass	Rubber
Zinc	Asbestos
Iron	Bakelite

STATIC ELECTRICITY

One of the fundamental laws of electricity is that *like charges repel each other and unlike charges attract each other.* Therefore, there is a force of attraction in the atom between the positive nucleus and the negative electrons revolving about the nucleus in the planetary elliptical orbits.

The word *static* means "standing still" or "at rest." Originally static electricity was considered electricity at rest, because experimenters thought that electrical energy produced by friction did not move. Now, however, a simple experiment can easily be performed to produce static discharges. If a dry comb is run vigorously through the hair several times, and a cracking or popping sound is heard, it is an indication that static discharges are taking place. Charges are first built up on the hair and the comb by the transfer of electrons from one to the other due to the friction between them. The discharge that follows is the rapid movement of electrons in the opposite direction from the comb to the hair as the charges neutralize each other. These discharges appear in the dark as tiny sparks.

Charged Bodies. In the experiment previously described, strands of hair may stand out at angles because the loss of electrons has caused the hair to become positively charged and like charges repel each other. The comb, on the other hand, has gained electrons and thus acquired a negative charge.

If the negatively charged comb is held near a small piece of paper, the paper will be attracted to it and will cling for a short time. The negative charge on the comb will repel free electrons on the paper to the far side leaving the side nearest the comb positively charged. The unlike charges on the comb and on the nearest side of the paper account for the attractive force that draws the paper into contact with the comb. During contact, some of the excess electrons move from the comb to the paper, thus giving the paper a negative charge. Since like charges repel each other, the paper is repelled from the comb. Thus, the paper is first attracted to the comb and then repelled by it.

Summarizing, a *charged body* is one that has more or fewer than the normal number of electrons. It may be either positively or negatively charged. A positively charged body is one in which some of the electrons have been removed from the atoms and there is a deficiency of electrons, or fewer electrons than protons. A negatively charged body is one in which there are more than the normal number of electrons in each atom—that is, there are more electrons than protons. A body in which there is an equal number of electrons and protons in each atom is an uncharged body.

Removing electrons from a body involves physically attaching them to another body and then moving the other body some distance away. The second body will have an excess of electrons and thus will be negatively charged. The first body will have a deficiency in electrons and thus will be positively charged. This principle can be illustrated by rubbing glass with silk. Some of the electrons are rubbed off the glass onto the silk, leaving the glass with a positive charge (deficiency of electrons) and the silk with a negative charge (surplus of electrons). So long as the silk and the glass are not brought into contact, they will retain the charges. However, when they are allowed to touch, the surplus of electrons on the silk will move onto the glass and neutralize the charge on the two bodies.

Coulomb's Law of Charges. It has been shown experimentally that charged bodies attract each other when they have unlike charges and repel each other when they have like charges. Thus, electrons and protons attract each other, electrons repel other electrons, and protons repel other protons. The forces of attraction or repulsion change with the magnitude of the charges and also with the distance between them. This relation is dealt with in the law of forces discovered by a French scientist named Charles A. Coulomb, which states that *charged bodies attract or repel each other with a force that is directly proportional to the product of the charges on the bodies and inversely proportional to the square of the distance between them.*

The charge on one electron or proton might be used as the unit of electric charge, but it would not be practical because of its very small magnitude. The practical unit of charge is the *coulomb*, which is equivalent to the charge on 6,280,000,000,000,000,000 electrons. A more convenient way of expressing this number is 6.28×10^{18} electrons.

Electric Fields. The space between and around charged bodies in which their influence is felt is called an *electric field of force.* The electric field requires no physical or mechanical connecting link; it can exist in air, glass, paper, or a vacuum. *Electrostatic fields* and *dielectric fields* are other names used to refer to this region of force.

Fields of force spread out in the space surrounding their point of origin and, in general, *diminish in proportion to the square of the distance from their source.* An example of the force of gravity is the field of force that penetrates the space surrounding the earth and acts through free space to cause all unsupported objects in that region to fall to the earth. Newton discovered the law of gravitation, which states that *every object attracts every other object with a force that is directly proportional to the product of the masses of the objects and inversely proportional to the square of the distance between them.*

Note the similarity between the law of gravity and the law of attraction of charged bodies. The gravitational fields hold the universe together, for with no gravitational field the planets, including the earth, would fly off into space instead of revolving around the sun. The moon would cease to revolve around the earth, and, because of the earth's rotation about its own axis, objects from its surface would fly out into space. Similarly, electrons revolving at high velocity around the positive nucleus of the atom are held in their orbits by the force of attraction of the positive nucleus. We conclude that a field of force must exist between the electrons and nucleus.

Relatively speaking, there are enormous spaces between the electrons and nucleus, even in the densest atoms. *For example,* if a copper penny were enlarged to the size of the earth's orbit around the sun (approximately 186,000,000 miles in diameter), the electrons in the penny would be the size of baseballs and the average distance between them would be about three miles.

Lines of Force. In diagrams, lines are used to represent the direction and intensity of the electric field of force. The intensity of the field (field strength) is indicated by the density (number of lines per unit area), and the direction of the field is indicated by arrowheads on the lines pointing in the direction in which a small test charge will move or tend to move when acted upon by the field of force.

A small test charge, either positive or negative, can be used to test the direction in which the force acts, because the force of the dielectric field will act on either. By agreement, a small positive charge is used when determining the direction of the field. The test shows that the direction of the field about a *positive* charge is *away* from the charge, because a positive test charge is *repelled;* and that the direction about a *negative* charge is *toward* the charge, because the positive test charge is *attracted* toward it. The direction of the field between the positive and the negative charges is from positive to negative, as shown in Fig. 3.

The electric field about like charges is shown in Fig. 4. It can be seen that the lines of force apparently repel each other. Note, in Fig. 3, the lines of force between the two charged bodies are not parallel. They bend outward at the center as if they were repelling each other. In Fig. 4, the lines of force located in the region between the two charges apparently repel each other.

In both illustrations the lines terminate on material objects and extend from a positive charge to a negative charge.

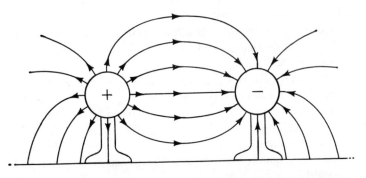

Fig. 3. Direction of electric field about positive and negative charges.

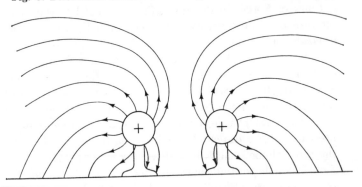

Fig. 4. Electric field between two positively charged bodies.

They are regarded as imaginary lines in space along which a real force acts. In both examples, the direction in which the force acts is that in which a small positive test charge placed in the field will move—that is, from the positive charge to the negative charge.

ELECTRIC CURRENT

Up to this point reference has been made to static electricity, or electricity at rest. The free electrons in a conductor are moving constantly and changing positions in a vibratory manner.

Electric current can be simply defined as the movement of free electrons within an electrical conductor caused by a potential difference between the ends of the conductor. Since the electron carries a negative charge, it will be attracted to the positive end of the conductor—i.e., to the point of higher potential. Therefore, an *electric current* actually moves in a direction from negative to positive.

Direct current is a current which flows in *one* direction only. A pure direct current has a constant magnitude (value) with respect to time (Fig. 5, A). A varying direct current varies in magnitude with respect to time (Fig. 5, B). When such variations occur at regular intervals, the direct current is called a *pulsating direct current* (Fig. 5, C).

Alternating current is an electric current which moves first in one direction for a fixed period of time and then in the opposite direction for an equal period of time, constantly changing in magnitude. From a zero value, alternating current builds up to a maximum in a positive direction, then falls off to zero value again before building up to a maximum in the opposite or negative direction, and then finally returning to zero. For this reason, alternating current may be further defined as a current which is constantly changing in magnitude (either building up or falling off) and periodically (at set intervals of time) changing direction. The shape of such alternating flow is that of a wave.

1. The most common and most important alternating current wave shape is the sine wave, so named after the

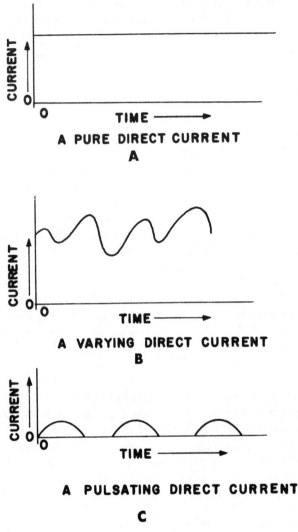

Fig. 5. Types of direct currents.

trigonometric function whose graphic pattern it follows. However, due to the presence of electrical noise (random voltages generated within an electric circuit), sine waves are often distorted. Consequently, many different kinds of alternating wave shapes can occur. These alternating wave shapes are referred to as complex waves. Figure 6 illustrates a sine wave and two types of complex waves.

2. In each instance, one cycle or complete pattern from positive to negative is shown (Fig. 6). Assuming that each wave shape starts at the same instant of time, it can be seen

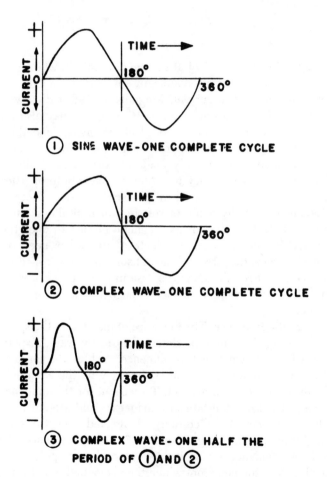

Fig. 6. Alternating-current wave shapes.

that the wave shape of (Fig. 6, C) completes its cycle in half the time of either (Fig. 6, A or B). Therefore, the frequency of (C) is twice that of (A and B).

3. It is apparent, then, that the proper recognition and definition of an alternating current depends upon:

(a) The wave shape of one cycle.

(b) The value at some specified point in the cycle.

(c) The length of time to complete one cycle (the period).

MAGNETISM

Magnetism is a power of attraction or repulsion which can be introduced most pronouncedly in certain metals by means of an electric current. Magnets formed in this way are referred to as artificial magnets. Magnets may also be classified as either permanent or electromagnets, depending upon their ability to retain their magnetic properties. Permanent magnets are usually made of steel or steel alloys, since *steel* has a tendency to *retain* its magnetic properties once it has been magnetized by an electric current. Electromagnets are usually made of soft iron or iron alloys, since *iron* tends to *lose* most of its magnetic properties once the magnetizing influence of an electric current has been removed. Hence, the effect of magnetism can be controlled by an electric current. Such is the case in many types of electric machines, wherein an electromagnet is a major component.

Magnetic Influence. The influence of magnets in the space surrounding them may be detected in many ways. Experiments have shown that this influence, such as the force of attraction and repulsion, varies inversely as the square of the distance from the magnet. To account for this influence a magnet is said to establish a magnetic field around itself which is represented pictorially by directed lines. Consider the permanent bar magnet of Fig. 7. For convenience, the ends of a magnet are arbitrarily referred to as poles. The north pole is the end from which the magnetic lines of force leave the magnet and the south pole is the end into which the magnetic lines of force reenter the magnet. The magnetic lines of force pass through the magnet from its south to its north pole. Lines of magnetic force are always closed loops.

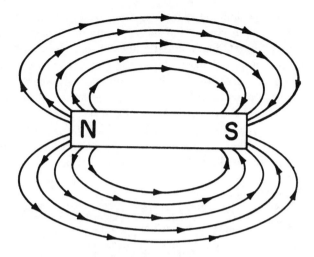

Fig. 7. Magnetic lines of force about a bar magnet.

Lines of Force. If a second bar magnet is placed near the first bar magnet (Fig. 8) so that the unlike poles are adjacent, a force of attraction results, and some of the magnetic lines of force from the first bar magnet are diverted toward the pole of the second bar magnet. However, if one magnet

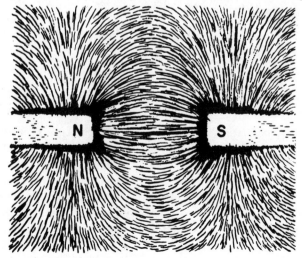

Fig. 8. Magnetic lines of force about two bar magnets of opposite polarity.

is reversed so that the like poles are now adjacent, a force of mutual repulsion will result, tending to separate the magnets (Fig. 9). The magnetic lines of force leaving from each of the adjacent like poles of the two bar magnets will tend to repel each other while simultaneously seeking to reenter the opposite end (opposite pole) of their respective magnets. From these observations it can be seen that unlike poles of magnets attract each other and like poles repel each other. This phenomenon of magnetic attraction and repulsion produces the torque action in many types of electric motors.

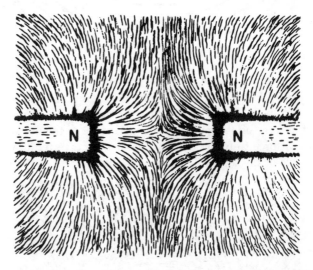

Fig. 9. Magnetic lines of force about two bar magnets of like polarity.

Electromagnetism. When an electric current is passed through a wire, a magnetic field is produced around the wire (Fig. 10). The direction of the magnetic lines of force, which form concentric circles around the wire, depends upon the direction of the electric current through the wire. The direction of the magnetic lines of force about a current-carrying conductor can be determined by using the left-hand rule for a current-carrying conductor. This rule states that if a current-carrying conductor is grasped in the left hand with the thumb pointing in the direction of current flow

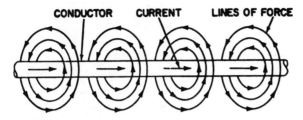

Fig. 10. Magnetic lines of force about a current-carrying conductor.

(negative to positive), the fingers will encircle the wire in the direction of the magnetic lines of force, as shown in Fig. 11.

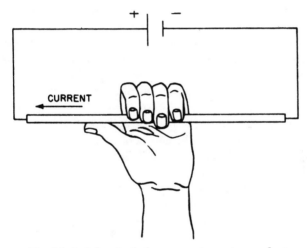

Fig. 11. Left-hand rule for a current-carrying conductor.

Magnetic field about a coil. The magnetic field resulting under the conditions shown in Fig. 10, even with high currents, is relatively weak. But if the electrical conductor is formed into a coil, a relatively stronger magnetic field is created with magnetic lines of force running through the center of and perpendicular to each turn of the coil, as shown in Fig. 12. The coil current sets up circular magnetic lines of force around each coil turn in accordance with the left-hand rule. Consequently, the magnetic lines of force

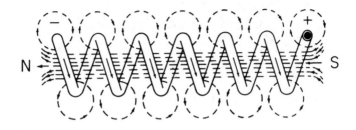

Fig. 12. Magnetic field about a current-carrying coil.

encircling the upper part of each coil turn in a counterclockwise direction and those encircling the lower part of
each coil turn in a clockwise direction. In the center of the
coil, all of the magnetic lines of force run in the same
direction, thereby aiding each other to produce a net positive effect. Between adjacent coil turns, the magnetic lines
of force cancel each other. Consequently, magnetic lines of
force travel the entire length of the coil in order to complete
their loops. This makes the coil behave as a magnet with a
north and a south pole. By using the left hand, again, the
north pole end of a current-carrying coil can be determined.
This time, if the coil is grasped so that the fingers of the
left hand encircle the coil in the direction of the current
flowing through the coil, the thumb will point to the north
pole of the coil.

Electromagnets. If a piece of magnetic material, usually
soft iron, is placed within a coil through which current is
flowing, the magnetic properties of the coil are tremendously
increased. This increase in magnetic strength is due to the
greater permeability of the soft iron. Permeability is the
ease with which magnetic lines of force pass through a substance. A coil wound around a core of magnetic material is
called an electromagnet. The coil may be wound with one
or more layers of wire from one end to the other and back,
providing, of course, that the current flows around the core
continuously in the same direction.

Magnetic field about a coil with an iron core. Figure 13 shows the effect that an iron core has on the magnetic lines of force surrounding a current-carrying coil. In Fig. 13, A, note that the lines of force passing through the coil are confined to the iron core. If the iron core is pulled partially out of the coil as shown in Fig. 13, B, the magnetic lines of force will be extended in order to enter the end of the iron core which is outside of the coil. Once the lines have established themselves in the core, they tend to shorten, thereby

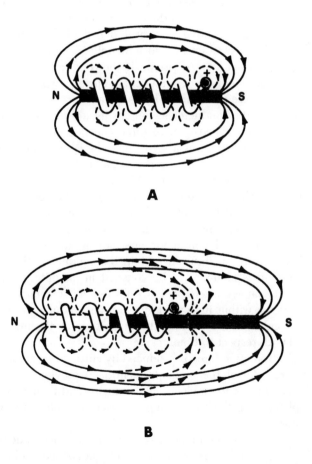

A

B

Fig. 13. The effect of an iron core on the magnetic field of a coil.

exerting a force on the core. This force tends to pull the core until its center coincides with the center of the coil, as shown by the dotted lines in the figure. This action has many practical applications in the various types of electrical controlling devices used in industry and the home.

Induced Electromotive Force. Having determined the existence of magnetic lines of force around any wire due to the flow of an electric current, one may conceive of establishing an electric current by means of a magnetic field. Such is the case when an electrical conductor is moved across a magnetic field. As the conductor cuts the lines of magnetic force, an electromotive force is induced in the conductor causing an electric current to flow only if the conductor is part of a closed loop or circuit. If the conductor is not part of a closed loop or circuit, then current will not flow through. There will still be, however, an electromotive force induced in the conductor as long as it cuts across magnetic lines of force. Electromotive force (emf) is defined as the force or pressure necessary to cause the flow of an electric current. An emf can still be present in a circuit without the flow of an electric current. It can also be induced in a stationary electrical conductor if the magnetic field is made to move so that its lines of force cut across the conductor. In each of these instances, such as where a magnetic field and an electrical conductor are moving relative to each other causing an induced emf in the conductor, the magnetic field is assumed to be constant in magnitude. However, a magnetic field which does not move relative to a conductor within its boundaries can also induce an emf in the conductor, by varying the magnitude of the magnetic field with respect to time.

Electromotive force is produced by cutting magnetic lines of force. All of the methods mentioned previously for inducing an emf in an electrical conductor find their practical application in the various types of electric motors and generators.

Figure 14 shows a loop wire revolving in a magnetic field. The magnetic field is created by field poles of the kind found in the most elementary type of electric machine. The ends

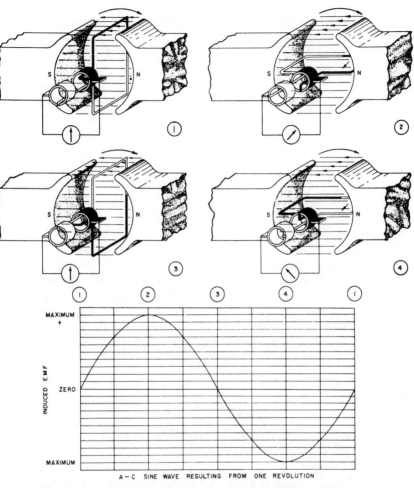

Fig. 14. Loop of wire rotating in a magnetic field. The encircled arrow simulates a galvanometer for indicating the strength and direction of the induced emf.

of the loop are connected to sliprings which revolve with the loop. Stationary brushes are used to collect the current from the rings and deliver it to an external circuit. In Fig. 14, A, the white conductor moves to the left while the black conductor moves to the right, and the induced emf's in both conductors may be added together. The total emf will depend upon the position of the loop in the field. This is true because only that portion of the motion perpendicular to the field is effective in producing emf. The wave shape which results from one complete revolution of the loop is also shown in the figure.

1. To trace the development of the wave shape, start with the loop as shown in Fig. 14, A. In this position, each conductor is moving parallel to the magnetic field, the loop is in a neutral position, and the generated emf is zero.

2. As the loop continues in a clockwise direction the emf increases, due to the loops cutting more lines in a given period of time, until it reaches a maximum, at which time the loop is parallel to the field (Fig. 14, B).

3. As the loop continues to rotate to the position shown in Fig. 14, C, the emf decreases until it is again zero.

4. If the loop is turned further through an angle of 90° in the same direction (Fig. 14, D), it will again be cutting lines of force at a maximum rate. However, the emf and resulting current will be reversed with respect to the loop. The reversal occurs because of the change in the direction in which the conductor passes through the field.

5. As the conductor continues to rotate, the generated emf decreases until, at the starting point, it is again zero. The wave shape of Fig. 14 is a sine wave.

Let us see what would happen if the sides of the loop of Fig. 14 were connected to split rings instead of sliprings, as in Fig. 15. *Note* that one brush is shown as being black and the other white.

1. Starting with the loop in a neutral position (Fig. 15, A), the generated emf rises from zero to a maximum at the position shown in Fig. 15, B. This corresponds to the first portion of the wave shape of B, Fig. 14. *Note* that the black brush is in contact with the black side of the coil.

2. As the loop continues to rotate in a clockwise direction, the induced emf again decreases to zero (Fig. 15, C), as it did in Fig. 14, C. However, the black brush is directly over the split portion of the ring, indicating that the white portion of the loop will be in contact with the black brush while the loop is rotated through the next 180° (Fig. 15, D and E).

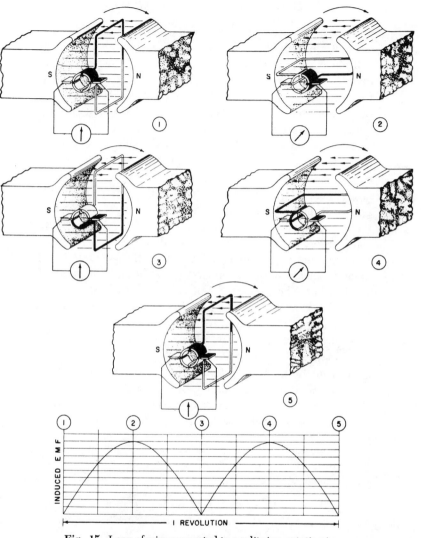

Fig. 15. Loop of wire connected to a split ring, rotating in a magnetic field.

3. As the loop is rotated through the remaining 180° to complete one revolution (Fig. 15, D and E), the generated emf once again rises to a maximum at the position shown in Fig. 15, D, and decreases to a minimum at the position shown in Fig. 15, E. However, since the black brush is al-

ways in contact with the side of the loop passing from left to right through the field, the current through the brush will always be in the same direction. This is shown by the reversal of current in the second half of the wave shape in Fig. 15. The action of collecting the induced current in the same direction is called commutation, and the split ring by which it is accomplished is called a commutator.

4. Commutation can be explained as follows. At the instant that each brush is contacting two segments on the commutator (Fig. 15, A, C, and D), a direct short circuit is produced. If an emf is generated at this time, a high current will flow in the short circuit, causing an arc, and damaging the commutator and the brushes. For this reason, the brushes must be positioned so that the short between the commutator segments occurs when the generated emf is zero. This position is referred to as the *neutral plane* of the brushes. The generated emf is zero, at the instant when the coil is not cutting magnetic lines of force. This instant occurs at Fig. 15, A, C, and E.

Figure 16 shows the wave shape generated by three separate loops which are connected to sliprings and are rotating in a magnetic field. *Note* that the voltage generated by the second and third coils follows the first by 120° and 240°, respectively. This displacement, or phase angle, results from the loops being mutually spaced at 120°; in other words,

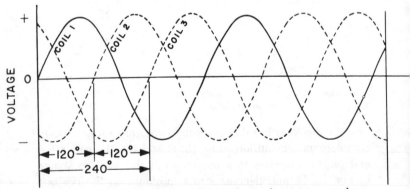

Fig. 16. Wave shapes of three separate loops of wires connected to sliprings.

the distance between the first and second loops is one third
the periphery of the armature core as is the distance between the second and third loops. The resultant wave shape
is similar to that of a three-phase generator.

Figure 17 shows the wave shape generated by four separate coils connected to an eight-segment commutator, rotating in the magnetic field established by two poles. A
comparison of Figs. 15 and 17 will show that the variation
between maximum and minimum values of emf decreases

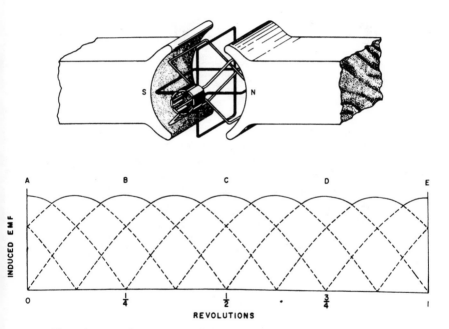

Fig. 17. Wave shape generated by four separate coils connected to an
eight-segment commutator.

with the addition of loops or coils. This variation, called
ripple, is present in all emf's generated in the manner described previously.

The maximum value of the emf is not affected by the
number of loops, but rather by the number of turns of wire

per loop, the field strength, and the speed of rotation. This is sometimes expressed as being the rate of change of flux linkages. Flux is merely another name for lines of magnetic force.

Left-Hand Rule for Generators. The relationship found to exist among the direction of the magnetic field, the direction of motion of the conductor, and the direction of the induced current is shown in Fig. 18. This relationship, referred to as "left-hand rule for generators," may be stated as follows: Extend the thumb, forefinger, and middle finger

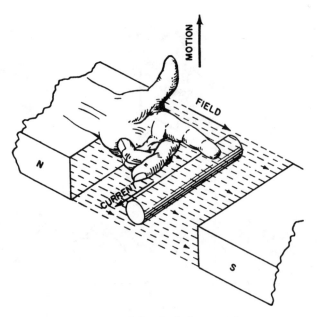

Fig. 18. Left-hand rule for generators.

of the left hand so that they are at right angles to each other. Place the forefinger in the direction of the magnetic field (north to south pole) with the thumb in the direction of the motion; then the middle finger will indicate the direction of current flow through the conductor.

Motor Effect. If current is generated in a wire moving through a magnetic field, one might logically assume it is possible to cause a conductor, located in a magnetic field, to move by passing a current through it. Such is the case. In fact, this action, called the *motor effect,* is the basis of all electric motors. Consideration of the conductor shown in Fig. 19 will aid in understanding how this comes about.

The conductor in Fig. 19 is assumed to carry an electric current coming out of the page. This current establishes lines of force about the wire in a clockwise direction according to the left-hand rule. With the magnetic field in the position shown, some of the lines which would normally pass through, or immediately beneath, the area occupied by the conductor may be thought of as being deflected over it. In doing so, they are stretched somewhat and crowded into the area above the conductor. The natural tendency of the lines to straighten exerts a force on the conductor which would tend to push it downward and entirely out of the magnetic field.

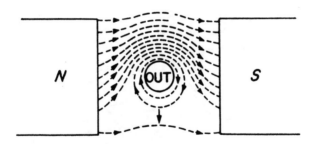

Fig. 19. Expulsion of current-carrying wire from a magnetic field.

Let us see what would happen if a loop of wire, rather than a single conductor, were placed in a magnetic field and current passed through it. Figure 20 shows a loop, with the direction of current flow indicated, located in a magnetic field. The direction of rotation of the lines of force set up by the current is clockwise about the left half of the loop and counterclockwise about the right half (left-hand rule). This distorts the lines of the magnetic field by deflecting the lines in such a manner that they pass over the left conductor and beneath the right conductor. The left half of the

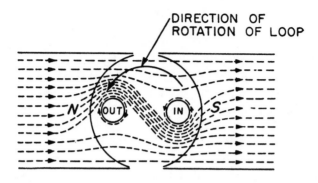

Fig. 20. Rotation of a current-carrying loop of wire in a magnetic field.

loop experiences a downward force, as shown in Fig. 19, while the right half is forced upward. If the loop were free to move about an axis located midway between the sides of the coil, it would rotate in a counterclockwise direction. If the direction of current flow through the loop were reversed,

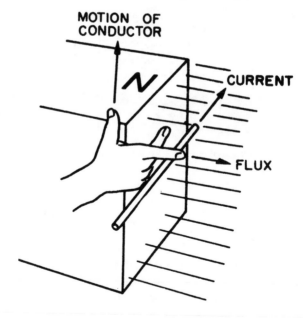

Fig. 21. Right-hand rule for motors.

the loop would tend to rotate in a reverse direction, or clockwise.

Right-Hand Rule for Motors. Since an electric motor performs the reverse function of a generator, in that it changes electrical energy into mechanical energy, the opposite hand, or right hand, is used in the motor rule (Fig. 21). The rule, referred to as the right-hand rule for motors, can be performed as follows. Extend the thumb, forefinger, and middle finger of the right hand at right angles to each other. Place the forefinger in the direction of the magnetic field (north to south pole), with the middle finger in the direction of current flow in the conductor; the thumb will then indicate the direction of motion. *Note* that, as in the case of the generator rule, the forefinger is associated with the field, the middle finger with the current, and the thumb with the direction of motion. When you know any two of the following—motion, magnetic field, or current—you can determine the direction of the third.

CHAPTER 2

Electrician's Tools and Equipment and How to Use Them

Electrician's Pocketknife. The blades of the electrician's pocketknife are used for prying, boring, cleaning crevices, and cutting and scraping wire (Fig. 1). The large cutting blade should be kept sharp, the other blades dull.

Fig. 1. Electrician's knife.

Screwdrivers. Screwdrivers should be used only for tightening or loosening slotted screws. The most common types of screwdrivers are the *standard* or ordinary, the *offset*, and the *Phillips*.

The *standard* screwdriver (Fig. 2, A) is available in lengths from 3″ to 18″ and with blades in sizes to fit the standard screw heads (Fig. 2, B).

The *offset* screwdriver can be used in corners where there is insufficient space to manipulate a standard screwdriver (Fig. 2, D). It has two blades: one in line with the shank or handle, the other at right angles to the shank.

The *Phillips* screwdriver is made with a blade to fit the Phillips cross-slot screws (Fig. 2, C).

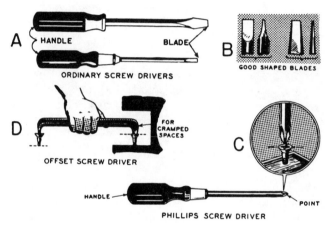

Fig. 2.

Wood Chisels. The *socket* and *tang* chisels (Fig. 3) are used to cut wood. Socket types of chisels are used for deep cuts and are driven with a heavy mallet. Tang chisels are used for light hand-driven work. The size of these chisels is determined by the distance across the blade.

Cold Chisels. The *flat* and *cape* cold chisels (Fig. 4) are generally used to cut, chip, or slot any metal that can be filed. Sizes of these chisels are determined by the width of their cutting edges.

Flat chisels are used to cut thin sheet metal, rods, and

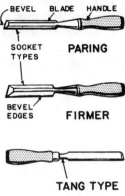

BEVEL   BLADE   HANDLE

SOCKET   **PARING**
TYPES

BEVEL
EDGES   **FIRMER**

TANG TYPE

Fig. 3. Wood chisels.

conduit, to chip flat surfaces, and to cut rivets or split nuts. Cape chisels are used to cut keyways, grooves, and slots.

Braces and Bits. The *plain brace* (Fig. 5, A, C) and the *corner brace* (Fig. 5, B) are used for boring holes in wood. Both braces have holding devices (*chucks*) that clamp the shank of the *auger bit* (Fig. 5, D). The point of the auger bit does the boring. Auger bit sizes are stamped on the shank, and auger bits are available for boring holes ¼″ to 1″ in diameter. *Gimlet bits* and *awls* are used to bore holes smaller than ¼″. Use the *expansive* or *Forstner bits* (Fig. 5, B) to bore holes larger than 1″.

Pliers. Pliers commonly used are the combination, the round nose, the flat nose, the long nose, and cutting pliers or nippers.

The slip joint of *combination* pliers permits the jaws to be opened wider at the hinge pin for gripping large diameters (Fig. 6, A). Combination pliers are available in lengths ranging from 5″ to 10″. Adjustable combination pliers are used to hold and bend flat or round stock.

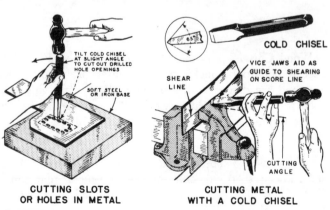

TILT COLD CHISEL
AT SLIGHT ANGLE
TO CUT OUT DRILLED
HOLE OPENINGS

SOFT STEEL
OR IRON BASE

**CUTTING SLOTS
OR HOLES IN METAL**

65°

COLD CHISEL

VICE JAWS AID AS
GUIDE TO SHEARING
ON SCORE LINE

SHEAR
LINE

CUTTING
ANGLE

**CUTTING METAL
WITH A COLD CHISEL**

Fig. 4.

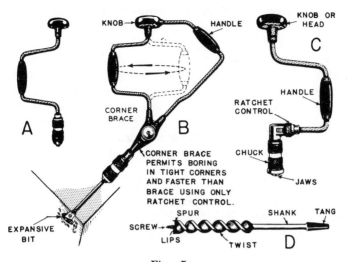

Fig. 5.

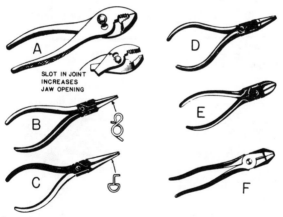

Fig. 6. Pliers.

The *round nose* and the *flat nose* pliers (Fig. 6, B, D) are used to bend or form thin metal or wire into various shapes (note bent wires in Fig. 6, B, C).

Long nose pliers (Fig. 6, C) are designed for use in inaccessible places.

The *diagonal cutting* pliers (Fig. 6, E) are used to cut soft wires, insulation (from electrical cables), and pins or similar fasteners, and also to remove or apply safety wire.

The *side cutting* pliers (Fig. 6, F) are used for cutting all types of electrical wire.

Wrenches. Types of wrenches used to tighten, twist, or remove nuts, bolt heads, or cap screws, or to grip and turn conduit, pipe, and other round rods are as follows: adjustable, socket, open-end, box, and pipe or Stillson wrench.

Adjustable wrenches have one adjustable jaw that can be opened or closed to fit the flats of nuts or bolt heads. The monkey wrench and the adjustable open-end wrench are two common types of such wrenches (Figs. 7 and 8). The method of using these wrenches is shown in Fig. 7.

Socket wrenches are designed for use in close or inaccessible places. Sockets are available in sets to fit standard-sized nuts. They fit into the ratchet handle.

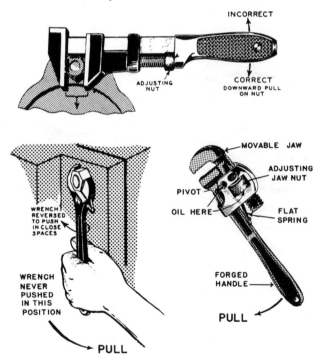

Fig. 7.

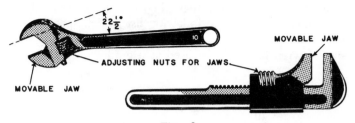

Fig. 8.

Open-end wrenches are solid, nonadjustable, with openings in each end (Fig. 9). They are used for working in close quarters and are made to fit standard-sized nuts on bolt heads. The sizes of the openings between the jaws determine the size of the wrench. The over-all length of the wrench is determined by the size of the openings.

Box wrenches and combination wrenches are also used for working in close quarters (Fig. 9). These wrenches can be used to loosen or to tighten a nut continuously with a minimum 15° swing of the handle as compared to a 60° swing of the standard open-end wrench. The box wrench head cannot slip off the nut.

Pipe or *Stillson* wrenches are used to turn conduit, pipe,

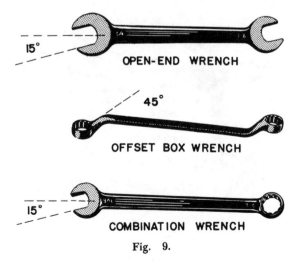

Fig. 9.

round rod, or smooth fittings (Fig. 7). They are made in a number of sizes. The following list, which specifies the maximum pipe or rod size that can be turned with each size of wrench, should be used as a guide:

WRENCH SIZE (in inches)	PIPE SIZE (in inches)
6	¼
10	⅜, ½
14	¾
18	1¼
24	1½, 2

Hammers. The *claw* hammer (Fig. 10) is used to drive nails, wedges, and dowels. The curved claws are used to pull out nails. Claw hammers are available in sizes from 4 oz. to 2½ lbs.

Ball peen hammers are made with hardened steel faces and fitted with a hard-wood handle (Fig. 11). The flat portion of the head of the ball peen hammer is called the face and the other end the peen. The most commonly used are the 6 oz., 12 oz., and 16 oz. hammers.

Handsaws. The *backsaw* (Fig. 12) is used to saw wood trim and for similar small jobs. It is a cross-cut saw with

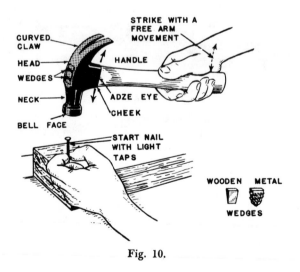

Fig. 10.

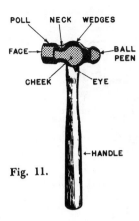

POLL NECK WEDGES

FACE→

BALL PEEN

CHEEK   EYE

←HANDLE

Fig. 11.

fine teeth (12 to 14 points per inch) and with a stiff steel band reinforcing the top edge.

The *keyhole* saw (Fig. 12) is used to cut holes and slots for electrical receptacles and accessories. *Interchangeable keyhole saws are available in several lengths and types of blades and with an adjustable handle.*

Hacksaws. Hacksaws with adjustable frames (Fig. 13) are used to saw conduit, metal pipe, and all types of metals. These frames can be adjusted to hold blades from 8″ to 16″ long. Hacksaw frames are provided with wing nuts to tighten the blade. Hacksaw flexible blades are used for sawing conduit, BX cable, similar hollow shapes, and light metals (Fig. 13). All-hard hacksaw blades are used for sawing brass, tool steel, cast iron, and similar heavy cross-section materials.

KEYHOLE SAW

BACKSAW

Fig. 12.

Center Punches. Center punches are pointed tools made of round or octagonal steel rods (Fig. 14). A center punch point marks the location of a hole that is to be drilled. Keep the octagonal type of center punch points taper-ground to an angle of about 90°, and keep the round type taper-ground to a 60° angle.

Hand, Breast, and Twist Drills. The *hand* drill and the *breast* drill (Fig. 15) are used for holding and turning drills. Holes up to ¼″ in diameter can be drilled in metal by hand with a *twist* drill (Fig. 16).

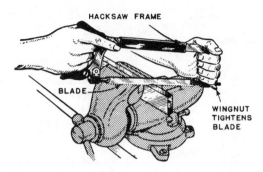

Fig. 13. Hacksaw and correct method of sawing work fastened in vice.

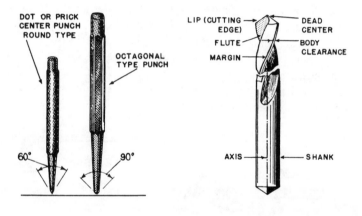

Fig. 14. Center punches.   **Fig. 16.** Twist drill.

For drilling, some materials require no lubricant, but others require a lubricant peculiar to their nature. Use the following list as a guide:

Tool steel—oil	Brass—dry
Soft steel—oil or soda water	Copper—oil
Wrought iron—oil or soda water	Babbitt—dry
Cast iron—dry	Glass—turpentine

The procedure for using a breast drill is generally the same as that for using a brace, excepting that a straight shank drill must be used. To apply pressure to the breast drill, push against the end of the tool with the body. The capacity of most breast drills is ¼″.

Electric Hand Drill. Figure 17 shows the various parts of a portable electric hand drill. Electric drills are used with twist drill bits for drilling holes in metal. The drill supplies the power for turning the twist drill, but the operator must exert the forward pressure to feed the bit into the work. In general, high speed and light feed are recommended.

Check the nameplate of a portable electric drill to make sure that the motor specifications agree with the supply line in voltage, phase, and cycles or kind of current.

Alcohol Torch. The automatic alcohol torch (Fig. 18) is a self-contained unit that creates its own pressure. The tank is filled about two-thirds with alcohol and sealed. The flame from the wick heats the jet tube, causing the liquid in the container to vaporize and expand. The expansion forces the alcohol vapor from the jet opening, where it is ignited to form a hot, light-blue flame.

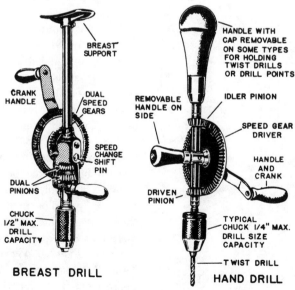

BREAST DRILL

HAND DRILL

Fig. 15.

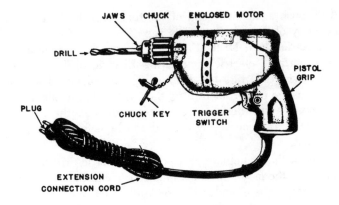

Fig. 17. Portable electric drill.

Soldering Irons. The two types of soldering irons are the *electric* and the *fire-heated* (Fig. 19). They are available in many weights and sizes. Electric soldering irons are built with internal heating coils and removable and in-terchangeable heads, thus allowing the use of various-shaped tips. Soldering irons are used for soldering electrical wires and fixtures, for soldering sheet metals together where the sheets are either lapped or locked together, and for other small soldering work. Soldering irons may be heated by a blowtorch, gas flame, or electricity. The point or working face of a soldering iron should be blunt. It serves merely as a heating device.

Fig. 18. Alcohol torch and its use.

Snake Fishing Wire or Tape. A snake made of tempered steel wire or rectang-

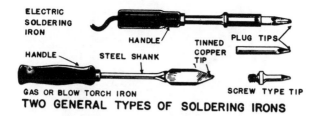

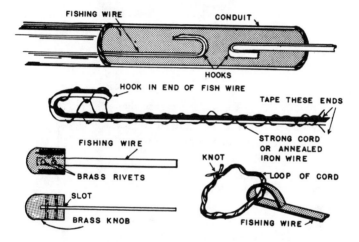

Fig. 19. Two general types of soldering irons.

Fig. 20. Fishing wires or tapes — methods of using.

Fig. 21. Test lamp.

Fig. 22.   Stock and dies.

ular cross section (Fig. 20) is used for fishing wire through wall spaces, under flooring, and through conduit. It is available in various widths and in coils of 50', 75', 100', 150', and 200'. Figure 20 shows methods of using the wire.

Test Lamps. There are many devices for testing electrical circuits, but the simplest and most effective is the test lamp set shown in Fig. 21. It has two test tips with insulated handles, a length of ordinary light cord, a male plug, a light socket, and a 23-volt, low-wattage bulb.

Stock and Die. Stocks and dies are used for cutting outside or male threads on rods, pipes, or bolts. The dies with standard threads are held for turning leverage in the stock. Figure 22 shows the complete assembly ready for cutting.

Plumb Bob. The plumb bob (Fig. 23) is used to establish vertical lines.

Fig. 23.
Plumb bob.

Fig. 24. Spirit level.

Spirit Level. The spirit level (Fig. 24) is used to determine the accuracy of horizontal or vertical lines. The frame, which may be of either wood or metal, contains two tubes that are partially filled with alcohol, leaving a small bubble in each. The tube for horizontal lines is at right angles to the one for vertical lines.

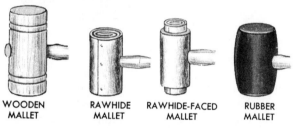

WOODEN MALLET   RAWHIDE MALLET   RAWHIDE-FACED MALLET   RUBBER MALLET

Fig. 25. Types of mallets.

Mallets. A *mallet* is a hammer-like tool made of hickory-wood, rawhide, or rubber (Fig. 25). It's used for pounding down sheet metal seams and for shaping sheet and strap metal, and it will not dent as a steel hammer would. You should always use a wood mallet to pound a wood chisel or a gouge.

C-Clamp. The *C-clamp* may be used for holding any kind of stock (Fig. 26). To replace a C-clamp screw or jaw, pry open crimped portion of swivel head and remove head from ball on end of screw, then turn the screw out of jaw (Fig. 26). When using new parts as needed, turn the screw into position through its base at end of jaw. Slip swivel head onto ball on end of screw and crimp head sufficiently to hold it on ball.

Files. *Files* are graded according to the spacing of their teeth. A coarse file has a small number of large teeth, and a

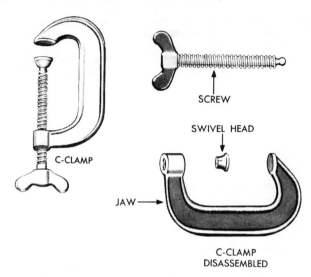

C-CLAMP

SCREW

SWIVEL HEAD

JAW →

C-CLAMP
DISASSEMBLED

Fig. 26.

smooth file has a large number of fine teeth. The coarser the teeth, the more metal will be removed on each stroke of the file. The terms used to indicate the coarseness or fineness of a file are rough, coarse, bastard, second-cut, smooth, and dead-smooth. The file may be either single-cut or double-cut. (*See* Fig. 27 for file terminology.)

Rasps. The *rasp* is similar to the file except for its coarse teeth raised by a triangular punch. It produces a rough cut and is used on wood, leather, aluminum, lead, and similar soft metals, for fast removal of waste material (Fig. 28).

Nippers. *Nippers* look like pliers, but are used only for cutting (Fig. 29). *Do not* try to use them for holding. Various types can be used for cutting wire, rod, nails, rivets, and bolts. For light work on soft metals you would use the nipper shown at Fig. 29, A. For heavier work, use the nippers shown at Fig. 29, B. They have replaceable blades, a strong joint, and a short fulcrum that provides plenty of leverage.

Reamers. *Reamers* (Fig. 30), being precision cutting tools for accurate sizing of holes, must be used, handled, and stored with every precaution to prevent denting, marring, or damage to the cutting edges. They are used to smooth

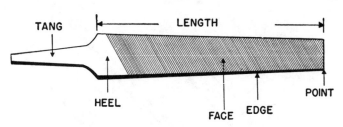

Fig. 27. File terminology.

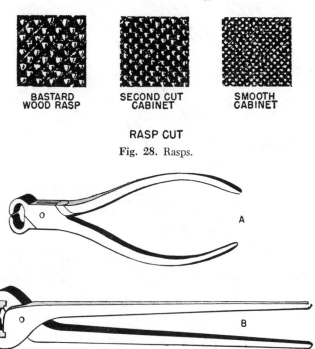

BASTARD WOOD RASP

SECOND CUT CABINET

SMOOTH CABINET

RASP CUT

Fig. 28. Rasps.

Fig. 29. Nippers.

and enlarge holes to *exact size*. An adjustable reamer should not be adjusted beyond the maximum size for which it was made.

Shears. There are many types of *shears* (Fig. 31) used for cutting sheet metal and steel of various thicknesses and shapes. All types of shears are used in the same manner. Place the cutting edge of the upper blade exactly on a guideline marked on the metal, and insert the sheet as far

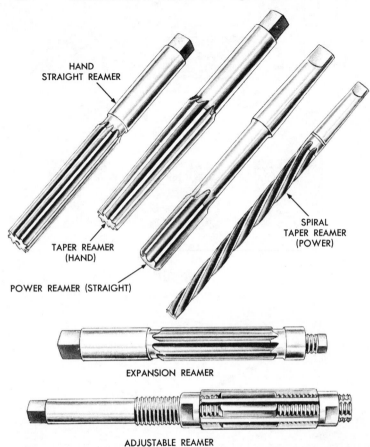

HAND
STRAIGHT REAMER

SPIRAL
TAPER REAMER
(POWER)

TAPER REAMER
(HAND)

POWER REAMER (STRAIGHT)

EXPANSION REAMER

ADJUSTABLE REAMER

Fig. 30.

back as possible between the blades. Hold the shears so that the flat sides of the blades are at all times perpendicular, or at right angles, to the surface of the work.

Rules. *Rules* are graduated measuring instruments usually made of metal or wood. Flexible steel rules are the most commonly used. The graduations indicate inches and fractions of an inch (or centimeters and millimeters).

The electrician will find the common *flexible rule* suitable for most purposes. Other types often used are narrow rules and hook rules, slide caliper rules (Fig. 32), and the straight edge. The *narrow rule* is convenient on work where an ordinary rule is too wide. The *hook rule* is valuable for measuring inside dimensions where the hook is an advantage. The *caliper rule* is used for measuring outside diameters quickly. The *straightedge* is like a rule except that it is longer and has no graduation marks. You cannot measure with a straightedge. It is used as a reliable guide

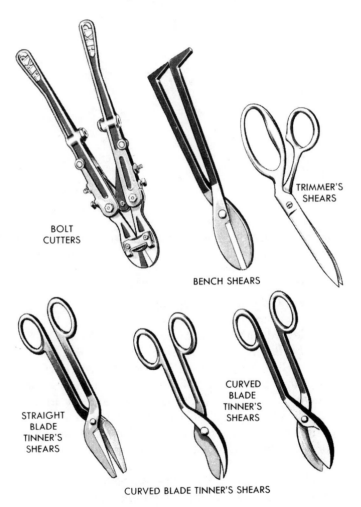

BOLT
CUTTERS

BENCH SHEARS

TRIMMER'S
SHEARS

STRAIGHT
BLADE
TINNER'S
SHEARS

CURVED
BLADE
TINNER'S
SHEARS

CURVED BLADE TINNER'S SHEARS

Fig. 31.

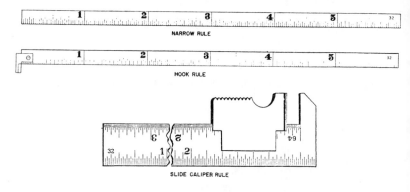

Fig. 32. Narrow, hook, and caliper rules.

for scribing long, straight lines and for checking flat surfaces. *Be sure* to keep it straight.

Once the most often used *6-inch* rule is understood, the use of other types becomes merely a question of applying them to the work at hand. Ordinarily, the four available edges of a 6-inch rule are graduated in 8ths and 16ths of an inch on one side and in 32nds and 64ths on the other. Figure 33 shows the two sides of such a rule. (*See* Fig. 34 for readings on a 6-inch rule.)

Calipers. *Calipers* are commonly made with a screw-and-nut adjustment, which puts tension on a spring to hold the setting desired. Other calipers are of the *firm-joint* (Fig. 35) *friction-holding* type. This type of caliper has thin flat legs and is excellent for measuring in narrow spaces. *Hermaphrodite calipers,* shown in the figure, are half caliper and half divider. They are used to measure and mark from an edge. The setting of the tool must be taken from a rule. You can use this caliper to mark parallel lines, locate centers for drilled holes, and to find the centers of pieces of round stock.

Micrometer calipers. The basic principle of the *micrometer caliper* is a very accurate screw which is free to move in an accurately threaded fixed nut. This provides a variable opening for measuring work held between one end of the

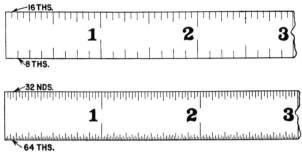

Fig. 33. Two sides of 6-inch rule.

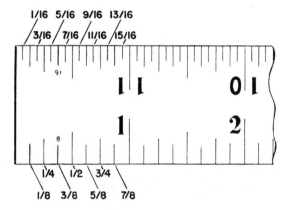

Fig. 34. Readings on 6-inch rule.

micrometer screw or spindle and a fixed contact face or anvil in the frame. (*See* Fig. 36 for sectional view of a micrometer caliper.) The graduated thimble rotates with the spindle and travels along a graduated barrel. The graduations conform to the pitch of the micrometer screw; hence, there is a graduation on the barrel for each revolution of the micrometer screw. The graduations on the beveled edge of the thimble accurately divide each revolution of the screw so that measurements can be made in .001″ or .01 mm. Readings to .0001″ are made on the ten-thousandth inch micrometer by using a vernier on the barrel.

The ratchet stop or convertible thimble furnished on some micrometer calipers automatically provides the correct measuring pressure, assuring consistent accuracy by eliminating variation in personal measuring "feel."

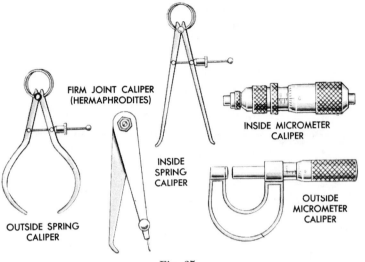

FIRM JOINT CALIPER
(HERMAPHRODITES)

INSIDE MICROMETER
CALIPER

INSIDE
SPRING
CALIPER

OUTSIDE
MICROMETER
CALIPER

OUTSIDE SPRING
CALIPER

Fig. 35.

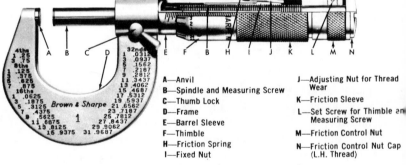

4ths		32nds	
4ths	.25	1	.0312
1	.25	3	.0937
2	.5	5	.1562
3	.75	7	.2187
8ths		9	.2812
1	.125	11	.3437
5	.625	13	.4062
7	.875	15	.4687
16ths		17	.5312
1	.0625	19	.5937
3	.1875	21	.6562
5	.3125	23	.7187
7	.4375	25	.7812
9	.5625	27	.8437
11	.6875	29	.9062
13	.8125	31	.9687
15	.9375		

Brown & Sharpe

A—Anvil
B—Spindle and Measuring Screw
C—Thumb Lock
D—Frame
E—Barrel Sleeve
F—Thimble
H—Friction Spring
I—Fixed Nut

J—Adjusting Nut for Thread Wear
K—Friction Sleeve
L—Set Screw for Thimble and Measuring Screw
M—Friction Control Nut
N—Friction Control Nut Cap (L.H. Thread)

Fig. 36. Sectional view of a micrometer caliper.

The clamp ring or clamp lever is a convenience in keeping the micrometer at a particular setting. A slight rotation of the knurled ring, or flick of the lever, locks the spindle. *Never* tighten the clamp when the spindle is removed. To do so injures the clamping mechanism.

Dividers. *Dividers* are used to measure dimensions between lines or points, to transfer to the work the dimensions taken from a steel rule, and to scribe circles or arcs. The contacts of dividers are sharp points at the ends of straight legs (Fig. 37). Close measurements are made by visual comparison. The range of a divider is restricted to the opening span of its legs, and when the legs make acute angles with the surface being worked on, the divider is less effective for scribing and similar uses.

Taps and Dies. *Taps* and *dies* are tools for cutting screw threads. *Taps* are used to cut internal threads; *dies* are used to cut external threads. Taps and dies are pigeonholed according to the type of thread they form—such as N.F. or N.C.—or according to the diameter of the screw formed, or to fit a hole that is tapped, or according to the number of threads to the inch.

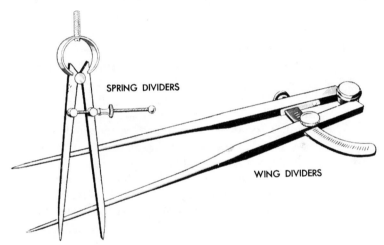

Fig. 37. Dividers.

Taper, plug, and *bottoming taps* for screw threads usually come in a set (Fig. 38). They make one size of thread. The *taper tap* may be used for internal threading where the work permits the tap to be run entirely through. The diameter of this tap gradually increases from or near the starting end.

When the taper tap cannot be run through the work, you will note that, near the bottom of the tapped hole, the diameter will be so small that the screw or bolt will not screw down as far as it should. In this case, run in a plug tap after the taper tap is removed. If you want full diameter threads all the way to the bottom of the hole, follow the plug tap with a bottoming tap which is the same diameter as its entire length.

Dies are made in a variety of forms (Fig. 38). When you are cutting threads with a die, the thread will be formed progressively, since the die is designed that way. Each cut produces a slightly deeper thread until the finished thread is obtained. Each die is made in one piece. The adjustment is made by splitting the die between two of its lands (cutting portions between grooves) and expanding or contracting it. Do not confine adjustable round split dies for thread cutting with hexagon rethreading bolt dies. Hexagon rethreading bolt dies have six cutting lands and are intended for dressing over bruised and rusty threads.

There are many forms of *wrenches* for turning taps, and *stocks* for turning dies. The diestock for adjusting round split dies, and the T-handled, and adjustable wrenches (Fig. 38), will enable you to work either in open or confined spaces.

The size drill for drilling a hole to be threaded is determined by the diameter of the tap minus 1½ times the depth of the thread used. A skilled worker can judge the size drill to use on rough work—a feat *you* can accomplish with enough experience. One fact you might remember is that taps and dies used in making the National Standard screw threads cannot be used in making the National Taper pipe thread.

If you use a lubricant in cutting threads, it will insure a smooth, clean-cut thread. Lard oil is generally preferred on steel, while kerosene is used on aluminum or aluminum alloy. A lubricant is not needed on brass. There are left-handed taps and dies for making left-hand threads.

Vises. *Vises* are used for holding work when it is being planed, sawed, drilled, shaped, sharpened, or riveted, or when wood is being glued. You will use a utility or pipe vise.

The *utility vise* (Fig. 39) is satisfactory for general work and is designed for a variety of uses. It has a small anvil

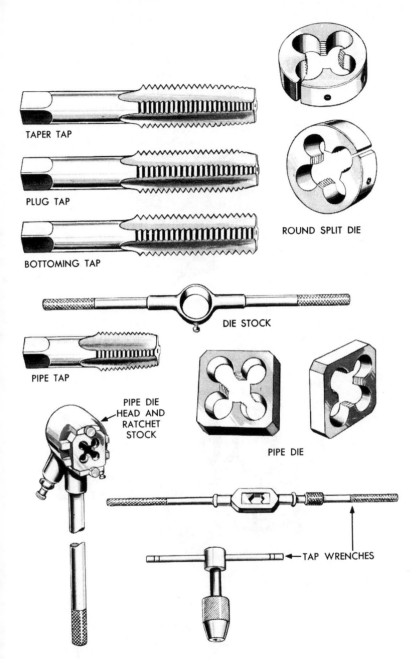

TAPER TAP

PLUG TAP

BOTTOMING TAP

ROUND SPLIT DIE

DIE STOCK

PIPE TAP

PIPE DIE HEAD AND RATCHET STOCK

PIPE DIE

TAP WRENCHES

Fig. 38. Taps and dies.

and anvil horn as part of the back jaw. The anvil surface is broken by a small hole into which the hardie fits. The hardie is the small tool shown with the utility vise.

A *pipe vise* (Fig. 40) is used to hold the pipe during the threading operation. The die shown in Fig. 40 is adjustable. It has a *guide clamp* which fits over the pipe and is tight-

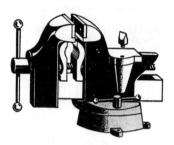

Fig. 39. Utility vise.

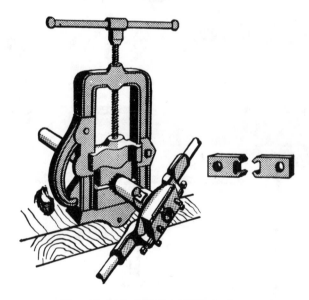

Fig. 40. Pipe vise, die, and die stock.

ened with a thumbscrew. This draws the die on the pipe as the die stock is revolved. The clamp also helps to get the threads straight. *Do not* forget to back the die up frequently to clear the chips. When it is necessary to cut internal threads, use a *pipe tap,* which cuts standard pipe threads with ¾-inch taper per foot. Pipe taps are fluted and are like common screw taps, except for the taper. They are used with tap wrenches.

Do not use vise jaws as a heavy anvil. There is danger of breaking jaws or battering inserts. Use an anvil for anvil jobs.

Cleaning. Wash grease and dirt from vise with dry cleaning solvent. Wipe dry with a cloth. Clean jaws with a wire brush. Pick chips from between serrations with a file scorer or a flattened iron wire.

Lubrication. Lubricate slide and worm lightly with preservative lubricating oil (special) or engine oil (SAE 10).

Drill Press. The *drill press* is arranged to hold and rotate the drill bit at the proper angle with the work. Drill press sizes range from the small bench type shown in Fig. 41 to huge multiple-spindle jobs weighing many tons. The smaller

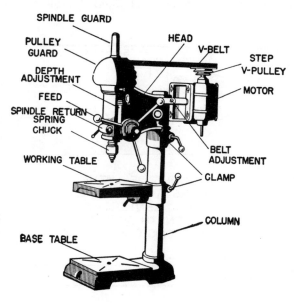

Fig. 41. Type of bench drill press.

drill presses usually have a separate motor which drives the drill spindle and chuck by means of a V-belt.

One advantage of the drill press over the portable electric drill is that you have *speed control*. Four grooves are usually found on the *cone step pulley* of the motor and four on the spindle pulley. The drill press shown in Fig. 41 has four speeds, and the belt is shown as adjusted for the highest possible speed. *Note* that it is on the largest-diameter groove of the *motor pulley*. This high speed—3,000 to 3,600 rpm— is suitable for small drills, but it is about 10 times too fast for a ½-inch drill. When you are in doubt, use the *slowest* speed. That means changing the belt to the *smallest-diameter* groove of the motor pulley.

The feed pressure is easily controlled on the drill press by means of a feed wheel with long handles. A *depth stop* is provided to stop the progress of a drill at a predetermined depth. That is important when you are drilling blind holes (those that do not go all the way through).

Engine Lathe. *Bench* or *engine lathes* are used for all forms of turning, external and internal grinding, facing,

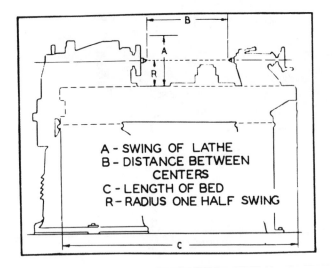

Fig. 42. Size and capacity of a lathe.

cutting-off, boring, and reaming operations, cutting external and internal threads, taper cutting, drilling, polishing, knurling, and similar operations They are available in several types and a wide variety of sizes, but they all embody the same basic principles of design and operation (Fig. 42).

Principal parts of the lathe. Figure 43 show the various parts of an engine lathe. Its principal function in the shop is the removal of material from revolving work by the use of suitably formed cutting tools made of hard and tempered steel or of an alloy metal such as carboloy, stellite, and the like.

Gas Oven. The *gas oven*, sometimes called a gas furnace, is the best device for heating soldering coppers (Fig. 44). These ovens have one, two, or three burners, on which the heat may be manually controlled by valves. Heated coppers are ready to use when they cause the gas flame to burn with a green color. Overheating must be avoided, as it softens the copper and burns away the film of "tin" on the surfaces of the point.

Extreme care should be taken in lighting the gas when a gas oven is in use. You will play safe if you twist up a piece of paper, light it, place it on or near the burner, *stand clear,* and then turn on the valve. Make sure that a valve does not leak when you turn it off.

Gas Blowpipe and Automatic Alcohol Torch. You can do *some* soldering jobs more easily and efficiently with a *direct flame* than with a soldering copper. The flame of a gasoline blowtorch can be used on certain large jobs, but on small jobs, you will have better luck with a *gas blowpipe* or an *alcohol torch.* These tools are *designed* for direct flame soldering.

The *gas blowpipe* looks like a welding torch's little brother, and it works on the same principle (Fig. 45). The blowpipe part of it is a tube within a tube. One tube furnishes the gas, either natural or manufactured, for the flame; the other tube supplies compressed air or oxygen.

The *automatic alcohol torch* (Fig. 45) works on the same principle as the blowpipe except that it burns alcohol and

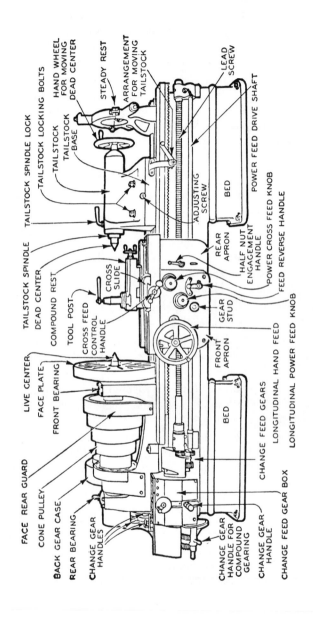

Fig. 43. Parts of an engine lathe.

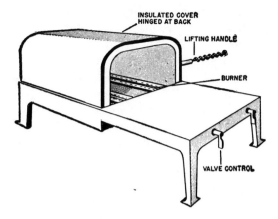

Fig. 44. Gas oven.

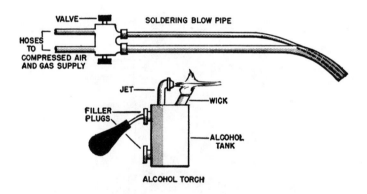

Fig. 45. Gas blowpipe and alcohol torch.

is a self-contained unit. The alcohol torch does its own "blowing." The burning wick heats the jet tube, causing the alcohol to vaporize. When a liquid is vaporized, it expands. This expansion continuously forces some of the alcohol vapor from the jet opening. There the vapor is ignited to form a hot, light blue flame.

Blowtorch. Where the metals are not flat surfaces in position for the use of a soldering copper, soldering is accomplished by playing the flames of a gasoline, kerosene, or alcohol blowtorch directly on the surfaces and then applying the solder cold in bar or wire form in small cross sections. The heated surfaces melt the solder.

The *gasoline blowtorch* (Fig. 46) is the one most commonly used in soldering. Its operation is simple. Fill the tank about two-thirds full of clean, unleaded gasoline. Operate the pump until sufficient pressure is built up in the tank to cause the gasoline to flow when the valve is opened.

With the valve open, liquid gasoline will flow from the jet of the torch and drip into the priming pan. When the pan is partly full, close the valve and ignite the gasoline with a match. The flame from this burning gasoline heats the perforated nozzle (or heating tube). When the nozzle is hot, open the valve slightly again, allowing the gasoline vapor which has been formed to flow from the nozzle. It

Fig. 46. Gasoline blowtorch.

burns with an almost colorless flame. By working the valve, you can adjust this flame to any desired intensity.

There is very little maintenance to the gasoline torch providing you use *only clean, clear, unleaded gasoline.* If you use leaded gasoline, a compound will form that will stop up the gasoline passages. The torch will be a source of trouble from then on, as it is almost impossible to clean these passages thoroughly.

Do not close the valve with too much force. Remember, the metal is hot and will contract when it cools, thus causing the valve to tighten up when cold. If you use too much force, it will be difficult to reopen the valve.

SPECIAL TOOLS AND EQUIPMENT

The tools and equipment peculiar to the repair of motors and generators are as follows.

Spring balance
Plug gauges
Sleeve-bearing arbors
Extractor and inserter set
Arbor press
Arbor press plates
Wire hook
Acetylene torch set
Coil winder drive and head
Commutator dresser

Insulation former
Wire gauge
Commutator stone dressing
  tool
Mica undercutting tool
Jackscrew and plate type
  bearing pullers
Hook type bearing puller
Bar type bearing puller
Air-gap feeler gauge

TEST EQUIPMENT

Electric meters and other test equipment necessary for the repair of electric motors and generators are as follows.

Circuit Tester.

Multimeter. The *multimeter* is a versatile piece of equipment that can be used for many purposes, such as checking voltages, both ac and dc, amperes, resistances, and continuity. It will not check ac amperes.

Watermeter, *a.c.* and *d.c.*

Clamp-on Ammeter. The split core multimeter, commonly called a *clamp-on ammeter,* is a portable piece of equipment that can be used to check ac voltages or ac amperes in ac circuits only.

Coil and Condenser Tester.

Power Factor Meter.

Frequency Meter. The *frequency meter* is used to check the 60 and 400 hertz (Hz) frequencies.

Electrical Analyzer. The *electrical analyzer* is designed for ac circuits and should not be used on dc. It consists of a voltmeter, ammeter, wattmeter, and power factor meter, together with two current transformers and the necessary switches suitably connected to facilitate the industrial testing of three-phase three-wire loads. The analyzer can also be used for measurements on single-phase and other polyphase circuits.

Megohmmeter. The *megohmmeter,* commonly called the megger, is used to check the quality of insulation around a wire. It can also be used to locate a grounded coil. It may be used to check for shorts between coils only if a high-scale model is used.

Tachometer. The *magnetic tachometer* is primarily a small electric generator coupled to a meter. The tachometer includes the generator, rpm indicator, 5-foot connecting cord, surface speed wheel attachment, and pointed and mushroom tips. To use the tachometer as a hand type, plug the generator into receptacle on indication, and twist to lock in position.

External Growler. The *external growler* is a versatile piece of equipment that can be used to test 2½- to 6-inch diameter armatures for opens, shorts, grounds, incorrect number of windings on the armature, and reversed coils.

Internal Growler. The *internal growler* is used to test 6- to 12-inch diameter ac stators for shorts between coil groups.

Oscilloscope. The *oscilloscope* is a solid-state portable instrument that combines small size and light weight with the ability to make precision wave form measurements. It is mechanically constructed to withstand the shock, vibration, and other extremes of environment associated with portability. A dc to 4 megahertz vertical system provides calibrated deflection factors from 0.01 to 20 volts/division (0.001 volt/division minimum with reduced frequency response). The trigger circuits provide stable triggering over the full vertical band width. The horizontal deflection system provides calibrated sweep rates from 1 second to 5 microseconds/division. An X10 horizontal magnifier allows each sweep rate to be increased 10 times to provide a maximum sweep rate of 0.5 microsecond/division in the 5 s position. X-Y measurements can be made by applying the vertical (Y) signal to the VERT INPUT connector and the horizontal (X) signal to the EXT TRIG or HORIZ INPUT connector (time/division switch set EXT HORIZ).

Pen Recorder or Oscillograph. *Pen recorders* are basically graphic dc millimeters designed for laboratory and field use where analog signals in the dc to 5-Hz range are to be recorded. An unusual feature is the straight line motion, especially important when a low-frequency, undistorted trace is required. The D'Arsonval galvanometers, coupled with unique pen systems, produce true sinusoidal wave forms for a sinusoidal input. In many applications, this will eliminate time-consuming data reduction and interpretation.

Ac and Dc Digital Voltmeter. The *digital voltmeter* is a compact, economical general purpose instrument with a high degree of accuracy and reliability. It features a servo-driven three-digit counter with overranging and combines in one instrument many virtues of both digital and analog voltmeters.

Transistor Tester. The *transistor tester* is a solid-state portable instrument used to check transistors and diodes present in solid-state control devices of motors and generators.

Ac Power Source (oscillator). The *power oscillator* is intended for use where requirements exist for ac signals over the frequency range of 0.1 to 100,000 hertz with a high degree of amplitude stability and purity of wave form. It is used as a source in ac voltage and current calibration work, galvanometer calibration, precise distortion measurements, vibration calibration and testing, filter and transformer testing, and other applications in the subsonic through ultrasonic regions. The power amplifier section has extremely stable gain, very low distortion and a high degree of regulation.

Electro-Hydraulic Actuator Test Stand. The *test stand* was designed basically as a governing system on generator sets using the LEH control unit. The test stand may be used to test and align other actuators provided the correct mechanical and electrical biases are available to the user. The stand has its own hydraulic system with three gauges and connections to provide hydraulic fluid to the actuator. It is an electrical system that requires 120 volts to provide the correct power to the actuator being tested.

CHAPTER 3

Wiring Materials and Procedures

When installing or replacing interior or exterior wiring, conform to local regulations and to the National Electrical Code.

CONDUCTOR SPLICES

Splicing is employed for connecting ends or pieces of two or more conductors or wires. Conductors should be securely spliced before they are soldered and taped. The three types of splices generally used are the Western Union, the tap or branch, and the rat-tail or pigtail splice.

Western Union Splice. The Western Union splice (Fig. 1)

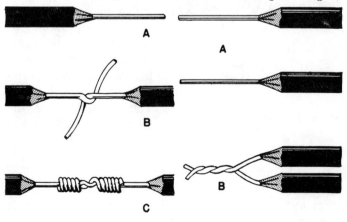

Fig. 1. Method of making W.U. splice.

Fig. 3. Procedure for making rat-tail or pigtail splice.

is used to lengthen or extend wires and also to join the ends of broken or cut wires. To make this splice, proceed as follows: Remove about 3″ of insulation from both ends of the wires (Fig 1, A). Scrape the ends until they are clean and bright, and twist the loose strands of wire tight. Cross the two bare ends of wire about ⅝″ from their insulation and twist the ends (Fig. 1, B). With pliers, hold the wires firmly together at their intersection, and wrap either end one or two turns around the other wire. Repeat this operation with the other end. Keep the coils of turns close together (Fig. 1, C), and continue to wrap the wires until you have made five or six turns close together on each side of the center. Leave about ¼″ space between the turns of the wire and the insulation, and cut off any remaining wire with pliers. Tighten and shape the joint with pliers; then solder and tape.

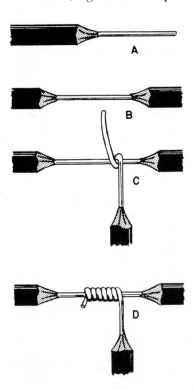

Fig. 2. Procedure for making tap or branch splice.

Tap or Branch Splice. In joining or splicing a main wire, a tap or branch splice is made as follows: Remove 1½″ of insulation in the main wire (Fig. 2, A). Scrape the exposed wires until they are bright and clean. Place the wires in position (Fig. 2, B). Hold the wires firmly in place with pliers and wrap the branch wires firmly around the main wire (Fig. 2, C). Make six turns of the branch wire on the main wire, leaving ¼″ space between the last turn of the tap and the insulation of the main wire. Cut off any excess length of branch wire. Tighten and shape the splice with pliers. Complete the joint by soldering and taping.

Rat-tail or Pigtail Splice. The rat-tail or pigtail splice is generally used for connecting lighting fixture leads and joining conductors in switch and outlet boxes. It should never be used where it may be subjected to any pull or strain.

A rat-tail or pigtail splice is made as follows: Remove 2″ of insulation from each wire end (Fig. 3, A). Scrape the wires until they are clean and bright; then place them parallel to each other. Make at least five tight twists (Fig. 3, B). Tape rat-tail and pigtail splices.

SOLDERING AND TAPING SPLICE

When soldering splices, use a wire solder with a rosin core. A common or electric soldering iron or an alcohol blow torch is used for soldering. To solder splices with a *blow torch,* proceed as follows: Apply solder only to portions of the splice or joint that come in contact with the wires. Use just enough solder to penetrate in all spaces between the loops and to cover the spliced wires completely. Before taping, allow the soldered splice to cool and set.

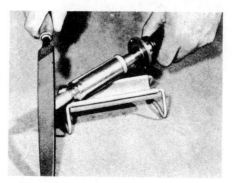

Fig. 4. Dressing the face of a soldering iron with a file.

Before using an electric soldering iron, tin or coat the copper tip with solder as follows: File the face of the tip clean and bright with a flat smoothing file while the iron is cold (Fig. 4), and turn on the current. Rub the rosin-core solder over the tip faces every 10 or 15 sec. while the soldering iron is heating. Continue until the solder melts and spreads smoothly and evenly over the faces of the tip (Fig. 5). Wipe the tip with a dry rag while the iron is still hot. This will produce a mirror-like even layer of molten solder on the tip of the iron.

If soldering irons are properly tinned and are not overheated, they will remain in good condition. If the iron faces become badly pitted because of overheating, file and retin the iron.

Fig. 5. Tinning the iron.

Fig. 6. Soldering a wire splice.

When soldering splices with a soldering iron, place the broadest tinned face of the iron flat against the surfaces to be soldered, and apply solder under the edge of the iron closest to the work. Apply enough solder to form a

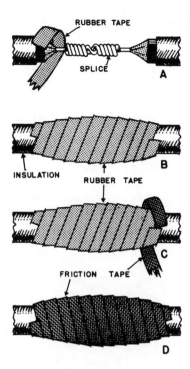

Fig. 7. Taping splices.

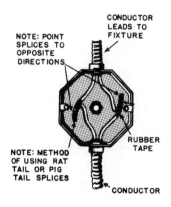

Fig. 8. Procedure for laying away splices.

heavy film between the working face of the iron and the surfaces being soldered. Move the iron slowly along the work, making sure that the applied solder melts, spreads, and penetrates properly. The correct method of soldering a wire splice is shown in Fig. 6. Allow the solder to cool and solidify before taping.

Taping. To tape a splice, proceed as follows: Apply a layer of rubber tape around the joint, overlapping the preceding turn by half the width of the tape (Fig. 7, A). Then wrap two layers of friction tape over the entire joint in a half-lap fashion (Fig. 7, B, C, and D). Reversing the direction, seal the end with one edge of the tape, and proceed toward the other end of the joint. The joint should be firm, not too bulky, and its ends should be properly sealed.

Laying Away Splices. Fixtures, receptacles, outlets, switches, and pull boxes that have limited space make the placing or laying away of splices difficult. Rat-tail or pigtail splices are generally used in such places. They are compact and are not subject to longitudinal strains under these conditions. They are laid away as shown in Fig. 8.

INTERIOR WIRING METHODS

Various methods can be used in the installation of interior wiring. Before beginning to install any interior wiring system, construct a wiring plan showing the location of switch boxes and convenience boxes and the appliance outlets and ceiling outlets. Then consult local authorities and electric companies to make sure that what you plan to do will be permitted. The following interior wiring methods are commonly used:

Open wiring.
Rigid metal conduit.
Flexible metal conduit.
BX armored cable.
Nonmetallic-sheathed cable.
Service entrance cable.

OPEN WIRING

Open wiring on insulators should be used only where installation costs must be kept to a minimum. It is usually installed under roofs and in basements and garages, outhouses, barns, poultry houses, and similar farm buildings.

Conductors (*wires*) must conform to local specifications covering adequate size, mechanical strength, insulation, and carrying capacity. (See Table 1A).

Installation. Support all open or exposed wiring on suitable knob or cleat porcelain insulators (Fig. 9). When wiring over flat surfaces, use rigid supports at least every 4½′ (Fig. 10, B). Conductors should be supported by a knob or cleat within 6″ of a tap. In barns or in similar buildings, where wire not smaller than No. 8 is used and where wiring will not be disturbed, the wires may be separated 6″ from each other. They can be run directly from timber to timber and supported from each timber only. Separators providing not less than 2½″ separation between parallel conductors should be installed at intervals not exceeding 4½′.

TABLE 1A

Conductor Sizes Based On Voltage Drop
(230-volt—1 Per Cent Drop)

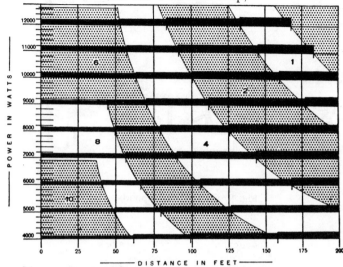

Use screws to mount knobs, and use nails or screws to mount cleats (Fig. 10, A). Nails or screws should penetrate framing or timbers to a depth one-half or more of the height of knobs, and through the full thickness of cleats. Insert composition washers between cleats and framing when using nails.

In dry locations and where voltages do not exceed 300 volts, parallel conductors should be separated 2½″ from each other and ½″ from the surface to which they are attached. They should be separated 4″ from each other and 1″ from the surface where voltages range from 301 to 600 volts. For damp or wet locations, and for all voltages, separate wires at least 1½″ from the surface and at least 4″ from each other.

Figure 9 shows methods used for securing or tying conductors to insulator knobs.

Do not dead-end or terminate open conductors at a lampholder or receptacle unless the last support is within 12″ of the lampholder or receptacle. Cleats are used to form a dead-end (Fig 10, A).

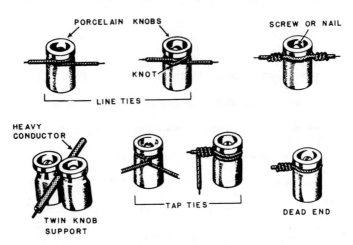

Fig. 9. Open wiring supports.

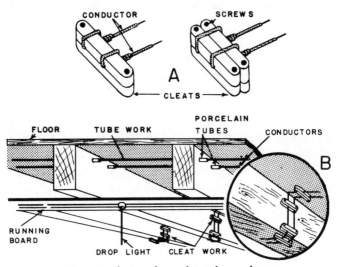

Fig. 10. Cleats and porcelain tubes or sleeves.

Noncombustible, nonabsorptive insulating tubes or bushings must be used to separate open wiring from direct contact with walls, floors, timbers, or partitions through which

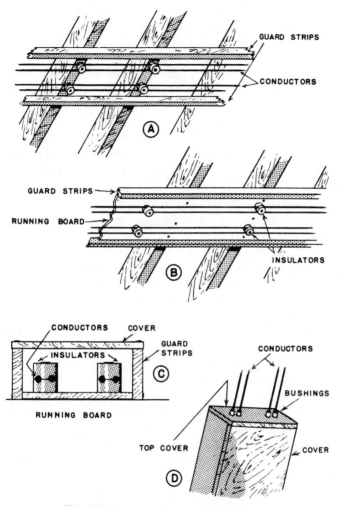

Fig. 11. Protection from mechanical injury.

the wires pass. If the bushings are shorter than the openings, noninductive waterproof sleeves should be inserted in each opening, and bushings should be slipped into the sleeves at both ends. Each conductor should be carried through a separate porcelain tube or sleeve (Fig 10, B).

Make certain that open conductors are at least 2″ from metallic conduit, piping, or other conducting material, and from any exposed lighting, power or signal conductors that are not separated by a continuous fixed nonconductor. If insulation tubes are used, their ends should be secured with friction tape. When open wiring is located close to water pipes or tanks, or in other damp locations, an air space should be maintained between the wiring and the pipes, tanks, or locations crossed. Always install open wiring over any moisture-condensing pipes.

Where open conductors cross ceiling joists and wall studs or are within 7' from the floor, protect them from mechanical injury by applying any of the following methods. (See Fig. 11.)

1. By placing guard strips (not less than ⅞" thick and at least as high as the insulating supports) on each side of and close to the wiring.

2. By placing a running board at least ½" thick behind the conductors with side protections. Extend the running boards at least 1" but not more than 2" outside the conductors. Protecting sides should be at least 2" high and at least ⅞" thick.

3. By making a boxing, as outlined above, and adding a cover that is separated at least 1" from the conductors within the enclosure. For vertical conductors on side walls, close the boxing and bush the holes through which the conductors pass.

4. By encasing the conductors in continuous lengths of flexible tubing, such as conduit or metal piping.

Unfinished Attics and Roof Spaces. Procedures for open wiring in unfinished attics or roof spaces are as follows: Run conductors through or on the sides of joists, studs, and rafters, except in attics and roof spaces where head room at all points is less than 3'. In accessible unfinished attics and roof spaces reached by a stairway or a permanent ladder, run conductors through bored holes in the floor or the floor joists. Protect them by fastening substantial running boards securely in place, extending them at least 1" on each side of the conductors. Where the conductors are carried along the sides of rafters, studs, or floor joists, running boards or guard strips are not required.

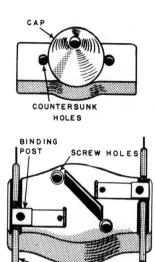

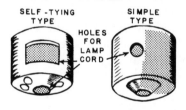

Fig. 12. Rosettes and ceiling buttons.

Rosettes, Switches, and Sockets. Rosettes or ceiling buttons are used to connect drop cords in open wiring to the branch circuits. Figure 12 shows a modern rosette with concealed protected contact lugs. Where there is considerable vibration, ceiling buttons are sometimes used instead of rosettes. Switches on open conductors should be mounted and supported as shown in the illustration.

RIGID METAL CONDUIT

There are two kinds of rigid metal conduit; the white and the black (Fig. 13). The white type is galvanized and used outdoors; the black one has a coating of enamel and is used for interior wiring. Both types have black insulating enamel baked on the inside of the conduit. The conduit is available in 10′ lengths threaded at both ends with one threaded pipe coupling screwed onto one end. Inside

Fig. 13. Black and white rigid steel conduit.

Fig. 14.   Nipples used for rigid conduit.

diameters range in sizes from ½″ to 6″.

Installation. Rigid metal conduit is moistureproof and fireproof and can be embedded in cement, brick, or plaster in partitions and walls. It can be used for both exposed and concealed indoor and outdoor wiring. Made of malleable metal, it can be bent if necessary, permitting easy insertion, removal, and replacement of conductors.

Conductors. According to the Standard Code, no conduit smaller than ½″ inside diameter should be used. (See Tables 1B, 2, 3, 4, and 5 for the minimum sizes and the number of conductors allowed.)

Conduit Nipples. Conduit nipples are short lengths of conduit threaded at both ends (Fig. 14) and are used to join fittings or to fill in short runs between conduit and fittings.

Conduit Fittings. Threaded conduit fittings are attached to conduit bends or elbows. They are available in a great variety of sizes and forms. One type is shown in Fig. 15. Conduit, boxes, or outlet fittings (Fig. 16) are installed without conductors or wires.

Bushings and Locknuts. Bushings and locknuts (Fig. 17) are screwed over the

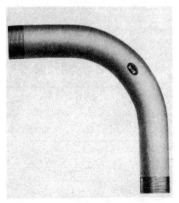

Fig. 15.   Elbow used for rigid conduit.

TABLE 1B

CONDUCTORS

Trade Name	Standard Type Letter	Maximum Operating Temperature	Special Provisions
Rubber-covered fixture wire.	RF-1	60 C. 140 F.	Fixture wiring. Limited to 300 V.
Solid or 7-strand.	RF-2	60 C. 140 F.	Fixture wiring, and as permitted in the code.
Rubber-covered fixture wire.	FF-1	60 C. 140 F.	Fixture wiring. Limited to 300 V.
Flexible stranding.	FF-2	60 C. 140 F.	Fixture wiring, and as permitted in the code.
Heat-resistant, rubber-covered.	RFH-1	75 C. 167 F.	Fixture wiring. Limited to 300 V.
Fixture wire, solid or 7-strand.	RFH-2	75 C. 167 F.	Fixture wiring, and as permitted in the code.
Heat-resistant, rubber-covered.	FFH-1	75 C. 167 F.	Fixture wiring. Limited to 300 V.
Fixture wire. Flexible stranding.	FFH-2	75 C. 167 F.	Fixture wiring, and as permitted in the code.
Thermoplastic-covered fixture. Wire—solid or stranded.	TF	60 C. 140 F.	Fixture wiring, and as permitted in the code.
Thermoplastic-covered fixture. Wire—flexible stranding.	TFF	60 C.	Fixture wiring.
Cotton-covered, heat-resistant fixture wire.	CF	90 C. 194 F.	Fixture wiring. Limited to 300 V.
Asbestos-covered, heat-resistant fixture wire.	AF	150 C. 302 F.	Fixture wiring. Limited to 300 V.
Code rubber.	R	60 C. 140 F.	General use.
Heat-resistant, rubber.	RH	75 C. 167 F.	General use.
Moisture-resistant rubber.	RW	60 C. 140 F.	General use and wet locations.
Moisture and heat-resistant rubber.	RH, RW	60 C. 140 F. 75 C. 167 F.	General use and wet locations. General use.
Moisture, heat-resistant rubber.	RHW	75 C. 167 F.	General use and wet locations.

TABLE 1B—*Continued*

CONDUCTORS

Trade Name	Standard Type Letter	Maximum Operating Temperature	Special Provisions
Latex rubber.	RU	60 C. 140 F.	General use.
Heat-resistant, latex rubber.	RUH	75 C.	General use.
Moisture-resistant latex rubber.	RUW	60 C. 140 F.	General use and wet locations.
Thermoplastic.	T	60 C. 140 F.	General use.
Moisture-resistant thermoplastic.	TW	60 C. 140 F.	General use and wet locations.
Mineral insulation (metal-sheathed).	MI	85 C. 185 F.	General use and wet locations with Type O termination fittings. Maximum operating temperature for special applications 250 C.
Thermoplastic and asbestos.	TA	90 C. 194 F.	Switchboard wiring only.
Varnished cambric.	V	85 C. 185 F.	Dry locations only. Smaller than No. 6 by special permission.
Asbestos and varnished cambric.	AVA	110 C. 230 F.	Dry locations only.
Asbestos and varnished cambric.	AVL	110 C. 230 F.	Wet locations.
Asbestos and varnished cambric.	AVB	90 C. 194 F.	Dry locations only.
Asbestos.	A	200 C. 392 F.	Dry locations only. Not for general use. In raceways only for leads to or within apparatus. Limited to 300 V.
Asbestos.	AA	200 C. 392 F.	Dry locations only. Open wiring. Not for general use. In raceways, only for leads to or within apparatus. Limited to 300 V.
Asbestos.	AI	125 C. 257 F.	Dry locations only. Not for general use. In raceways only for leads to or within apparatus. Limited to 300 V.

TABLE 1B—*Continued*

CONDUCTORS

Trade Name	Standard Type Letter	Maximum Operating Temperature	Special Provisions
Asbestos.	AIA	125 C. 257 F.	Dry locations only. Open wiring. Not for general use. In raceways, only for leads to or within apparatus.
Paper.		85 C. 185 F.	For underground service conductors, or by special permission.
Slow-burning.	SB	90 C. 194 F.	Dry locations only. Open wiring; and in raceways where temperatures will exceed those permitted for rubber-covered or varnished cambric-covered conductors.
Slow-burning, weatherproof.	SBW	90 C. 194 F.	Dry locations only. Open wiring only.
Weatherproof.	WP	80 C. 176 F.	Open wiring by special permission where other insulations are not suitable for existing conditions.

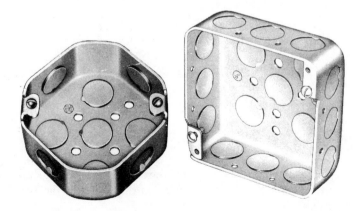

Fig. 16. Two types of outlet boxes used for rigid conduit, EMT, and flexible conduit.

TABLE 2

NUMBER OF CONDUCTORS IN CONDUIT OR TUBING

Rubber-covered, Types RF-32, R, RH, and RU Thermoplastic, Types TF, T, and TW. For more than nine conductors, see Table 5.

Size AWG MCM	Number of Conductors in One Conduit or Tubing								
	1	2	3	4	5	6	7	8	9
18	½	½	½	½	½	½	½	¾	¾
16	½	½	½	½	½	½	¾	¾	¾
14	½	½	½	½	¾	¾	1	1	1
12	½	½	½	¾	¾	1	1	1	1¼
10	½	¾	¾	¾	1	1	1	1¼	1¼
8	½	¾	¾	1	1¼	1¼	1¼	1½	1½
6	½	1	1	1¼	1½	1½	2	2	2
4	½	1¼	°1¼	1½	1½	2	2	2	2½
3	¾	1¼	1¼	1½	2	2	2	2½	2½
2	¾	1¼	1¼	2	2	2	2½	2½	2½
1	¾	1½	1½	2	2½	2½	2½	3	3
0	1	1½	2	2	2½	2½	3	3	3
00	1	2	2	2½	2½	3	3	3	3½
000	1	2	2	2½	3	3	3	3½	3½
0000	1¼	2	2½	3	3	3	3½	3½	4
250	1¼	2½	2½	3	3	3½	4	4	5
300	1¼	2½	2½	3	3½	4	4	5	5
350	1¼	3	3	3½	3½	4	5	5	5
400	1½	3	3	3½	4	4	5	5	5
500	1½	3	3	3½	4	5	5	5	6
600	2	3½	3½	4	5	5	6	6	6
700	2	3½	3½	5	5	5	6	6	......
750	2	3½	3½	5	5	6	6	6	......
800	2	3½	4	5	5	6	6	......	......
900	2	4	4	5	6	6	6	......	......
1000	2	4	4	5	6	6	......	......	......
1250	2½	5	5	6	6	......	......	......	......
1500	3	5	5	6	......	......	......	......	......
1750	3	5	6	6	......	......	......	......	......
2000	3	6	6	......	......	......	......	......	......

* Where a service run of conduit or electrical metallic tubing does not exceed 50 feet in length and does not contain more than the equivalent of two quarter-bends from end to end, two No. 4 insulated and one No. 4 bare conductors may be installed in one-inch conduit or tubing.

conduit end entering outlet and box. Bushings are necessary to protect the conductor from abrasion unless the box or fitting gives this protection.

Instructions for Cutting, Threading, and Bending Conduit. Conduit can be cut or bent to any required angle or length. The hacksaw is used for cutting conduit. The dies or tools are used for threading conduit.

Hickeys (Fig. 18) are used to bend conduit. Place the end of the conduit in which the bend *is not* to be made firmly against a wall. Make a chalk mark at the point where the bend is to be made. Slip the head of the hickey over the conduit within a few inches of the chalk mark. Place the left foot on the part of the conduit that is not to be bent (3″ or 4″ from this mark). Grasp the handle of

TABLE 3
CIRCUIT

NUMBER OF CONDUCTORS IN CIRCUT OR TUBING

LEAD-COVERED TYPES RL AND RHL—600 V.

Size AWG MCM	Number of Conductors in One Circuit or Tubing											
	Single-Conductor Cable				Two-Conductor Cable				Three-Conductor Cable			
	1	2	3	4	1	2	3	4	1	2	3	4
14	½	¾	¾	1	¾	1	1	1¼	¾	1¼	1½	1½
12	½	¾	¾	1	¾	1	1¼	1¼	1	1¼	1½	2
10	½	¾	1	1	¾	1¼	1¼	1½	1	1½	2	2
8	½	1	1¼	1½	1	1¼	1½	2	1	2	2	2½
6	¾	1¼	1½	1½	1¼	1½	2	2½	1¼	2½	3	3
4	¾	1¼	1½	1½	1¼	2	2½	2½	1½	3	3	3½
3	¾	1¼	1½	2	1¼	2	2½	3	1½	3	3	3½
2	1	1¼	1½	2	1¼	2	2½	3	1½	3	3½	4
1	1	1½	2	2	1½	2½	3	3½	2	3½	4	5
0	1	2	2	2½	2	2½	3	3½	2	4	5	5
00	1	2	2	2½	2	3	3½	4	2½	4	5	5
000	1¼	2	2½	2½	2	3	3½	4	2½	5	5	6
0000	1¼	2½	2½	3	2½	3	3½	5	3	5	6	6
250	1¼	2½	3	3	......	......	......	......	3	6	6	......
300	1½	3	3	3½	......	......	......	......	3½	6	6	......
350	1½	3	3	3½	......	......	......	......	3½	6	6	......
400	1½	3	3	3½	......	......	......	......	3½	6	6	......
500	1½	3	3½	4	......	......	......	......	4	6	......	......
600	2	3½	4	5	......	......	......	......	......	......	......	......
700	2	4	4	5	......	......	......	......	......	......	......	......
750	2	4	4	5	......	......	......	......	......	......	......	......
800	2	4	5	5	......	......	......	......	......	......	......	......
900	2½	4	5	5	......	......	......	......	......	......	......	......
1000	2½	4½	5	6	......	......	......	......	......	......	......	......
1250	3	5	5	6	......	......	......	......	......	......	......	......
1500	3	5	6	6	......	......	......	......	......	......	......	......
1750	3	6	6	......	......	......	......	......	......	......	......	......
2000	3½	6	6	......	......	......	......	......	......	......	......	......

The above sizes apply to straight runs or with nominal offsets equivalent to not more than two quarter-bends.

Fig. 17.   Rigid conduit locknut and bushing.

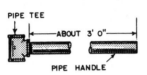

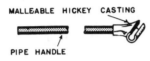

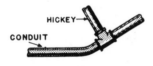

Fig. 18. Hickeys and method of using.

the hickey in both hands and draw it toward the body. Do not attempt to make a 90° bend in one operation, but gradually move the hickey forward or backward as required. Slip a piece of pipe larger than the conduit over the straight portion and stand on it. Remove the hickey, slip the handle behind the bent portion of the conduit, and pull toward the body.

A single unit run of conduit between outlet and outlet, between fitting and fitting, or between outlet and fitting should not have more than four quarter-bends, including all bends located at the outlet or fitting.

Methods of Securing and Supporting Conduit. Pipe straps are used to secure or support conduit (Fig. 19). They are available in various sizes

Fig. 19.   Rigid conduit pipe strap.

to fit standard conduit. Wood screws are used to secure straps on wooden surfaces. Rawplugs or machine screws should be used on masonry.

TABLE 4

Number of Conductors in Conduit or Tubing

More than Nine Conductors, Rubber-Covered Types RF-32, R, RH, RW, RU, Thermoplastic, Types TF, T, and TW, When Specially Permitted by the Code*.

Size	Maximum Number of Conductors in Conduit or Tubing						
AWG	¾"	1"	1¼"	1½"	2"	2½"	3"
18	12	20	35	49	80	115	176
16	10	17	30	41	68	97	150
14	....	10	18	25	40	59	90
12	....	....	15	21	35	50	77
10	....	....	13	17	29	41	64
8	....	....	....	10	17	25	38
6	....	....	....	....	....	15	23

* More than nine conductors are permitted in a single conduit for conductors between a motor and its controller and similar applications.

TABLE 5

Combination of Conductors

(Per Cent Area of Conduit or Tubing)

	Number of Conductors				
	1	2	3	4	Over 4
Conductors (not lead-covered)	53	31	43	40	40
Lead-covered conductors	55	30	40	38	35
For rewiring existing raceways for increased load where it is impracticable to increase the size of the raceway because of structural conditions	60	40	50	50	50

TABLE 6

Radius of Conduit Bends

Size of Conduit (Inches)	Conductors Without Lead Sheath (Inches)	Conductors With Lead Sheath (Inches)
½	4	6
¾	5	8
1	6	11
1¼	8	14
1½	10	16
2	12	21
2½	15	25
3	18	31
3½	21	36
4	24	40
5	30	50
6	36	61

TABLE 7

SPACING OF CONDUIT SUPPORTS

Conduit Size, Inches	Number of Conduits in Run	Location	Maximum Spacing of Supports, Feet
		Horizontal Runs	
½, ¾	1 or 2.	Flat ceiling or wall.	5
½, ¾	1 or 2.	Where it is difficult to provide supports except at intervals fixed by the building construction.	7
½, ¾	3 or more.	Any location.	7
1 and larger.	1 or 2.	Flat ceiling or wall.	6
1 and larger.	1 or 2.	Where it is difficult to provide supports except at intervals fixed by the building construction.	10
1 and larger.	3 or more.	Any location.	10
		Vertical Runs	
½, ¾	....	Exposed.	7
1, 1¼	....	Exposed.	8
1½ and larger.	....	Exposed.	10
Up to 2.	....	Shaftway.	14
2½ and larger.	....	Shaftway.	28
Any.	....	Concealed.	10

The Electragist Standards published by the National Electrical Contractors' Association include the standards shown in Table 8, which can be used as a guide for spacing conduit supports.

Conductors or wires are pushed or fished through conduit with a fishing wire or tape.

FLEXIBLE METAL CONDUIT

Flexible metal conduit is made of a single strip of flexible steel spirally wound (Fig. 20). The spiral is interlocked, producing a round flexible cross section with a high degree of mechanical strength. It is available in lengths of 25' to 250', depending on the inside diameter of the conduit. This makes it suitable for use in a finished house where it is necessary to rewire the entire system or to add one or more circuits. Flexible metal conduit is easier and simpler to

Fig. 20.　Single-strip flexible conduit.

handle than rigid conduit and can be used for installing interior wiring in the same manner as rigid conduit.

Flexible steel conduit can be easily fished through holes in walls, ceilings, studs, and floors. Manufacturers of flexible steel conduit furnish fish plugs for small sizes of conduit up to and including ¾″ inside diameter. Cut the conduit to the required length with a hacksaw and remove all burrs with an ordinary file. Then screw the fish plug into the end to be pulled through openings that were made for the concealed wiring. Attach the fishing wire to the fish plug and pull conduit through, outlet to outlet.

Conduit should be properly fastened at all openings. In some industries, the type of pipe strap shown in Fig. 19 can be used; in other cases, elbow clamps, available in various sizes, are suitable. Use iron plates where flexible conduit passes through slots or openings in floor joists, beams, or studs to protect it from nails. Elbow fittings are unnecessary, for the elbows are formed with the conduit.

Exposed flexible conduit should be clamped with cable clips or single-hole straps to the wired surface. Use two-hole conduit straps for large sizes of flexible conduit. Space supports in the same manner as that specified for rigid conduit. When joining flexible conduit to rigid conduit, use coupling.

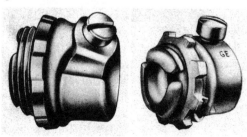

Fig. 21.

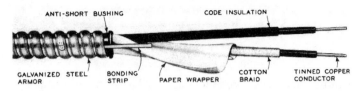

ANTI-SHORT BUSHING   CODE INSULATION

GALVANIZED STEEL   BONDING   COTTON   TINNED COPPER
ARMOR   STRIP   PAPER WRAPPER   BRAID   CONDUCTOR

Fig. 22.   BX armored cable.

Switch boxes or outlet boxes must be installed at all switches or outlets. Run the conduit continuously from outlet to outlet and fasten securely to boxes provided with adequate bushings. Figure 21 shows the connector used for connecting flexible steel conduit to an outlet box. The bolt of the connector should be clamped to the armor to insure a good electrical connection.

The same sizes and numbers of wires or conductors specified for rigid conduit are used for flexible conduit. Inside diameters are approximately the same as in the rigid type.

BX OR ARMORED CABLE

Modern armored cable, called BX cable (Fig. 22) is a complete assembly of flexible armor and insulated wires that can be installed in one operation. A flame-retardant and moisture-resistant fiber protects the conductors and is colored so that the wires may be easily identified. Insulating wires provide a water-repellent protection between the conductors and the armor. An interlocking, galvanized steel strip covers the entire wire assembly, providing protection, flexibility, and a continuous ground. The ACL leaded-type or the unleaded AC-type of BX cable are available for various sizes of wires used in systems of not more than 600 volts. Use the unleaded type only in dry locations and the leaded in both wet and dry locations. The armor will have permanent low electrical resistance if a bonding strip of tin, copper, or aluminum is used with the armor in sizes 14 and 12 AWG (Fig. 22). BX cable is subject to exacting tests prescribed by the Underwriters' Laboratories (UL)

Fig. 23.

before it is approved and shipped from the factory and is available for two- and three-phase systems.

Installation of BX Cable. BX cable can be run through walls and floors and down existing partitions without use of a fishing wire. When this cable is used, entire sections of walls and floors do not have to be torn up, for it is necessary to cut only the required openings. The general wiring methods with BX cable are the same in new as in old houses, with the exception that the necessary openings must be made in old houses.

Switch Boxes. All switch boxes and outlet boxes should be installed before the cable is run. BX cable boxes are made of specially designed metal and are grounded by means of the cable armor. They should not be secured on laths, but should be fastened to a joist, or a stud, or a backboard so that they are not pulled loose when fixtures

Fig. 24.   Extension ring outlet box used for conduit, EMT, and flexible conduit.

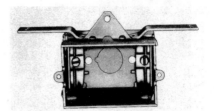

Fig. 25. Sectional switch box with side brackets.

Fig. 26.

are hung and adjusted. Figure 23 shows a large box used for either a switch or a receptacle when more than two cables are to be led into the outlet. Boxes used with BX cable are shown in Figs. 24 and 25. The box shown in Fig. 26 is mounted with brackets and is secured with nails. It is quickly installed at exactly the right depth in relation to the plaster line (Fig. 26). Figure 27 shows another method of supporting a box screwed to a board mounted between two studs. When a ceiling outlet box is between joists, a supporting device is required (Fig. 28). Move the box along the supporting bar hanger

Fig. 27.

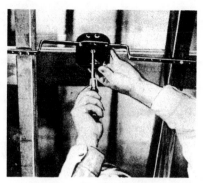

Fig. 28.

Fig. 29.   Fig. 30.

(that provides free access to all knockouts in the box) to the right spot and fasten it.

Method of Running Cable. After all the boxes have been installed, the next step is to run the cables. Select the most direct route from one box to another. Holes must be bored where the BX cable crosses joists or studs. Use a brace with an extension bit (Fig. 29), and bore a hole somewhat larger than the cable. When the holes have been bored, the end of the cable should be drawn out from the center of the coil in a counterclockwise direction (Fig. 30) and threaded through from one outlet to the next. If the cable is pulled in this manner, it will come out smoothly, easily, and without kinking. All cable runs must be continuous from outlet to outlet (Fig. 31).

The cable should be stapled at intervals of not more

than 4½′ (Fig. 32) and as near as possible to the center of the stud or joist. When it is necessary to support a loop in the cable, the intervals should be shorter. Staple the cables securely to studs not more than 12″ from each box. Various types of fasteners and staples are used; one type is shown

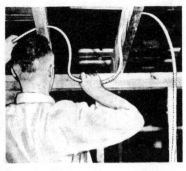

Fig. 31.

in Fig. 33. At each outlet, allow about 8″ of additional cable for the conductors inside the box before cutting. When cutting the cable, hold the saw as shown in Fig. 34 and cut part way through and across a convolution of the armor. Break the armor, and saw through the conductors.

Method of Connecting Cable. Each box has knockouts that provide for the cable entrance (Fig. 35). Knockouts are removed by a tap of the hammer and a twist of the pliers. BX cables must be firmly secured to the box with connectors (Fig. 36). Prepare the cable by cutting partially (about 8″ from the end) through a convolution of the armor (Fig. 37). Break the armor and slip it off, exposing the bonding wire (Fig. 38). Then snip off all but

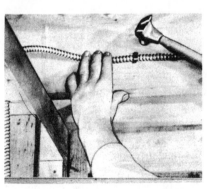

Fig. 32.

Fig. 33.

Fig. 34.

Fig. 35.

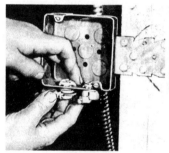

Fig. 36.

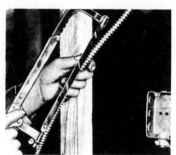

Fig. 37.

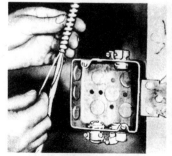

Fig. 38.

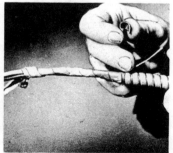

Fig. 39.

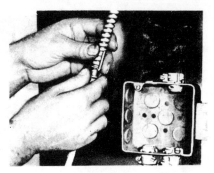

Fig. 40.

a couple of inches of the bonding wire and bend it back against the armor (Fig. 39). Now the cable is ready for the insertion of the insulating bushing (Fig. 40).

The conductors of the cable are fed into the box through the connector in the knockout, and the armor is securely clamped into the connector (Fig. 41). Some boxes are designed with built-in cable clamps that eliminate the need for connectors. However, connectors such as those shown in Fig. 41 are generally preferred for new installations. The switch box shown in Fig. 42 requires a cover to bring the opening flush with the plaster line and to support the switch.

Fig. 41.   Fig. 42.

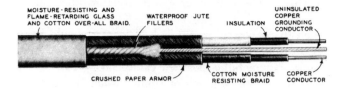

Fig. 43. Nonmetallic-sheathed cable (PVX) with ground wire.

NONMETALLIC-SHEATHED CABLE

Nonmetallic-sheathed cable consists of two or three rubber- or thermoplastic-insulated wires that are bound together and covered with a cotton-bound paper sheath. The entire assembly is further protected with an outer braid treated with a moisture- and heat-resisting compound (Figs. 43 and 44). It is used for exposed work and for some types of concealed work. It can be run or fished in existing holes of concrete blocks, tile walls, or other types of construction where the walls are not exposed nor subject to excessive moisture.

Installation of Nonmetallic-Sheathed Cable. For exposed work, nonmetallic-sheathed cable must be supported directly on walls or ceilings by means of approved staples, straps, or similar fittings. Typical staples and straps are shown in Fig. 45. The cable should be secured at intervals not exceeding 4½' and within 12" of every outlet or fitting. If the cable is exposed, it must follow the contour of the house finish or of the running boards. If it is necessary to

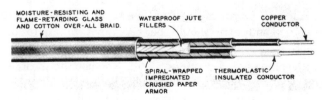

Fig. 44. Nonmetallic-sheathed cable (PVX) without ground wire.

protect the cable, use either guard strips or other suitable means.

When the cable passes through a floor, it should be enclosed in rigid conduit or pipe extending at least 6″ above the floor. Nonmetallic-sheathed cable can be run through holes bored in studs, joists, or similar wood timbers without any additional protection. Bore holes slightly larger than the cable, near the center of the timber.

If the cable is run at angles with joists in unfinished basements, secure conductors (two No. 6 or three No. 8) directly to lower edges of the joists. Smaller conductors should be run through holes bored in the joists. When cables are run parallel to joists, they should be secured to their sides or faces.

Fig. 45.

If the cable is run within 7′ across the top of floor joists, or across the face of rafters or studding in accessible attics, it should be protected by guard strips the height of the cable. If the attic is not permanently accessible, the cable should be protected only within 6′ of the edge of the attic entrance. Guard strips or running boards are not required if the cable is run along the sides of rafters, studs, or floor joists.

Splices or *joints* should not be used during installation of nonmetallic-sheathed cable, excepting in outlet boxes or junction boxes. *Bends* in nonmetallic-sheathed cable should be made with a radius less than five times the diameter of the cable, and care should be taken not to injure the protective covering of the cable.

Outlet Boxes and Fittings. With nonmetallic-sheathed cable, approved boxes or fittings must be used at all outlets or switch locations and the cable must be attached with a

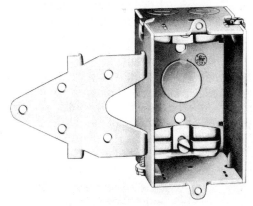

Fig. 46. Sectional switch box with standard bracket and built-in clamp used for nonmetallic cable.

built-in clamp or a squeeze connector (Figs. 46 and 47). In localities where outlet boxes must be grounded, use a special nonmetallic-sheathed cable with an uninsulated grounding wire. Skin 6″ or 8″ of the grounding wire before securing the cable to the outlet box. Methods of attaching the cable vary with the type of box; specific directions are furnished by the manufacturer. Several approved types of porcelain outlets are available for surface mounting. A

Fig. 47. Sectional switch box with mounting ears used for nonmetallic cable.

special clamp is provided inside these outlets that permits the cables to be brought in from both ends.

SERVICE ENTRANCE CABLE

In the home a service entrance cable (Fig. 48) is used for branch circuits and for supplying current to electric ranges, water heaters, or other heavy-duty appliances. On the farm this cable is used principally as a feeder to supply current to other buildings from the master service.

Installation of Service Entrance Cable. Types SE and ASE service entrance cable may be used for interior wiring. Type SE does not have a metallic armor and is installed in the same way as nonmetallic-sheathed cable. Type ASE has an interlocking metal armor, and its installation is similar to that of armored or BX cable. Service entrance

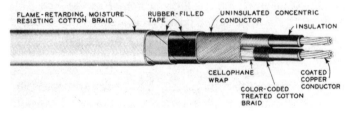

Fig. 48. Service entrance cable.

cable with rubber-covered or thermoplastic conductors may be used in either concealed or exposed wiring systems. The type having an uninsulated grounded conductor may be used for circuits supplying ranges, water heaters, or other appliances, provided that it has a nonmetallic outer covering and that the current does not exceed 150 volts to ground. Service entrance cable is supported by single-hole cable straps. These are available in various sizes. Special screws connect service entrance cable to switch boxes or outlets.

EXTERIOR WIRING

For exterior wiring, materials and procedures are the same as those used for interior wiring, unless climatic con-

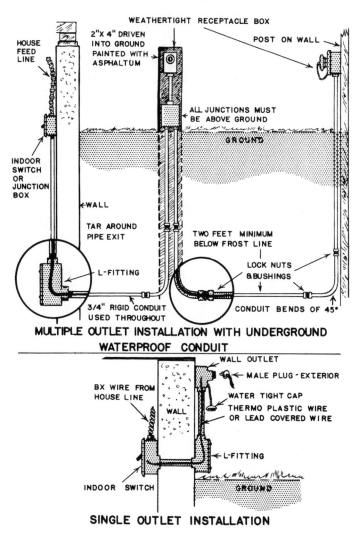

WEATHERTIGHT RECEPTACLE BOX

2"x 4" DRIVEN INTO GROUND PAINTED WITH ASPHALTUM

POST ON WALL

HOUSE FEED LINE

ALL JUNCTIONS MUST BE ABOVE GROUND

GROUND

INDOOR SWITCH OR JUNCTION BOX

WALL

TAR AROUND PIPE EXIT

TWO FEET MINIMUM BELOW FROST LINE

L-FITTING

LOCK NUTS & BUSHINGS

3/4" RIGID CONDUIT USED THROUGHOUT

CONDUIT BENDS OF 45°

MULTIPLE OUTLET INSTALLATION WITH UNDERGROUND WATERPROOF CONDUIT

WALL OUTLET

BX WIRE FROM HOUSE LINE

MALE PLUG - EXTERIOR

WATER TIGHT CAP

THERMO PLASTIC WIRE OR LEAD COVERED WIRE

WALL

L-FITTING

INDOOR SWITCH

GROUND

SINGLE OUTLET INSTALLATION

Fig. 49.

ditions and changes in either weather or humidity have to be considered. Consult local authorities.

The safest and most practical method of installing exterior wiring is to use rigid conduit. All wire or cable used in underground runs or in conduit must be in unbroken lengths. All splices should be made in specially designed weathertight receptacle boxes located above the ground. Where the exterior wiring is to be extended from the house into the grounds and connections are to be made in various parts of the grounds, the conduit and wiring should be buried at least 36″ below the frost line. Weatherproof or waterproof types of sockets should be used. Procedures for several typical exterior installations are shown in Fig. 49.

CHAPTER 4

Adequate Wiring

Better wiring in homes is more important today than ever before. In many homes the wiring systems, planned mainly for lighting, are inadequate for the numerous appliances now in use. Operating appliances on low voltage causes reduced efficiency, wasted current, and excessive repairs or replacements of appliances. Too low voltage may also slow down electric motors or make them stall and burn up. Every homeowner should carefully check the wiring system and make installations required for safety and economy.

THE ELECTRIC SERVICE ENTRANCE

Electricity reaches the home through the electric service entrance. So that full voltage may be supplied, it is important to have a large enough service entrance. The two basic parts of a service entrance are the service entrance wires and the service entrance equipment.

Service Entrance Wires. The service entrance wires are connected to the outside power supply. In many areas these wires are brought in through a pipe or conduit. Sometimes they are combined into a cable (Fig. 1, A). If three service entrance wires are necessary, they may be combined into a flat, oval-shaped cable (Fig. 2, A). Many services are equipped with only two wires, which provide 115-volt service. (Although the standard service voltage ranges from 110 volts in some areas to 130 volts in others, regular voltage is referred to here as 115-volt. Appliances are rated within

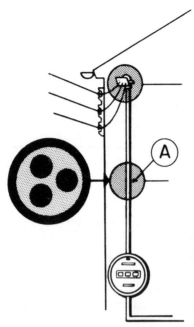

Fig. 1.

this range and are designed to operate at 110 to 130 volts.)
A third wire is required for large appliances such as ranges,
water heaters, and clothes dryers; this provides both 115-
volt and 230-volt service. Capacity may range from 70 amps.
to 115 amps. The size of the service entrance wires determines the amount of electricity that can be brought into
the home at any one time. All wire sizes are designated by
numbers (Fig. 3).

Service Entrance Equipment. The main switch and fuse,
or the main circuit breaker, is the junction point from which
electricity is dispatched to various parts of the house and
from which the main flow of current can be disconnected.
The common types of service entrance equipment are as
follows:

The fuse type with the main switch and the main fuses
contained in a box (Fig. 4, A).

The fuse type with the main switch and the main fuses in
the same box as the branch circuit fuses (Fig. 4, B).

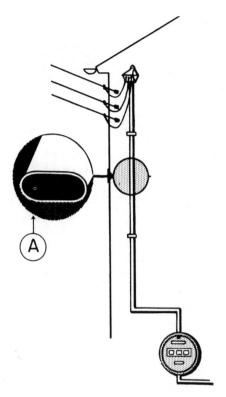

Fig. 2.

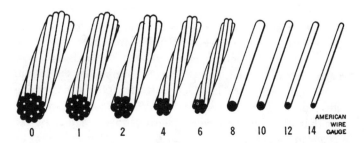

Fig. 3. Actual diameters of typical sizes of copper conductors or wires.

SERVICE ENTRANCE EQUIPMENT

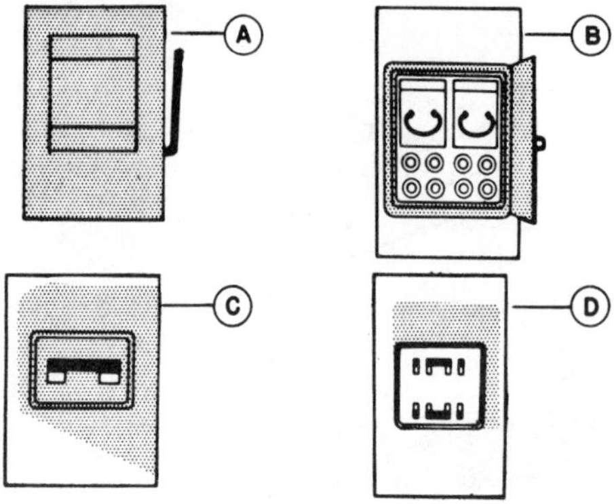

Fig. 4.

The circuit breaker type with the main circuit breaker in a separate box (Fig. 4, C).

The circuit breaker type with the main breakers in the same box as the branch circuit breakers (Fig. 4, D).

Adequate electric service depends upon the capacities of the service entrance wires and equipment. If either is too small, the addition of a new major electrical appliance such as an electric clothes dryer, an electric range, or even a dishwasher may necessitate replacing the entire service entrance with larger wires and equipment.

Recommended capacities for service entrance wires and equipment have been worked out by electrical engineers after a long and careful study of average electrical consumption. In most cases, the capacities of service entrance wires and equipment recommended in Table 8 will be large enough. A double check should be made to allow space for additional equipment or appliances. National Home Builders Association now calls 100 amps. service "adequate." But the Washington Water Power Company says that 100 amps. service is already obsolete. Home power consumption is doubling every 10 years. The definition of "adequate" is changing.

Size of the Electric Service Entrance. The following

table and illustration (Fig. 5) provide a rough guide toward learning the capacity of the electric service entrance in an

TABLE 8

Checking the Size of Electric Service Entrance Capacities

Mechanism for fuel-fired heating plant	800 watts
Dishwasher-waste disposer	1500 watts
Waste disposer alone	500 watts
Automatic washer	700 watts
230-volt electric clothes drier	4500 watts
Electrostatic air cleaner	60 watts
Home freezer	350 watts
Water pump	700 watts
Built-in bathroom heaters	
How many? each	1000-1500 watts
Room air conditioner (¾ ton)	1200 watts

	30 AMPERES May be only 120 Volt	60 AMPERES 240 Volt	100 AMPERES 240 Volt
	OBSOLETE	**MINIMUM**	**ADEQUATE**
If Service Entrance Equipment looks about like any of these	TYPICAL 30 AMP. FUSE TYPE MAIN SWITCH / TYPICAL 30 AMP. COMBINATION MAIN BREAKER AND BRANCH CIRCUIT PANEL	TYPICAL 60 AMP. FUSE TYPE COMBINATION MAIN SWITCH AND BRANCH CIRCUIT PANEL / TYPICAL 60 AMP. COMBINATION MAIN BREAKER AND BRANCH CIRCUIT PANEL	TYPICAL 100 AMP. FUSE TYPE COMBINATION MAIN SWITCH AND BRANCH CIRCUIT PANEL / TYPICAL 100 AMP. MAIN BREAKER
The home has a basic electrical capacity of	Probably 3,600 Watts	14,500 Watts	24,000 Watts
It will supply	Lighting and a few plug-in appliances. Name of the major appliances listed above could be added.	Lighting and plug-in appliances — Electric Range — Water Heater. Would have to be enlarged before any of the major appliances listed above could be added.	Lighting and plug-in appliances PLUS any of the major appliances listed above, without changing the Electric Service Entrance wires or equipment.

Fig. 5.

existing house. Both fuse-type and circuit breaker equipment are shown. One or the other, but not both, will be found in actual installations.

When building a new house, note that the proper size for the electric service entrance (based on an estimate of present and future requirements) should be part of the building specifications. In an existing house, checking the capacity of the electric service entrance is sometimes complicated for those who are not familiar with the design and all makes of service entrance equipment. If necessary, enlist the help of the local electric power supplier who will be glad to cooperate. If contemplating the purchase of a newly constructed house, consult the contractor who did the electrical work for the capacity of the electric service entrance.

BRANCH CIRCUITS

After electricity has passed through the electric service entrance, it enters the branch circuit box, and from there it is carried throughout the house by a number of smaller wires (branch circuits). The number of branch circuits in the electric system and the size of the wires in these circuits determine how efficiently electricity will be utilized in the home.

If there are not enough branch circuits (with wires of proper size) to accommodate all the electrical equipment and lamps, electricity is wasted as the wires become overheated and the voltage is reduced. Overloaded circuits are clearly indicated when the lights dim as the refrigerator or another appliance starts; when the iron, toaster, or other heating appliances operate slowly; or when motors seem sluggish and have difficulty in starting.

Each branch circuit is protected by a fuse or a circuit breaker, located in the branch circuit box (Fig. 6, A). When a branch circuit becomes dangerously overloaded, a fuse blows or a circuit breaker trips open.

In an overloaded or faulty circuit a fuse of the right size automatically stops the flow of electricity from the branch circuit before the circuit wires become overheated. Electricity will remain *Off* in that circuit until the cause of the difficulty has been removed and the fuse replaced.

The circuit breaker (Fig. 6, B) performs the same function as the fuse. When an overloaded or a short circuit

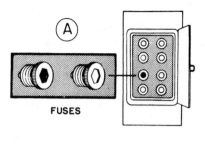

FUSES

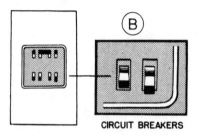

CIRCUIT BREAKERS

Fig. 6.

occurs in the branch circuit, the handle of the current breaker automatically jumps to the *tripped (Off)* position. Electricity will not flow over that circuit until the cause of the difficulty is remedied and the circuit breaker is reset.

When a fuse or circuit breaker goes out, disconnect lamps and appliances which were in use at the time. If a fuse has blown, open the main switch in the service entrance to cut off the current. Replace the blown fuse in the branch circuit box with a new one of proper size (they screw in and out like light bulbs). If after replacing a branch fuse, lights do not go on, check fuses in main switch box. Make sure hands are dry and stand on a dry board. When the fuse has been replaced, close the main switch to restore service. If a circuit breaker has opened, push the handle to the extreme *OFF* position and then return it to the *ON* position.

An electric system should contain three kinds of branch circuits: general purpose, small appliance, and individual circuits, each wired with a wire of correct size and having a fuse or circuit breaker of the proper size. (See Table 9, which also shows the number of watts that can be connected simultaneously on a single circuit.)

Houses erected prior to 1945 still rely entirely on one or two general purpose circuits for all lighting and appliance use, and some that have been built since then have in addition only one or more small-appliance circuits. Even new houses being built today are without adequate circuit requirements. Tables 8 and 9 list appliances usually connected to general purpose or appliance circuits and show how easy it is to overload a single general purpose circuit, or even an appliance circuit, when several household tasks are being performed at the same time. Automatic washing machines, electric water heaters, dishwashers, or ranges should have individual circuits that will permit them to operate independently of each other.

To make sure that all the lamps and appliances are getting the necessary amount of electricity and that the branch circuits will not be overloaded, consult Table 9, which gives directions for budgeting the use of lights and appliances so that no more wattage than the branch circuits can carry will be used at one time. If necessary, install additional circuits with wire large enough to correct the situation.

When an adequate number of branch circuits cannot be installed at the time the house is being built, the branch circuit box should be made large enough to accommodate extra branch circuits which may be added at some future time.

The efficiency and convenience of the electric system can be stepped up by using a No. 12 wire for all general purpose and small-appliance circuits. If a No. 12 wire is installed throughout, and if local ordinances permit, 20 amp. fuses or circuit breakers may be used on general purpose circuits, thereby increasing the capacity of each circuit to 2300 watts. When a No. 14 wire is used in general purpose conduits, the use of 15-amp. fuses or circuit breakers is mandatory, thus limiting the circuit's capacity to 1750 watts. Actual comparative sizes of both wires are shown in Fig. 7.

TABLE 9

DIRECTIONS FOR BUDGETING LIGHTS AND APPLIANCES

Branch Circuits	Service	Requirements	Number of Watts Connected Simultaneously
General Purpose Circuits Lamp Radio Vacuum cleaner	Lights all over the house and convenience outlets everywhere except in the kitchen, laundry and dining areas.	15-Amp. Fuse or Circuit Breaker A 20-Amp. fuse or circuit breaker may be used if a No. 12 wire is used in the circuit. This allows..................	1750 watts 2300 watts
	No. 14 wire, requiring a 15 amp. protection is found in most homes. No. 12 wire is larger and is therefore recommended for present-day electrical systems for general purpose and small appliance circuits.		
Small Appliance Circuits Coffee maker Refrigerator Television Toaster	Only the convenience outlets, no lights, in the kitchen, laundry and dining areas, where portable appliances are most often used.	20-Amp. Fuse or Circuit Breaker	2300 watts
	No. 12 wire is the smallest which may be used in small appliance circuits.		
Individual Circuits Air Conditioner Automatic Washer Range Water Heater	One piece of major electrical equipment each.	Various sizes and types of fuses and circuit breakers, depending on the rating of the appliance.	Nothing more than the appliance served by each circuit.
	An individual circuit seldom becomes overloaded, as no other appliance is connected to it.		

ADEQUATE SWITCH CONTROLS

The switches in the electric system not only cut off or restore the flow of current to lights and convenience outlets but also contribute to safety.

Multiple switch control is the control of a single light source from two or more locations. Switches are placed on the wall or on the latch side of a doorway (sometimes in groups of two or three) when several lights are to be controlled from one spot. Closet lights are usually controlled by door switches that are installed in the door jamb and operate with the opening and closing of the door.

Master or remote control is another switching con-

Fig. 7. Actual size wires.

Fig. 9. Special purpose outlet.

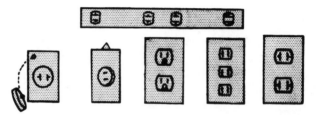

Fig. 8. Convenience outlets.

venience. With a master control switch in any room, lights located anywhere in the house or on the grounds can be controlled.

ADEQUATE ELECTRIC OUTLETS

Any opening that gives access to electricity can be considered an *outlet*, even if only a lighting fixture or a piece of electrical equipment is visible. Unsightly and unsafe tangles of extension cords are evidence that the home does not have enough outlets. Extension cords were never intended to be used in place of branch circuits.

Convenience outlets are available in several forms, the most common of which is the *standard duplex convenience outlet*. *Triplex* or *quadruplex outlets* as well as a number of outlets assembled on a strip or molding are also available. For some locations, there are convenience outlets combined with a switch. *Weatherproof outlets* have protective caps and should always be used in open porches or in any exposed location. *Hanger outlets* for electric clocks permit the clocks to be mounted high on a wall with the cord concealed. For safety, a new type of outlet designed to receive a *three-prong plug* provides a connection which grounds metal parts of appliances. (These outlets are shown in Fig. 8.)

Special purpose outlets (Fig. 9) are the openings through which certain major appliances are connected to their individual branch circuits. The electric range is the only major appliance that has a specially designed *plug-in outlet*. Other major appliances are connected to individual circuits through a standard convenience outlet or are connected directly to the wiring system. Openings for fixed lights are also called outlets, although they are seldom regarded as such because they are concealed behind the lighting fixtures. Table 10 lists minimum requirements as to the number and placement of outlets, lights, and switches. Consult the local supplier of electric power or an electrical contractor about the Standards of Wiring Adequacy approved by the local authorities.

THE ADD-A-CIRCUIT METHOD

An isometric view of a five-room, one-family house built twenty-five or so years ago is shown in Fig. 10. It is a modern home in all respects except the wiring. When built, it was equipped with one circuit, ten lights, five switches, and two convenience outlets. Fig. 11 shows the same house rewired with eight regular circuits, seven special circuits, twenty-one lights, twelve switches, thirty convenience outlets, and four special outlets.

To be electrically modernized, a house should be rewired from top to bottom. However, a complete rewiring job

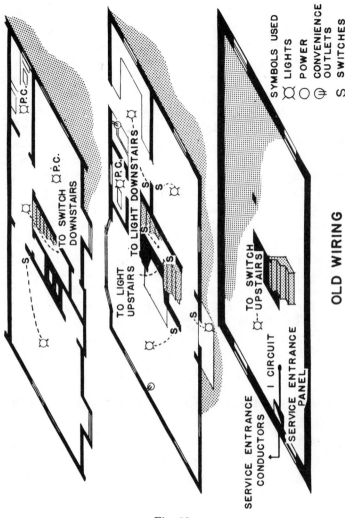

SYMBOLS USED

LIGHTS
POWER
CONVENIENCE OUTLETS
SWITCHES

P.C.

TO SWITCH DOWNSTAIRS

TO LIGHT DOWNSTAIRS

P.C.

S

S

TO LIGHT UPSTAIRS

S

S

OLD WIRING

TO SWITCH UPSTAIRS

SERVICE ENTRANCE CONDUCTORS

1 CIRCUIT

SERVICE ENTRANCE PANEL

Fig. 10.

may not be practical. The construction of the house may make it impossible to install all the needed additional circuits and outlets. The budget may not permit the necessary expenditures in one operation. Under such circumstances,

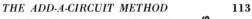

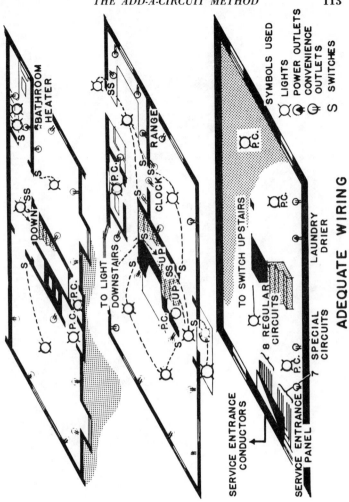

Fig. 11.

the wiring system can be improved sufficiently by simply adding one or, if needed, two or three circuits to the system.

The add-a-circuit method, developed by General Electric, is a practical and feasible system of remedying deficiencies in a wiring system. It provides convenience outlets and conductors of the correct size where most needed. The

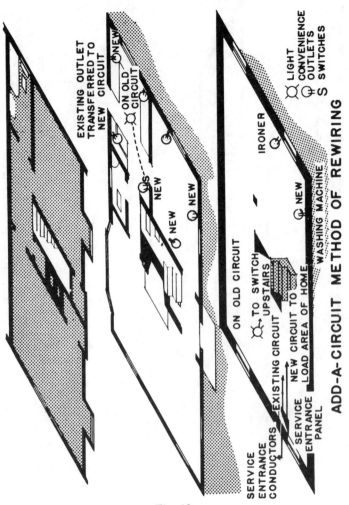

Fig. 12.

additional circuit (Fig. 12) or circuits can be installed at a low cost and with little damage to walls and finishes.

Method of Rewiring. Figure 12 shows an isometric view of the house illustrated in Figs. 10 and 11 with the addition of only one circuit to the old wiring system. Eight outlets have been installed on the circuit directed to the kitchen, dining room, and laundry area. Number 12 wire is used. This circuit permits the use of many appliances independently of all other parts of the house. Most appliances in the kitchen, dining room, and laundry draw large amounts of current; therefore, the first added circuit should be installed in this area. Heavy conductors should be used so that as many appliances as desired can be run at the same time. There should be enough conveniently located outlets so that one does not have to disconnect any appliance in order to use another appliance.

To install this circuit, run a line of cable from the service entrance panel to the outlets (Fig. 12). If the location of the outlets is properly planned, the cost of installing such a circuit will be less than the average cost of redecorating one or two rooms. Before proceeding with the installation, however, check the service entrance to see that it is large enough for the additional circuit. In houses wired for limited electrical service, use service entrance conductors and panels that match the rest of the wiring system. In most houses, the service entrance will have enough capacity in the conductors and panel to accommodate the extra circuit, but in some houses it may be necessary to add a small, inexpensive fuse panel and perhaps a new panel or new service entrance conductors, or both.

The add-a-circuit method of rewiring provides additional outlets and conductor capacity for small appliances. An additional electric range, water heater, or similar major appliance consumes a large amount of electricity and requires special circuits.

REMOTE-CONTROL WIRING SYSTEM

The G-E remote-control wiring system can be adapted to control 125-volt power circuits by low-voltage relays in an isolated circuit. In this system, the 125-volt lines are connected to all outlets in the same way that they are connected in any conventional wiring system, with the following exception (see Fig. 13): instead of connecting the switches in the 125-volt lines, one must connect the new control switches in a low-voltage isolated circuit, using relays directly in the outlet boxes to turn the current *On* or *Off*.

TABLE 10

MINIMUM REQUIREMENTS FOR OUTLETS, LIGHTS, AND SWITCHES

	Convenience Outlets	Special Outlets	Permanent Lighting	Switches
Living room, bedrooms, and general living areas.	1 at least every 12′, placed so that no point along floor line of usable wall space is more than 6′ from an outlet.	Required for FM radio. Television.	1 Ceiling light (2 in long narrow rooms). Wall; cove or valance lighting may be substituted.	
Dining areas.	1 for each 20′ of usable wall space, with one located near hostess' chair.		1 Ceiling light over table.	
Kitchen.	1 for each 4′ of work counter. 1 for refrigerator.	Required for Electric range. Dishwasher. Disposer. Home freezer. Clock. Ventilating fan.	1 Ceiling light. 1 over sink. Others over work counters as needed.	1 on latch side of each frequently used doorway.
Laundry.	1 at washing area. 1 at ironing area. 1 for hot plate.	Required for Automatic washer. Electric drier.	1 at washing area. 1 at ironing area.	Rooms with entrances more than 10′ apart should have multiple control switches.
Utility room.	1 at workbench. 1 near furnace.	Required for Fuel-fired heating equipment. Electric water heater.	1 for each enclosed space. 1 at workbench. 1 near furnace.	
Bathrooms.	1 adjacent to mirror.	Required for Built-in heater.	1 on each side of mirror. 1 in enclosed shower compartment.	
Entrances.	1 weatherproof near front entrance.		1 at front. 1 at rear.	1 inside front. 1 inside rear.
Hallways.	1 for each 15′ of hallway.		1 at least; 2 in long halls.	
Stairways.			1 each at head and foot.	1 each at head and foot.
Closets.			1 for each closet over 3′ deep.	
Porches, terraces, patios.	1 weatherproof for each 15′ of usable outside wall.		1 for each 150 sq. ft. of enclosed porch.	

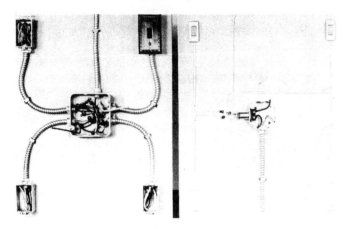

Fig. 13.   Left illustration shows the conventional wiring system for
one outlet controlled from four points. Right illustration
shows remote-control wiring system. Note the heavy,
three- and four-conductor cable necessary for the conventional wiring system, also the amount of space that
this cable takes in the outlet and switch boxes. The
same four-point control with the remote-control system is obtained with lightweight wire, smaller outlet
boxes, and neat, two-way push-button switches.

All outlets or several selected outlets can be switched from
one central point by means of a master station control.
Garage lights, outdoor lights, and convenience outlets in
the kitchen and laundry can be controlled from any part
of the house.

Components.   Components used in the remote-control
wiring system are: the relay, transformer, remote-control
wires, and remote-control switches.

Relay.   A remote-control relay (Fig. 14) operates on 24
volts. It is approved by the Underwriters' Laboratories for
ratings of ⅓ h.p., 15 amps., 125 volts a.c., or 5 amps. and
250 volts a.c.; it is also approved for use with tungsten
filament lamp loads. This relay installed in the outlet box
of the lighting fixture or convenience outlet actually switches
the 125-volt circuit. It can be mounted in a standard outlet
box or switch box that has ½″ knockouts. The barrel of the
relay is inserted through one of the knockouts from the

Fig. 14. Remote-
control relay.

inside of the box; thus the 125-volt end is inside the outlet or switch box; the low-voltage end is outside the outlet or switch box. The outlet or switch box acts as the barrier between the two systems. The type of relay shown in Fig. 14 is a split-coil solenoid-operated switch. When the *On* coil is energized from the control switch, it closes the 125-volt switch contacts on the relay. When the *Off* coil is energized, it opens the switch contacts on the relay.

Transformer. The transformer supplies the necessary low-voltage power (24 volts) that operates the remote control relays. Figure 15 shows the transformer designed for use with the G-E remote control wiring system. It is rated at 35 volt-amps. and is approved by the Underwriters' Laboratories. Only one transformer is necessary in the average home. It should be mounted on a universal-type box cover

Fig. 15. Transformer.

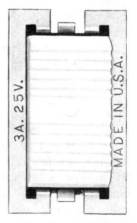

Fig. 16.   Remote-control wire conductors.

with screw slots and holes to fit standard ¾″, 3½″, or 4″ octagon outlet boxes or standard 4″ square outlet boxes. The 10″ pigtail leads used for connecting the 125-volt, 60-cycle power supply should extend through the mounting box cover, thus permitting direct connection to the power supply in the outlet box, and it also acts as a support for the transformer. Additions or alterations to the control circuits can be made easily because the low-voltage locking-type screw terminals are mounted on top of the transformer. The transformer must be connected to a branch circuit not over 15 amps.

Remote-Control Wires. Thermoplastic-insulated wires (Fig. 16) are available with two or three conductors and are rated for 30-volt service. The conductors are made of solid copper and are covered with $\frac{1}{65}$″ thermoplastic in-

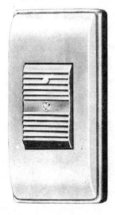

Fig. 17.   Remote-control switch for surface mounting.

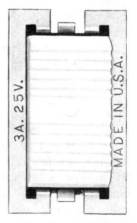

Fig. 18.   Remote - control switch for flush mounting

sulation. The three-conductor wire is thin, flat, and light in weight and has a double rib on one of the outer conductors for circuit identification. The wires can be fastened to studs, joists, walls, ceilings, and moldings; they may be used in new construction without any raceway protection. In old construction, snake the wires into the walls or staple

Fig. 19.   Remote-control switch strap
and plate.

them to the surface of wallboard. In plaster walls, the wire can be placed into a shallow groove and covered with plaster.

Remote-Control Switches. There are two types of remote-control switches. One switch is used for surface mounting (Fig. 17), and the other switch is used for flush mounting (Fig. 18). They are both rated at 3 amps., 25 volts a.c. Three of these switches can be installed in the same space required for a conventional tumbler switch. Any number of switches can be used.

Surface-mounted switches are used on wallboard construction or on thin partitions. They require a special switch box because they are not deep enough for a box of standard size. Flush-mounted switches can be installed in a conventional switch box in one-gang, two-gang, or three-gang combinations. A mounting strap is furnished with each wall plate for one, two, or three switches (Fig. 19). Both surface

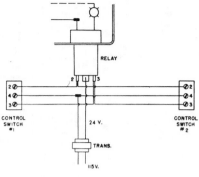

Fig. 20.

switches and flush-mounted switches operate on the same principle. They are a single-pole, double throw, momentary-contact, normally open push-button type. By depressing the end of the switch button (marked with a raised dot on the flush-mounted switch), the *On* coil of the remote control relay is actuated and the circuit is turned on. When the *Off* end of the switch button is touched, the other coil in the relay operates, turning the circuit off.

Both switches have binding screws that are marked to

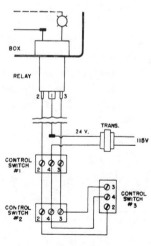

Fig. 21.

identify the circuits. Remove the back plate of the surface-mounted switch and screw it directly into the switch box, wallboard, or other wall surface. When the walls are ready for the switches, snap the switch to the back plate, thus automatically seating it in position. The flush-mounted switch is held in position by means of the wall-plate strap in the switch box. When the installation is complete, screw the wall plate into place.

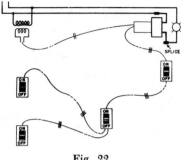

Fig. 22.

Installation. Figure 20 shows two typical installations. Note the amount of space that the heavy three-conductor and four-conductor cable takes in the conventional outlet and switch boxes in comparison with the four-point remote-control system. The latter has lightweight wire, smaller outlet boxes, and neat two-way button switches.

Figures 21, 22, and 23 illustrate some specific wiring applications.

Because the remote-control wiring system uses low voltage, it is not subjected to the same code restrictions as a conventional wiring system. The restrictions imposed upon low-voltage systems by the National Electrical Code are as follows: The *line side* of the transformer must be installed in accordance with the Code specifications. The *load side* of the transformer must have open conductors at least 2″ from any light or power circuit unless they are protected by a continuous nonconductor, such as flexible tubing or

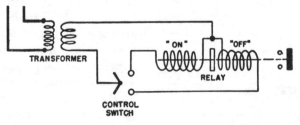

Fig. 23.

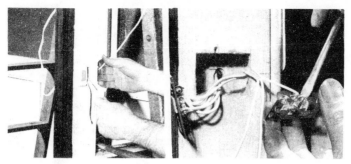

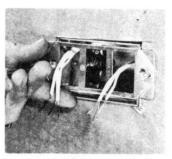

Fig. 24. Installing low-voltage wire for surface-mounted switch installation.

Fig. 25. Connecting low-voltage wires to black plate of remote-control switch.

Fig. 26. Mounting wall plate strap to standard switch box before installing flush-mounted remote-control switches.

Fig. 27. Installing switch with three large binding screws.

Fig. 28. Inserting flush-mounted switch into position.

Fig. 29. Completed installation.

porcelain tubing, in addition to insulation on the wire. In vertical runs the conductors must be separated from light and power circuits by at least 2″, or the conductors of either system must be encased in noncombustible tubing. If control circuits extend beyond one house, and are so run that they may accidentally come into contact with light or power circuits operating at a potential exceeding 300 volts, they must also comply with the Code.

Figures 24 to 29 show the various installation details.

CHAPTER 5

Batteries and Transformers

Batteries are devices that convert chemical energy into electrical energy. The two general types are dry cell batteries and storage batteries.

DRY CELL BATTERIES

The chemical in a dry cell is in paste form. As shown in Fig. 1, the container is a cylinder of zinc which, in addition to holding the ingredients of the cell, is itself a vital element in the making of electricity. A carbon post with a metal cap is in the center of the cell (Fig. 2) and forms the positive electrode which collects the current and conducts it to the outside circuit.

Current is created by the solution of zinc through action of the chemicals in the electrolyte (Fig. 3). As this action

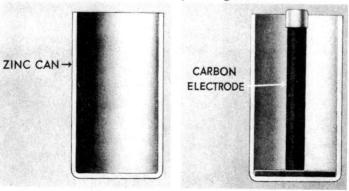

ZINC CAN →

CARBON
ELECTRODE

Fig. 1.   Fig. 2.

takes place, hydrogen atoms are released. Generally, the more zinc, manganese dioxide, and electrolyte chemicals put into a cell, the longer it will last. As the cell is used, the zinc is consumed and gets thinner (Fig. 4), and interaction between the zinc and the electrolyte chemicals produces reaction products, accumulation of which tends to retard the action of the cell. There is a gradual stop in the *working voltage* as illustrated by the *discharge curve* shown in Fig. 5.

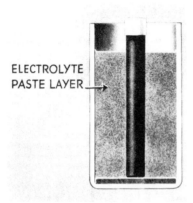

ELECTROLYTE
PASTE LAYER

Fig. 3.

ZINC
CONSUMED

Fig. 4.

Voltage. To cause water to flow in a pipe, pressure is necessary. Hydraulic pressure, which depends upon the *head* (or depth) of water, is expressed as *feet of water*. Similarly, electric pressure is required to cause a flow or current of electricity through a conductor.

Electrical pressure is called *electromotive force* (e.m.f.) or voltage. The unit is a *volt*. It is the pressure required to cause one coulomb per second (one amp.) to flow through a conductor having a resistance of one ohm.

The e.m.f. of voltage of a dry cell is approximately 1.5 volts. This is independent of the size of the cell. Two cells of different size may be compared to two tanks of water of different volume but of the same height (Fig. 6).

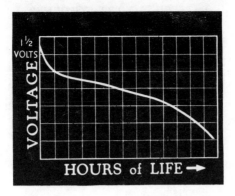

Fig. 5.

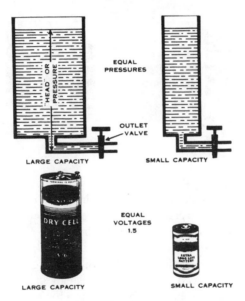

Fig. 6.

In these two tanks the pressures at the outlet valves are equal because the height of water above them is the same in both. The large tank, however, can discharge more water than the small one. Similarly, a large dry cell can deliver more electricity than a small one.

Series Connection. The pressure of a single tank of water may be multiplied by adding tanks, one above the

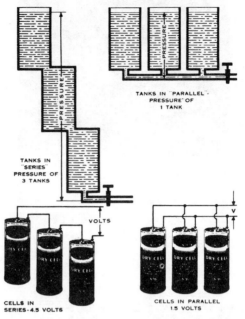

Fig. 7.

other, to increase the height or head of water at the outlet (Fig. 7). Dry cells arranged in the same manner as these tanks are said to be in *series*. Three dry cells connected in series, as shown in the illustration, have three times the voltage of a single cell.

Parallel Connection. At the right in Fig. 7, three tanks are shown placed side by side instead of one above the other. Here the pressure is the same for all three tanks as it would be for only one. Dry cells arranged in this manner are said to be in *parallel*. Three dry cells connected in parallel, as shown in the illustration, have the voltage of only one cell.

OHM'S LAW

Ohm's law expresses the electrical relation between voltage (E), resistance (R), and current (I). This law states that in two circuits subjected to equal voltages the current

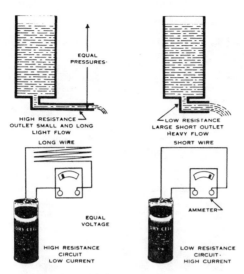

Fig. 8.

will be proportionately greater in the circuit with lower resistance. The law also states that in circuits with equal resistance the current will be directly proportional to the voltage applied. In other words, a high resistance or a low voltage yields a low current; and a low resistance or a high voltage yields a high current.

The first relation is graphically shown in Fig. 8. Here two tanks of water of equal height, thus having equal pressure at the outlets, are shown discharging through a small tube (with high resistance) and through a larger diameter outlet tube (with low resistance). Naturally the flow through the smaller tube is less than the flow through the larger tube. Two dry cells are also shown in the diagram, both of the same voltage. For one of these cells the circuit has a high resistance, and the ammeter shows a low current flowing through it. For the other (right-hand) cell the circuit has a low resistance, and the ammeter shows a higher current passing through the circuit.

The second relation is shown in Fig. 9. The outlet tubes in the two tanks of water are the same in size and, therefore,

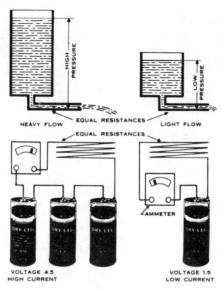

Fig.   9.

have the same resistance to the overflow of water. The left-hand tank is deep and exerts a high pressure at the outlet, whereas the other is not so deep and consequently creates a lower pressure at the outlet. Therefore the flow is heavier from the high-pressure tank. The two electric circuits shown in the illustration have the same resistance. At the left, the three dry cells are connected in series with 4½ volts of pressure, and the ammeter shows a high current. At the right, there is a single dry cell with 1½ volts, and consequently, the current is only one-third as much as in the other circuit.

The relations described show the current to be directly proportional to the voltage and inversely proportional to the resistance. This principle can be expressed by means of the following formula:

$$I \text{ (amperes) } = \frac{E \text{ (volts)}}{R \text{ (ohms)}}$$

It is evident that if the voltage and the resistance are in-

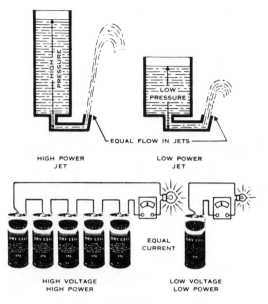

HIGH POWER JET LOW POWER JET

EQUAL FLOW IN JETS

HIGH VOLTAGE HIGH POWER EQUAL CURRENT LOW VOLTAGE LOW POWER

Fig. 10.

creased by equal amounts, the current will remain the same.

Power. Power is the rate of doing work. Electric power increases with additional current. Figure 10 shows two lamps drawing equal currents. The left-hand circuit connects a lamp of 7.5 ohms operating on five dry cells in series. It is operated with 7.5 volts. From Ohm's law the current is computed as follows:

$$I = \frac{7.5 \text{ volts}}{7.5 \text{ ohms}} = 1 \text{ ampere}$$

The right-hand circuit connects a lamp of ⅕ ohms which can be operated on a single dry cell of 1.5 volts and also takes 1 amp. It is obvious that the lamp at the left takes more power (7½ watts) and gives more light (candle power) than the lamp at the right (1½ watts).

Electric power, therefore, increases with voltage and with current and is proportional to their product. The unit of electric power is the watt. One watt is equal to a current of one amp. flowing under a pressure of one volt, and the formula is:

$$W \text{ (watts)} = E \text{ (volts)} \times I \text{ (amperes)}$$

Service. When the switch in an electric circuit is closed and the current starts to flow from the cell to operate some device, the terminal voltage will drop slowly until the cell is exhausted. If the same current is drawn from each of two cells of different sizes, the smaller one will be exhausted before the larger one, and its terminal voltage under these conditions will drop more rapidly. For the longest service of the cell, the current should be drawn from it intermittently (instead of continuously) and with the longest possible rest periods between the periods of drain. This method not only extends the withdrawal of electricity over a longer period but also allows recuperation to take place and thus increases the available energy.

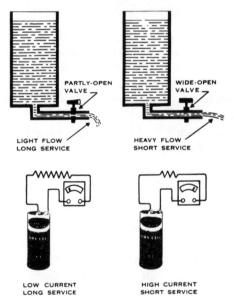

PARTLY-OPEN VALVE

WIDE-OPEN VALVE

LIGHT FLOW
LONG SERVICE

HEAVY FLOW
SHORT SERVICE

LOW CURRENT
LONG SERVICE

HIGH CURRENT
SHORT SERVICE

Fig. 11.

A dry cell will last longer with a low current output than with a high output. The left-hand cell in Fig. 11 is connected to a high-resistance circuit, and the ammeter shows a low output. In the right-hand circuit the situation is reversed. Often a dry cell will last longer than would be expected from its low rate of output, but efficiency is actually increased by a low output. If the flow of electricity is cut in half, for instance, the output of the cell will be more than doubled.

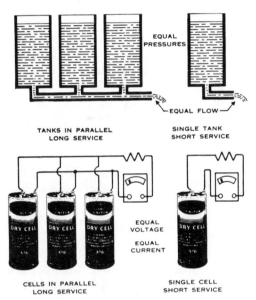

EQUAL PRESSURES

EQUAL FLOW

TANKS IN PARALLEL
LONG SERVICE

SINGLE TANK
SHORT SERVICE

EQUAL VOLTAGE

EQUAL CURRENT

CELLS IN PARALLEL
LONG SERVICE

SINGLE CELL
SHORT SERVICE

Fig. 12.

Connecting dry cells in parallel divides the total battery drain among them and makes them last longer than a single cell (Fig. 12). For this reason, particularly with a high output, it is often desirable to connect cells in parallel. When a dry battery has insufficient voltage, the apparatus to which it is connected ceases to operate satisfactorily (Fig. 13). The minimum voltage required is called the *cut-off-voltage*. The effect of the cut-off is not dependent upon the rate at which the current has been drawn. The lower the cut-off of the device using the cells, the more

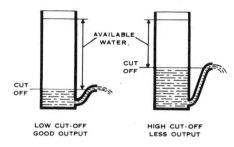

Fig. 13.

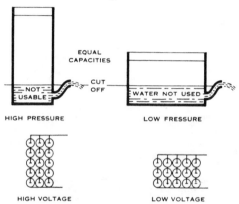

Fig. 14.

completely can the battery be used. For best results the cut-off voltage should be kept as low as possible.

Dry batteries can be arranged so that even with the same number of cells one will have a higher voltage than the other. The batteries shown in Fig. 14 illustrate such an arrangement. The battery at the left has higher voltage because it has more cells in series. The battery with the higher voltage will deliver a larger proportion of its energy down to a given cut-off. In the high voltage battery the cut-off voltage per cell in series will be lower.

CYLINDRICAL CELL "B" BATTERIES

As previously stated, a single dry cell has a nominal voltage of only 1.5 volts. For many purposes higher voltages are required, and it becomes necessary to connect a number of single cells in series. In the case of radio sets it has been necessary to provide 67½ volts or even more for the "B" circuit, and as a result "B" batteries have been evolved with voltages to 22½, 45, and 67½ volts and made up of 15, 30, and 45 cells in series. This is illustrated below, in Fig. 15.

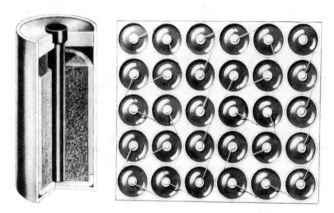

Fig. 15. A conventional 45-volt "B" battery.

"LAYER-BILT" BATTERIES

In the "Layer-Bilt" battery, flat cells are laid one against the other. The construction of this battery is shown in Fig. 16. Note that in contrast to the cylindrical batteries, the space between the cells has been eliminated and the space around the cells has been reduced. The zinc is protected by the carbon element. Moreover, the flat cell construction eliminates all soldered connections between the cells. The cells are connected directly to one another.

"MINI-MAX" BATTERIES

When the portable radio was invented there was a need for a highly efficient "B" battery of smaller size than had previously been required. Battery designers and engineers developed the battery shown in Fig. 17.

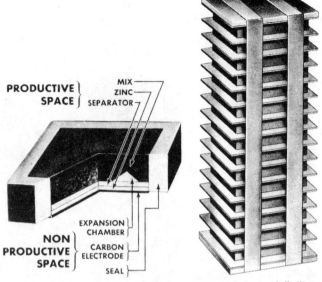

PRODUCTIVE SPACE — MIX, ZINC, SEPARATOR

NON PRODUCTIVE SPACE — EXPANSION CHAMBER, CARBON ELECTRODE, SEAL

Fig. 16. Two stacks of cells make a 45-volt "Eveready" "Layer-Bilt" "B" battery.

Fig. 17.

Fig. 18.   Fig. 20.

The materials in a cell of such a battery are the same as in a conventional dry cell, but are flat instead of cylindrical. A complete cell consists of four parts as shown in Fig. 18:

The Zinc—Number 2
The Separator—Number 3
The Mix Cake—Number 4
The Carbon—Number 5

Fig. 19.

Carbon (Number 1) is a part of the next cell below, and zinc (Number 6) belongs to the next cell above. In the construction of this battery the zinc of one cell and the carbon of the adjacent one are combined into a single plate called a *duplex electrode.*
Each unit in such a battery consists of a duplex electrode, a separator, and a mix cake placed close together (Fig. 18). The projection on the mix cake represents points of contact with the carbon of the next unit above.

An elastic envelope is put around the four sides of each

unit (Fig. 19), enclosing it to prevent moisture loss and the passage of electrolyte from cell to cell. Fifteen of these units are assembled into a stack and bound tight together between stiff end plates with strong tape (Fig. 20). Two such stacks connected in series deliver 45 volts. Placing two stacks side by side in a container and attaching a socket terminal complete the battery.

USAGE AND DETERIORATION OF DRY CELLS

Dry cells deteriorate with age, no matter how little they may be used. Deterioration results from slow chemical reactions and moisture changes in the cell. These effects are referred to as *shelf* deterioration. They gradually reduce the output of the cell. But cells of good quality and reasonable size should remain usable for two years or more if not subjected to improper conditions of storage. (Small cells have shorter shelf-life than large ones. The shelf-life of very small cylindrical cells may be only a few months.)

Dry cells may deteriorate rapidly if short-circuited or subjected to continuous heavy drains. Where the total current required is excessive for a single series of cells, additional strings of cells in parallel should be added to the battery in order to distribute the drain more widely.

Dry cells operate best at 70° F., which is normal room temperature. A high temperature of 100° F. or over may destroy the cells, or, at any rate, the resulting excessive loss of moisture and the accelerated chemical action may hasten the rate of shelf deterioration conversely. Low temperatures are also objectionable, for they retard chemical action within the cell. A temperature of 10° F. below zero will probably make the cells inoperative. However, exposure to low temperature will not cause permanent damage, for even cells which have been frozen may be warmed and restored to their original condition.

STORAGE BATTERIES

A storage battery is a means of storing chemical energy. Two metals are suspended in a solution (electrolyte) so that they cannot touch each other. If the solution can attack these metals chemically, there will be electric pressure or voltage between them which can be measured with a voltmeter. If wires from the two metals are connected outside the solution, current will flow and discharge will take place. A modern type of storage battery is shown in Fig. 21. The two metals are sponge lead (for the negative plates) and lead peroxide (for the positive plates). These plates are kept apart by separators. The electrolyte is a mixture of pure water and sulphuric acid.

During discharge the active material of the plates combines with the acid of the electrolyte. During charge this acid is withdrawn from the plates and returned to the

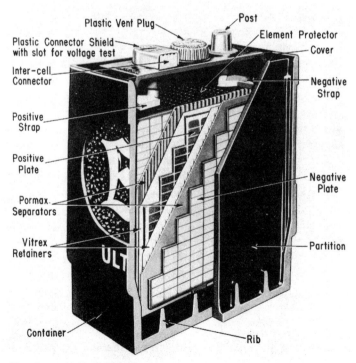

Fig. 21. Cutaway view of a cell of a 3-cell 6-volt automobile battery showing parts and construction.

electrolyte. When all the acid has been withdrawn, the cells are fully charged (Fig. 22).

The voltage in a storage battery depends on the materials selected for plates and electrolyte. All storage batteries

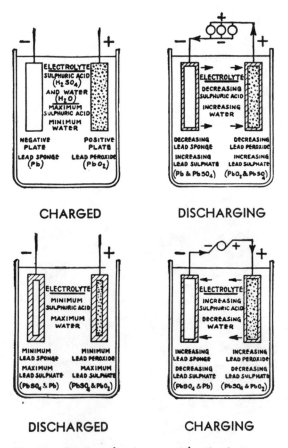

Fig. 22. Diagram showing essential action in storage battery.

using lead plates and acid electrolyte have a voltage of approximately 2 volts per cell on open circuit. The voltage is somewhat higher than this when the battery is being charged, and lower when it is being discharged.

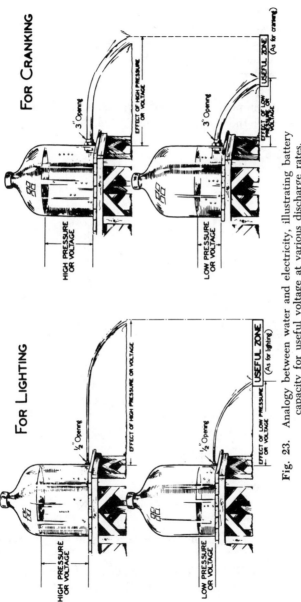

Fig. 23. Analogy between water and electricity, illustrating battery capacity for useful voltage at various discharge rates.

In most automobile batteries with 6 volts there are three cells. A battery with 6 cells has 12 volts. Regardless of its size each cell has only 2 volts.

Capacity. Capacity is the term used for the quantity of electricity which can be taken from a storage battery for a particular unit of time. The capacity of a battery, sometimes called its electrical size, is measured either in ampere hours or in amps. of current delivered continuously for a definite number of hours or minutes. Capacity depends on how well the battery is constructed, how large its plates

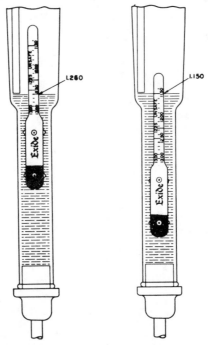

Fig. 24.   Hydrometer   **Fig. 25.**   Hydrometer
reading 1.260.   reading 1.150.

are, and how many plates are assembled in each cell (Fig. 23).

Specific Gravity. The passage of acid from the plates into the electrolyte strengthens it. The strength of the electrolyte

is measured by means of a float called a hydrometer (Fig. 24). The lower the specific gravity in a battery, the farther the hydrometer sinks in the electrolyte (Fig. 25).

Measuring the specific gravity, that is, taking a hydrometer reading, discloses the battery's state of charge. For automobile batteries the following readings indicate approximate charge values:

1.250-1.270—full charge
1.200-1.220—half charge
1.130-1.150—discharged

On the float in the type of hydrometer shown in Fig. 26, the scale reading below 1.225 (half charge) is colored red and marked *unsafe*. In this way an unsatisfactory state of charge is readily noted. The syringe automatically regulates the amount of electrolyte that must be drawn into the syringe barrel to obtain readings. After each reading the electrolyte should be returned to the cell from which it was taken. Hydrometer readings should be made at intervals of two to four weeks and always before water is added to the battery. Reading dates should be recorded.

The full charge specific gravity of most new batteries (with the electrolyte level ⅜" above the separators and temperature at 77° F.) is nominally 1.260 (1.250 to 1.270). However, batteries manufactured for use exclusively in tropical climates where

Fig. 26.
Hydrometer syringe.

water seldom freezes are designed so that they will be fully charged at a specific gravity of 1.210. This special design reduces the harmful effects of sulphation caused by high gravity electrolytes in above normal temperatures.

During the normal life of a battery, there is a slight decrease in its specific gravity. Loss of electrolyte when water is being added, for instance, reduces the full charge specific gravity. Electrolyte may be spilled if too much water is added or if the vent plugs are not kept tight.

CARE AND OPERATION

A storage battery is used to supply electricity for starting engines (as in automobiles or boats), to light lamps, to work the ignition when an engine is not running or when it is running at so low a speed that the generator is *not cut in,* i.e., is not providing electricity. At ordinary speeds the generator is *cut in* and supplies the required electricity. Some of the electric current is returned to the battery to replace the current that had been supplied by the battery when the generator was idle.

Overloading of Battery. The battery system in the automobile maintains the battery automatically at or near full charge when the car is in use. However, as with any storage system, if more energy is taken out than is put in, the reservoir for storage ultimately becomes empty (Fig. 27). For example, if the car owner drives less than the normal minimum for which the battery system has been designed, the battery will gradually lose specific gravity because of insufficient charging. Conversely, in a battery connected to an excessive number of appliances, there will be complete discharge.

Adding Water. The space for water is between the top of the separators and a point $3/8''$ above the separators (Fig. 28). If the water level is higher than $3/8''$ above the separators, electrolyte will be lost through the vent plugs, the battery will get wet, and the metal parts adjacent to the battery will corrode. (If the cells flood or sputter electrolyte, the level is too high and some of the electrolyte should be removed.)

Every cell can be filled automatically to the correct level by means of the self-leveling syringe shown in Fig 28. The nozzle of this syringe has a small hole about ⅜″ above the tip. When the tip rests on the tops of the separators, any

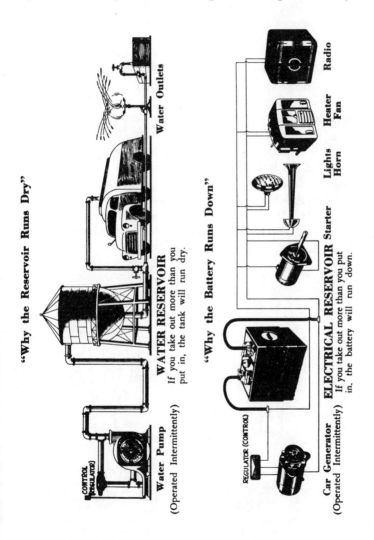

Fig. 27.   Analogy between water tank and generator.

water above the hole and the nozzle is sucked out until the water in all cells is at the correct level.

For batteries having a bull's eye cover, water is added as shown in Fig. 29.

How often water must be added to a battery will depend upon the mileage covered, the generator charge rate, and the weather. For most cars (with a mileage of 1,000 to 2,000 miles a month) water need be added only about once monthly. If more frequent additions are required, this may indicate that the rate is too high and that there is an overcharge with excessive wear on the plates. In cold weather water should be added, if needed, before the car is to be used. In this way the water will mix with the electrolyte; otherwise, the water will remain on top and freeze if the temperature drops below 32° F., the freezing point of water.

All cells of a battery should take the same amount of water. However, if one cell requires more water than others, make sure that the vent plugs are tight and that there is no leakage at the top. Check for leaks in the jar or case of

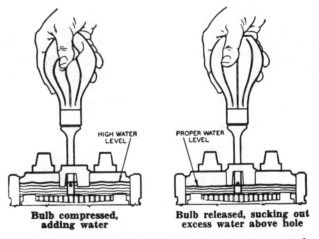

Bulb compressed,   **Bulb released, sucking out**
adding water   **excess water above hole**

Fig. 28. How the Exide filling and self-leveling syringe works.

the battery. Also examine the floor and underside of the case.

In most places in the United States, drinking water may be added to storage batteries, but distilled water should be used wherever there is any question of impurities. Water

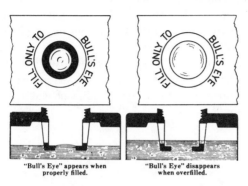

"Bull's Eye" appears when properly filled.　　"Bull's Eye" disappears when overfilled.

Fig. 29. Watering level for "Bull's Eye" covers.

from a tap or spigot should be allowed to run a few moments to remove accumulations of particles in the pipes. Glass, earthenware, rubber, or wooden receptacles that are thoroughly clean may be used for storing or transporting water. Metal vessels, except lead ones, should never be employed for this purpose. Furthermore, nothing but water should be added to the electrolyte.

Low and High Temperatures. Low temperatures temporarily reduce battery capacity. Modern batteries are available with capacity ratings at 80° F. and zero° F. Since these temperatures are often encountered, it is important to know the corresponding capacity ratings of the batteries.

The change from warm to cold weather just about doubles the starting torque requirements of an engine. Since battery capacity is reduced at low temperatures, it is wise to install only a battery of the proper size and to keep it charged. The battery cables, breaker points, spark plugs, ignition cables, and carburetor should be kept in good condition. A volatile fuel, an efficient engine, and the correct dilution will help to increase the efficiency of the battery and also, will make it easier to start the engine in cold weather.

Continued or frequently recurring high temperatures will shorten the life of the wood separators and plates. Ventilation should be provided for the battery compartment to keep the temperature below 110° F.

Freezing Points. The electrolyte has the highest specific gravity when the battery is fully charged. The freezing point of electrolyte at a full charge specific gravity of 1.275 is about 85° F. below zero. As a battery becomes discharged, its specific gravity drops because the electrolyte contains a larger amount of water; consequently, its freezing point approaches the freezing point of water, as shown in the following table:

SPECIFIC GRAVITY	FREEZING POINT
1.275	− 85 degrees F.
1.250	− 62 degrees F.
1.225	− 35 degrees F.
1.200	− 16 degrees F.
1.175	− 4 degrees F.
1.150	+ 5 degrees F.
1.125	+13 degrees F.
1.100	+19 degrees F.
Water (1.000)	+32 degrees F.

Removal of Battery. To remove a battery, disconnect the cable terminals and loosen the holding device. Note the location of the positive post before lifting the battery out, for it will be necessary to install the new battery in the same position. The positive post, which is larger than the negative, may be painted red or may be marked POS or +. Push back the cables and lift the battery together with the battery carrier out of the cradle. Clean the terminals. Check the cables to make sure they are in good condition.

Installation of New Batteries. Make sure that the battery cradle is clean and in good condition. Raise the new battery carrier and set it into the cradle so that the positive post will be in the position specified by the manufacturer. The battery should seat firmly and evenly with the holding device so that it makes a snug fit.

Apply a thin film of terminal grease or vaseline to the terminals. For batteries having grease ring seal nuts, fill the terminal post rings with grease. Attach the terminals temporarily to the battery posts, with the positive terminal connected to the positive post and with the negative terminal connected to the negative post. If the battery has been connected correctly, turning on the lights will cause the dashboard ammeter to register a *discharge*. If this occurs, the installation is correct. Tighten the connections and wipe off surplus grease.

Cleaning of Batteries. The top and sides should be wiped frequently with a rag that has been moistened with water. This should be done when water is being added. Once a year the battery should be removed from its cradle and the entire outside surface should be cleaned. A rag dipped in bicarbonate of soda (1 lb. of soda per gal. of water) should be used for this purpose. The surface should then be rinsed with water. The battery compartment may be cleaned in the same way and painted with asphaltum paint. This solution (or ammonia) will neutralize any electrolyte that may have spilled onto the outside of the battery. If applied immediately when acid is spilled on clothing, woodwork, or cement, it will similarly neutralize the electrolyte and overact harmful effects.

Note that vent plugs must always be kept tight and in place and that the small hole in the vent plugs must be kept open.

Caution. The operation of a battery produces gases which may be explosive if ignited. Lighted matches or burning cigarettes, sparks, and inflammatory liquids must never be allowed near a battery. Finally, it is necessary before touching the battery or vent plugs to discharge any static electricity generated on one's person or clothing. This can be done by touching a grounded metal object not attached to the battery.

TRANSFORMERS

Transformers are built to do just one job—change voltage and current values. They are designed to perform as efficiently as possible. Cores are laminated, winding turns are shortened, and artificial cooling is used. All these make the transformer the most trouble-free and efficient device used by electricians. Transformers get their name from the kind of work they do—they *transform electrical* power.

Transformers do not increase or decrease power. Remember that power is voltage multiplied by current. And when a transformer changes voltage it also changes current. If you *increase* the voltage, the current is *decreased.* One goes up and the other goes down—exactly enough to keep the power *output* the same as the power *input.* The *voltage* and *current* of a circuit *are changed by transformers.* But the *total power* remains the same.

Transformers require little care and maintenance because of their simple, rugged, and durable construction. The efficiency of transformers is high, and it is probably because of this fact that transformers are responsible for the more extensive use of an alternating current than direct current. The conventional constant-potential transformer is designed to operate with the primary connected across a constant-potential source and to provide a secondary voltage that is substantially constant from no load to full load.

Various types of small, single-phase transformers are used as component parts of equipment. In many installations transformers are used on switchboards to step-down the voltage for indicating lights. Low-voltage transformers are included in some motor control panels to supply control circuits or to operate overload relays. Other common uses include low-voltage supply for special signal lights, and high-voltage ignition circuits for automatic oil burners.

Instrument transformers include *potential,* or *voltage, transformers,* and *current transformers.* Instrument transformers are commonly used with a-c instruments when high voltages or large currents are involved.

Electronic circuits and devices employ many types of transformers to provide necessary voltages for proper electron tube operation, interstage coupling, signal amplification, and so forth. The physical construction of these transformers differs widely.

The *power-supply transformer* used in electronic circuits is a single-phase constant-potential transformer with one or more secondary windings, or a single secondary with several tap connections. These transformers have a low volt-ampere capacity and are less efficient than large constant-potential power transformers. Most power-supply transformers for electronic equipment are designed to operate at a frequency of 50 to 60 cycles per second. Aircraft power-supply transformers are designed for a frequency range of from 400 to 1,600 cycles per second. The higher frequencies permit a saving in size and weight of transformers and associated equipment.

CONSTRUCTION

The ordinary transformer has two windings insulated electrically from each other. The windings are wound on a common magnetic circuit made of laminated sheet steel. The principal parts are: (1) The *core*, which provides a circuit of low reluctance for the magnetic flux; (2) the *primary winding*, which receives the energy from the a-c source; (3) the *secondary winding*, which receives the energy by mutual induction from the primary and delivers it to the load; and (4) the *enclosure*.

When a transformer is used to step up the voltage, the low-voltage winding is the primary. Conversely, when a transformer is used to step down the voltage, the high voltage winding is the primary. It is common practice to refer to the windings as the primary and secondary rather than the high- and low-voltage windings.

The principal types of transformer construction are the *core-type* and *shell-type* shown in Fig. 30, A and B. The cores are built of thin stampings of silicon steel. Eddy currents, generated in the core by the alternating flux as it cuts through the iron, are minimized by using thin laminations and by insulating adjacent laminations with insulating varnish. Hysteresis losses, caused by the friction developed between magnetic particles as they are rotated through each cycle of magnetization, are minimized by using a special

grade of heat-treated grain-oriented silicon-steel laminations.

In the *core-type* transformer, the copper windings surround the laminated iron core. In the *shell-type* transformer
the iron core surrounds the copper windings. Distribution
transformers are generally of the core type; whereas some
of the largest power transformers are of the shell type.

If the windings of a core-type transformer were placed
on separate legs of the core, a relatively large amount of
the flux produced by the primary winding would fail to

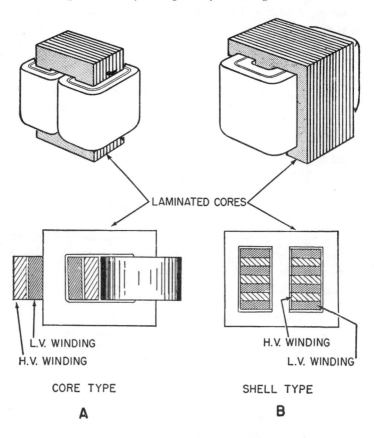

Fig. 30. Types of transformers.

link the secondary winding, and a large leakage flux would result. The effect of the leakage flux would be to increase the leakage reactance drop in both windings. To reduce the leakage flux and reactance drop, the windings are sub-divided and half of each winding is placed on each leg of the core. The windings may be cylindrical in form and placed one inside the other with the necessary insulation. The low-voltage winding is placed with a large part of its surface area next to the core, and the high-voltage winding is placed outside the low-voltage winding in order to reduce the insulation requirements of the two windings. If the high-voltage winding were placed next to the core, two layers of high-voltage insulation would be required, one next to the core and the other between the two windings.

In another method, the windings are built up in thin flat sections called pancake coils. These pancake coils are sand-wiched together, with the required insulation between them.

The complete core and coil assembly are placed in a steel tank. In commercial transformers the complete assembly is usually immersed in a special mineral oil to provide a means of insulation and cooling.

Transformers are built in both single-phase and polyphase units. A three-phase transformer consists of separate in-sulated windings for the different phases, wound on a 3-legged core capable of establishing three magnetic fluxes displaced 120 in time phase.

VOLTAGE AND CURRENT RELATIONS

The operation of the transformer is based on the principle that electrical energy can be transferred efficiently by mu-tual induction from one winding to another. When the primary winding is energized from an a-c source, an alter-nating magnetic flux is established in the transformer core. This flux links the turns of both primary and secondary, thereby inducing voltages in them. Because the same flux cuts both windings, the same voltage is induced in each turn of both windings. Hence, the total induced voltage in each winding is proportional to the number of turns in the winding.

EFFICIENCY

The efficiency of a transformer is the ratio of the output power at the secondary terminals to the input power at the primary terminals. It is also equal to the ratio of the output to the output plus losses, that is efficiency =

$$\frac{\text{output}}{\text{input}} = \frac{\text{output}}{\text{output} + \text{copper loss} + \text{core loss}}$$

The ordinary power transformer has an efficiency of 97 to 99 per cent. The losses are due to the copper losses in both windings and the hysteresis and eddy-current losses in the iron core.

The copper losses vary as the square of the current in the windings and as the winding resistance. In the transformer being considered, if the primary has 1,100 turns of number 23 copper wire, having a length of 1,320 feet, the resistance of the primary winding is 26.9 ohms. If the load current in the primary is 0.5 ampere, the primary copper loss is $(0.5)^2$ X 26.9 = 6.725 watts. Similarly, if the secondary winding contains 110 turns of number 13 copper wire having a length of approximately 132 feet, the secondary resistance will be 0.269 ohm. The secondary copper loss is $I_2^2 R_2$, or $(5)^2$ X 0.269 = 6.725 watts, and the total copper loss is 6.725 X 2 = 13.45 watts.

The core losses, consisting of the hysteresis and eddy-current magnetic flux in the core, are approximately constant from no load to full load, with rated voltage applied to the primary.

POLARITY

Standard markings have been adopted for transformer terminals to insure that the instantaneous polarity of the induced voltages of the windings will be known. This knowledge is necessary in paralleling transformers or in connecting them in wye or delta in 3-phase circuits and in properly metering power, current, and voltage. These markings are called *polarity markings.*

Polarity markings do not indicate the internal voltage stress in the windings but are useful only in making external connections between transformers.

SINGLE-PHASE CONNECTIONS

Single-phase distribution transformers usually have their windings divided into two or more sections. When the two secondary windings are connected in series their voltages add.

In the series connection, care must be taken to connect the coils so that their voltages add.

In the parallel connection, care must be taken to connect the coils so that their voltages are in opposition.

THREE-PHASE CONNECTIONS

Power may be supplied through 3-phase circuits containing transformers in which the primaries and secondaries are connected in various wye and delta combinations. *For example,* three single-phase transformers may supply 3-phase power with four possible combinations of their primaries and secondaries. These connections are: (1) primaries in delta and secondaries in delta, (2) primaries in wye and secondaries in wye, (3) primaries in wye and secondaries in delta, and (4) primaries in delta and secondaries in wye.

If the primaries of three single-phase transformers are properly connected (either in wye or delta) to a 3-phase source, the secondaries may be connected in delta.

If the three secondaries of a live transformer bank are properly connected in delta and are supplying a balanced 3-phase load, the line current will be equal to 1.73 times the phase current. If the rated current of a phase (winding) is 100 amperes, the rated line current will be 173 amperes. If the rated voltage of a phase is 120 volts, the voltage between any two line wires will be 120 volts.

CHAPTER 6

Use of Electricity in Your Home

Once we have become accustomed to electricity in our homes, most of us are inclined to take it as a matter of course. It seems so natural to turn on the light or to plug in a vacuum cleaner that we forget the following important points about electricity and electrical appliances:

Safety. How can we control electricity properly, and avoid danger of shock and fire by unsafe use? Careless handling, faulty wiring, poor materials, and bad workmanship all add up to danger.

Efficiency. How can we use electricity to get the best returns for the money we spend for electric service?

Economy. How can we *manage* not to waste electricity? Most families economize by being careful about *turning off the lights*. Actually, we are being unthrifty when we overload circuits, use appliances not suited to our present wiring, or invest in second-rate materials, such as cords, which use more current—and add to our electric bills.

This is the electrical age. Electricity is a cheap transported form of power. It can be used in small amounts to run a razor, light our house, bring us music and world events over the air, or connect us by telephone with our friends. It can be used in large amounts to run the motors in our factories, produce metals from ore, power streamlined trains, and for hundreds of other industrial applications.

Electric current consists of a movement or flow of electricity and may be divided into two general classes: direct current (d.c.) and alternating current (a.c.). Direct current always flows in the same direction. In alternating current the direction is reversed at regular intervals.

USE ELECTRICITY SAFELY

Water is a conductor of electricity. Electricity always seeks the shortest and easiest path to the ground. If the human body completes the circuit, shock is naturally the result.

Most everyone has received slight shocks from electrical equipment without harm. This gives a false sense of security. Actually, such shocks are a warning that immediate repairs are needed. In every home, it is wise to use precautions, especially in places where water or dampness can cause fatal shock, if conditions are right.

In bathrooms. If you have a pull chain light fixture over, or near, the wash bowl, make sure that it has an insulating link. When remodeling an old bathroom or installing a new one, a wall switch near the door is safer than a pull-chain at the fixture. Make sure that all cover plates for wall switches and outlets are of insulating material, such as bakelite. Locate the electric switch at least an arm's length (preferably) from the lavatory and bathtub. Finally, never touch any electrical appliance while in the bathtub.

When using an electrical appliance near a water faucet, keep far enough away from the faucet to prevent your touching the appliance with one hand and the faucet with the other.

When handling electrical equipment, do not touch water pipes, faucets, or radiators with the other hand. Also make sure your hands are thoroughly dry when handling appliances already plugged into the outlet.

Keep appliances and cords *in good repair* all the time. Defective appliances in need of repair are always dangerous. *For example,* a wire connection in your vacuum cleaner may have loosened. While using the cleaner, you take hold of a metal floor lamp in which a bare wire touches the metal frame. Such a contact completes an electrical circuit, and may give you a dangerous shock. Even a slight shock from an electrical appliance is a warning that inspection and prompt repair is needed.

ELECTRIC WASHING MACHINES

Electric washing machines present special problems of safety. Usually, motors and switches are well insulated. However, if these parts become worn, the insulation will be weakened. A worn cord may cause a wire to come in contact with the metal frame. The washing machine will then become "alive" and because water is a good conductor of electricity, a fatal shock may be the result.

All washing machines should be grounded. As with other electrical appliances, it is a good practice to check your washer periodically for defective wiring.

WIRING

Wiring either a new or remodeled home for electricity is usually a job for an experienced electrician. However, knowing some general facts about electricity in the home will help the homemaker to plan the wiring layout more efficiently.

To begin with, it is helpful to compare your home electric system with your water system. Both come to your home from a central outside source—the water through pipes, the electricity through wires. Both require certain conditions for smooth and efficient operation.

As a homemaker, you have doubtless had the experience of turning on a faucet and getting only a thin trickle of water, or perhaps a stream of water that runs slowly. The same thing may happen in connection with your electric system. Lights may flicker or burn dimly. Or you may plug in an appliance, blow a fuse and temporarily disrupt your electrical service.

The most important thing to remember at the start is to give electricity a chance to do its best work for you by providing adequate wiring. Just as your water system requires pipes large enough to carry water to different parts of your house, so your electric system needs wires of the proper size to convey electricity to all rooms where you want lighting, or where you will use electrical appliances.

Whether remodeling an old house or building a new one, planning ahead for future electrical needs is as important as thinking ahead on plumbing. Nine chances out of ten you will be adding this or that convenience—an electric iron or range, a TV set, or a freezer. Providing enough circuits at the start saves money in the long run.

USE ELECTRICITY EFFICIENTLY

The following three important guideposts to good electrical wiring are easy to remember.

1. The size of the *service entrance wires.*
2. The size of the *service entrance switch.*
3. The number and kinds of *circuits.*

We will take up each in turn, beginning with the *service entrance* wires which bring the electricity from the power line into your home.

The size of the service entrance wires determines the amount of electrical power that can be brought into your house at any one time. Consequently, these wires must have enough carrying capacity to provide all the power needed. For most average houses, three No. 6 wires are usually recommended. However, three No. 4 wires (the next size larger) are more efficient, assuring enough power for both present and future needs.

The size of the service entrance switch is determined by the electrical equipment you now have, or plan to have in the future.

For three No. 6 wires a 60-ampere switch is commonly used. This will provide enough power for both lights and major appliances, including a range, water heater, iron, washer, freezer, and the like.

For three No. 4 wires you will need a 90-ampere switch. These larger wires and the larger switch are especially needed if you have appliances that require more than the usual power, or if you plan to invest in these later.

In some houses, No. 2 service entrance wires with a 100-ampere switch are used. This, however, is usually in connection with the completely electrified home with fully modernized equipment in both kitchen and workroom. Two-family homes where the equipment load is heavy may also find this combination more efficient.

TYPES OF CIRCUITS

In today's electrified homes, the following three kinds of branch circuits are needed. Few older homes have enough circuits to carry the average electrical load.

1. General-purpose circuits
2. Appliance circuits
3. Individual circuits

General-purpose circuits are designed to serve all lights, as well as convenience outlets in all rooms except the kitchen, the dining areas, and the workroom. Low-wattage appliances (500 watts or less) can be served by these circuits. This includes most floor and table lamps, radios, electric clocks, vacuum cleaner, sewing machine, and the like. The wattage load on a general-purpose circuit, however, should not exceed 1725 watts at one time. (*See* Table 11.)

TABLE 11

WIRE AND FUSE SIZES
(Wattage Load that Can Be Safely Carried)

	Wire Size (American Wire Gauge)	*Fuse Rating (Amperes)*	*Voltage of Circuit*	*Total Load in Watts*
General Purpose Circuits ----------	No. 14	15	115	1725
Appliance Circuits ----------------	No. 12	20	115	2300
(toaster, iron, roaster, etc.)	No. 10	30	115	3450
Individual Circuits ---------------	No. 10	30	230	6900
(range, water heater, drier, etc.)	No. 8	40	230	9200
	No. 6	55	230	12650
Service Entrance Wires -----------	No. 4	70	230	16100
	No. 2	95	230	20950

Provide at least one general-purpose circuit for each 500 feet of floor space. *For example,* a house having 2000 square feet of floor space needs a minimum of four general-purpose circuits. One or two extra circuits are also advised as a provision for future needs.

To serve the kitchen and dining areas of a home and the all-important workroom, appliance circuits are needed for toaster, roaster, electric iron, washer, and the like. As the term implies, such circuits should serve appliances only. The National Electrical Code requires that every house should have at least one appliance circuit. However, for greater efficiency, it is well to provide two or more of these special circuits, thereby avoiding an overload if extra appliances should be purchased later.

Every home also needs one or more *individual* circuits. For satisfactory and efficient use, an individual circuit is recommended for each major piece of electrical equipment, such as a range, water heater, home freezer, automatic heating plant, automatic washer, clothes dryer, dishwasher, waste disposer, built-in bathroom heater, and the like. The number of individual circuits in your home will depend on the type of electrical equipment you already have, and what you plan to buy.

LONG CIRCUITS

Because of loss of electrical energy, it is desirable to *avoid long circuits.* When using No. 14 wire, the outlets should be not more more than 25 feet from a fuse box or panel. With a No. 12 wire, the distance to an outlet should not be more than 35 feet.

Occasionally, in large houses, it is necessary to have longer branch circuits. In such cases, the circuits may be shortened by installing fuse boxes or circuit breaker panels. These panels should be supplied by a feeder wire of suitable size from the *service entrance switch.* Such a feeder wire should be at least a No. 10 size.

"OVERLOADING" PROBLEM

The importance of installing enough electric circuits in a new or remodeled house is best illustrated by the following explanation of "overloading." Perhaps you want to use your 1000-watt iron and 1350-watt roaster *at the same time.* Both are on the same circuit which in most houses consists of two No. 14 wires protected by a 15-ampere fuse. The two appliances, both heating at the same time, "overload" the wire and "blow" the fuse.

In such a situation, the blown fuse is often replaced with a larger 20- or 25-ampere fuse. The wire will now try to carry enough current for both the iron and the roaster. As a result, both appliances will slow down, take longer to do the job, and use more electricity. At the same time, the wires may become overheated and create a fire hazard.

Providing enough circuits to take care of the appliances you already have, as well as those you hope to have, makes for more efficient use of your electrical services.

SIZE OF WIRES

For household use, electric wires come in the following sizes: No. 14, 12, 10, 8, 6, 4, 2. The larger the number, *the less the load. For example,* No. 14 electrical wire carries only a 1725-watt load, whereas a No. 2 wire, mainly used for *service entrance wiring,* can carry up to 20,950 watts. Homemakers will note that wire sizes are like thread sizes —the higher the number, the finer the thread.

In most homes, No. 14 (American Wire Gauge) is commonly used for general-purpose circuits. For greater efficiency, however, No. 12 should be more widely used. For appliance circuits, the National Electrical Code permits the use of No. 12 wires. In most cases, a No. 10 wire will serve appliances more efficiently, especially where several are used at one time.

The size of wire for individual circuits serving a single piece of equipment, such as a water heater, range, clothes dryer, and so on, depends on the wattage of the equipment, and the distance it is located from the *entrance switch.* (*See* Table 11.)

CORDS

When buying new cords, choose the proper type for your particular use. Choose lamp cords for lamps and low wattage appliances; asbestos insulated cords for heating appliances; heavy-duty cords for tools; and rubber-covered cords for all electrical equipment that will be used in damp or wet places.

UL Label on Cord. Some years ago, manufacturers of electrical products asked the Underwriters' Laboratories, Inc., a recognized authority on safety, to work out ways by which the public could know which cords comply with safety standards. As a result labels are attached to cords and cord sets which meet UL standards. An enlarged sketch of the bracelet-type label is shown in Fig. 1. It is *yellow* and is attached at five-foot intervals on all types of approved flexible cords.

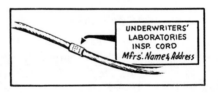

Fig. 1.

A *green* label is attached to a length of cord which has an attachment plug assembled to one end with bare wires on the other end. The green label may be either a bracelet wrap-around label, or a "flag" type label.

The "flag" type label consists of a strip of paper that is about one-third inch wide with an adhesive back. It is wrapped around the cord once and the two loose ends are pressed together, which then projects from the cord as a flag. This flag bears the words "Inspected—Underwriters' Laboratories, Inc."

A *blue* label (shaped like a washer) placed on cord sets indicates that the plug and cap as well as the cord have been approved by the UL.

Cord sets may also be labeled with either a *blue* "flag" type or a *blue* bracelet-like label. These labels have the same meaning as the "doughnut" or washer shaped label.

TYPES OF CORDS

Electrical cords come in different sizes. The two sizes most commonly used in homes are Nos. 16 and 18. No. 16 is larger than No. 18 and carries more electricity. No. 18 is satisfactory for lamps and will do for low-wattage appliances. However, a No. 16 cord is more efficient for low-wattage appliances, and is a *must* for high-wattage appliances.

Some cords get more wear than others. Your ironing cord is an example. Strongly made, durable ironing cords are a worthwhile investment. The same may be said for other cords that must stand up under continuous and hard usage. Cords with cotton or rayon outer coverings can only be used where it is dry.

Brief information about different kinds of cords follows. (*See* Fig. 2.)

Type SP, rubber, or SPT, thermoplastic covered (Fig. 2, A) is recommended for lamps, clocks, radios, and other small appliances. Avoid using on appliances that produce heat (irons, toasters, radiant heaters, and the like) or those which receive hard usage. They are available in black, brown, or ivory colors and in sizes 18 and 16.

Type HPD heater cord (Fig. 2, B) is the only cord recommended for appliances that give off heat (irons, toasters, waffle makers, and the like). They are asbestos-covered wires with cotton, rayon, or rubber outer covering. Size 16 is recommended; Size 14 is also available.

Type SV vacuum cleaner cord (Fig. 2, C), either rubber or thermoplastic outer covering, may be used in damp places and where given hard wear. This cord is available in Size 18 only.

Type SJ junior hard service cord (Fig. 2, D) should be used for home appliances, food mixers, vacuum cleaners, washing machines, and the like. It is also recommended for extension cords, except for appliances that produce heat. This cord is covered with either rubber or thermoplastic, and comes in Sizes 18 and 16.

Type S hard service cord (Fig. 2, E) is the most durable of all cords. Recommended for large motors and for use in workshops and garages where cords get hard wear, this cord is covered with rubber or thermoplastic, and is available in Sizes 16, 14, 12, and 10.

EXTENSION CORDS

Well-planned wiring minimizes the need for extension cords. When extension cords are needed, follow these suggestions for safety.

1. For irons, use an extension cord made from Size 16, Type HPD heater cord.

2. For lamps, radios, and other low-wattage appliances, use rubber or thermoplastic covered extension cords. Avoid cotton or rayon covered cords.

3. Make extension cords no longer than necessary. The longer the cord, the greater loss in electricity.

ATTACHMENT CAPS

On most cords, caps are molded around the cord (Fig. 3, A). These are more durable. When buying new caps, consider the style shown at Fig. 3, B, or a type of cap with cord grips, as shown at Fig. 3, C. Caps like this are more easily and firmly gripped and have spring-action prongs which hold them more securely in the outlet.

Avoid small, fragile bakelite caps that break easily and are difficult to remove from tight-fitting receptacles.

REPAIRING CORDS

Watch your cords carefully, keep them in good repair, and make all repairs at once. Whenever outside covering

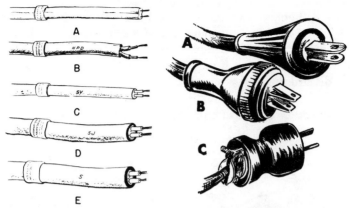

Fig. 2.   Fig. 3.

breaks down, wrap with electrical tape. Always replace a badly worn cord as soon as possible. Spliced cords are not considered safe and should be used only in emergencies.

Frayed, worn ends, as shown in Fig. 4, are a common ailment of household cords. Most homemakers can learn to repair them.

First, turn the cap toward you so you can see the inside. This should be covered by a protector disc (a fiber insulator). Remove this disc by prying gently with a small screwdriver. Then proceed as follows.

1. Loosen screws using screwdriver (Fig. 5). (Do not remove screws completely.) Pry out each *conductor* from around its screw.

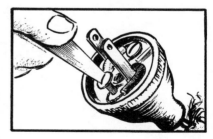

Fig. 4.   Fig. 5.

2. Slip cap down on cord. Cut worn part of cord back to where rubber insulation is firm (Fig. 6). Be sure to *cut evenly.*

3. Remove two inches of outer covering (Fig. 7). Examine rubber insulation around each conductor. If this splits easily, cut off more cord. Repeat until you find firm rubber insulation. If the rubber insulation keeps on splitting, discard the cord.

4. If outer covering is cotton or rayon, wrap coarse thread or electrical tape around end to prevent fraying, as shown at Fig. 8, A.

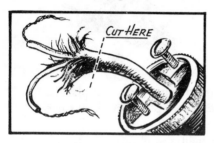

Fig. 6.

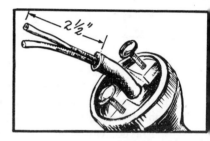

Fig. 7.

5. If cap opening is large enough and rubber insulation is in good condition, make an Underwriter's knot, as shown in Fig. 9. Otherwise, omit knot. The Underwriter's knot is especially good for a drop light, or wherever the cord must support a weight. In this case, apply the knot at both the ceiling and socket end.

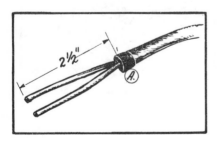

Fig. 8.

Fig. 9.

6. Bring each conductor around its adjoining prong in a clockwise direction and measure to edge of screw head. Cut strands evenly to within ½ inch from this point. Remove ½ inch of rubber insulation, being careful not to cut the tiny wire strands (Fig. 10). If strands are dark, scrape with dull edge of knife or polish with emery cloth. Twist the wire strands of each conductor to the right to make a firm end.

7. Bring each twisted end around its adjoining prong and screw in a clockwise direction. Fasten screws and replace fiber disc.

HOW TO CONNECT APPLIANCE PLUGS

Cords used for flat irons, electric plates, roasters, and the like are known as *heater* cords. They have cotton outer coverings and asbestos-covered conductors in order to withstand heat. Cords of this type will last longer if the plugs at each end are interchanged occasionally. To repair the plug (Fig. 11) at the appliance end, proceed as follows.

1. Be sure cord is disconnected from wall outlet. Take the plug apart by removing screws or spring clips (Fig. 12).

Fig. 10.   **Fig. 11.**

2. Loosen the screws that hold the conductors. Do not remove screws (Fig. 12). Cut off cord until you find firm rubber insulation. *Be sure* to cut cord evenly.

3. Wrap electrical tape or strong thread around cord to prevent cotton outer covering from fraying. (*See* Fig. 13, A.)

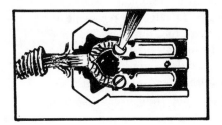

Fig. 12.

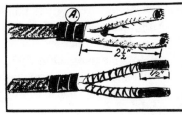

Fig. 13.

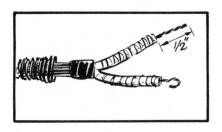

Fig. 14.

4. Remove ½ inch of the asbestos from the end of each conductor (Fig. 14). Wrap remaining asbestos with strong thread to keep in place.

5. Remove ½ inch of rubber from the end of each conductor. Make sure ends are cut evenly. Twist wire strands of each conductor firmly to the right and form into curve as shown in illustration. Fit around screws in a clockwise direction. Fasten screws firmly with screwdriver. *Caution:* Be sure the conductors and spring fit into the right grooves.

Fit the two parts of the plug together and replace screws or clips.

USE ELECTRICITY ECONOMICALLY

Electricity is measured and sold by the kilowatt-hour (kwh). The kilowatt-hour is the unit of energy measured by the meter. The *watt* is the actual unit of power, but since it is so small the kilowatt (1000 watts) is used. To figure the cost of operating an appliance, it is necessary to know its *wattage rating*. This is usually stated on the name plate of the appliance.

When figuring the cost, the homemaker needs to know the power (watts) and length of time the power is used (hours). *For example,*

$$\frac{\text{Watts X hours used}}{\text{divided by 1000}} = \text{kilowatt hours.}$$

For example, if an iron using 1000 watts, according to the name plate, is operated continuously for five hours, it would use five kilowatt-hours.

$$\frac{\text{divided by 1000}}{\text{100 watts X 5 hours}} = \frac{5000}{1000} = \text{kilowatt-hours}$$

Kilowatt-hours multiplied by the rate per kilowatt-hour equals the cost of operating: Five kilowatt-hours times four cents per kilowatt-hour equals 20 cents. In the case of electric irons and some other appliances, where there is automatic heat control, the current is not on all of the time. The cost figured may then be cut one third to one half.

NAME PLATES

A name plate (or stamping) on each piece of equipment tells what kind of current (alternating current, a.c., or direct current, d.c.), frequency of current (cycles), voltage (v), or pressure the appliance was made to operate on, and watts or amperes used by the appliance.

The ampere is the measure of rate of flow of electricity like gallons per minute in a water system. When the number of amperes required by a piece of equipment is indicated on the name plate, it can be changed to watts by using this formula:

Volts X amperes X power factor = watts.

For example, an appliance with the above name plate would be figured as follows: 1.8 amperes X 115 volts X 1.00 = 207 watts. The power factor is 100 per cent (1.00) for heating equipment and incandescent lights, and about 50 per cent (0.5) for small motors.

Alternating current (a.c.) is the kind that is found in our homes. Direct current (d.c.) is the kind that is stored in the car battery and still exists in a few older industrial areas.

Read the name plate on each piece of equipment and observe the kind of current on which it is made to operate. Those marked a.c. only must be used on alternating current. Those marked d.c. only must be used on direct current. Those marked a.c. or d.c. can be used on either.

Voltage. The volts on the name plate give the voltage on which the appliance was made to operate. Common lighting and appliance voltage ranges from 110 to 120. For heavy appliances and motors of one half horsepower and larger, the common voltage ranges from 220-240.

Wattage. The watts indicated on the name plate indicate the power the appliance needs. When purchasing equipment, this must be considered unless the house has been wired with large enough wire and enough circuits to supply enough electricity. Therefore, it is important to plan circuits that will carry the desired amount of power.

Motor Appliances. The name plate on appliances with motors usually states horsepower (hp). This may vary from 1/20 to ½ hp. A motor does not operate at 100 per cent efficiency. In most household appliances, the efficiency of small motors is usually from 30 per cent to 60 per cent.

The operating cost of motor-driven appliances usually runs considerably less than heat appliances.

Table 12 indicates the approximate wattage of different appliances. Figures cannot be exact because various makes and sizes differ in amount of power required.

For other appliances, see rating on name plate.

FUSES

A *fuse* is a "safety valve" that prevents the flow of more current than the circuit should carry. In this way, circuits

TABLE 12

WATTAGE OF APPLIANCES

Wire Size (American Wire Gauge)	Fuse Rating (Amperes)
No. 14	15
No. 12	20
No. 10	30
No. 8	40
No. 6	55
No. 4	70

are protected against overloading. In some new homes, circuit-breakers (described later in this chapter) are replacing fuses. Both, however, are a protection that add to the safety and efficiency of your electrical system.

CAUSES OF BLOWN FUSES

Some of the *causes of blown fuses* are (1) overloading the circuit by plugging in too many appliances, (2) a short circuit caused by a worn-out cord or faulty plug connection, and (3) starting a motor. Electric motors require three to seven times more electricity to start than they do to operate. A washing machine started in gear may use just enough additional current to blow a fuse.

Use of right size fuse. The importance of the right size of fuse used to protect the circuit cannot be stressed enough for safety. When a fuse blows out, never replace it with a larger size fuse or a penny. Actually, this permits an overload that is not only inefficient but dangerous. Too often, the danger of overloading is not understood. Excessive heat resulting from overloading a circuit causes the insulation to harden, crack, and break away. This may cause a fire.

First, determine the size of wire in your circuit. This may be done quite simply by comparing the size of the bare wire with that of various common nails used around a home. (Compare with body of nail and not nail head.) Make this comparison at your fuse box, or wherever wires are attached to switches or outlets. *Be sure to open your main switch* so current will be off while you are taking these measurements. Next, refer to Table 13 to find the size of fuse you need for that circuit.

TABLE 13

FUSE/RATING (AMPERES)

Wire Size (American Wire Gauge)	Fuse Rating (Amperes)	Diameter (bare) Inches	Nails about the size of electrical wire
No. 14	15	0.064	3-penny finishing nail
No. 12	20	0.081	3-penny common nail
No. 10	25	0.102	8-penny finishing nail
No. 8	35	0.128	8-penny common nail
No. 6	50	0.162	16-penny common nail
No. 4	70	0.204	30-penny common nail

CHANGING A FUSE

To change a fuse proceed as follows (*See* Fig. 15).
1. Disconnect appliance.
2. Open switch.
3. Stand on a dry board.
4. Replace fuse.

Fig. 15.

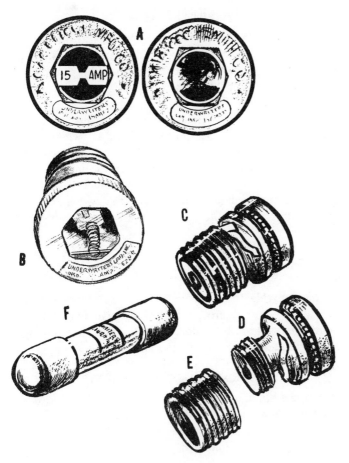

Fig. 16.

TYPES OF FUSES

Figure 16 shows the types of fuses: (A) Common plug fuse, left; blown plug fuse, right. (B) Fusetron, a time-lag fuse designed to prevent "blowing" when motors start. It fits standard socket. (C) shows fustat and adapter ready to screw into fuse socket. (D) shows fustat without adapter. (E) shows an adapter needed with fustat so it will fit fuse socket and need not be replaced when changing a fuse. (F) shows a cartridge fuse for main switch.

CIRCUIT-BREAKERS

Instead of fuse-box a circuit-breaker panel board is sometimes used. Instead of fuses, circuit-breakers give complete protection and are tamper-proof. There is nothing to replace. After trouble has been corrected, your lights go on again at a flip of the switch.

For safety and convenience, circuit-breakers are now widely recommended. (*See* Fig. 6, Chap. 4.) These offer automatic protection against overloads. On the circuit-breaker panel board are a series of switches, one for each of your circuits. When a short circuit or an overload occurs, the circuit is automatically opened and the switch changes position. After the trouble has been corrected, simply reset the switch.

Install high quality circuit-breakers which will trip quickly without sticking. Cheap circuit-breakers are less reliable.

Fustats provide the same protection as circuit-breakers at less cost, but they are not quite as convenient. Fustats are tamper resistant, have delayed action, and the adapter prevents the use of pennies or too large a fuse.

ELECTRICAL CODE REQUIREMENTS

If you are building a new home, remodeling an old one, or making changes in your wiring system, you would do well to know what your state and local codes specify. These codes *must* be followed.

CHAPTER 7

Electric Motors

Electric motors provide power in one of the safest forms known. They are economical and highly efficient, and they can be operated with the push of a button or the flick of a switch. Their initial cost is relatively low in proportion to the power they develop. Electric motors are capable of starting a reasonable load, withstanding temporary overloading, and have the advantage of being automatically and remotely controlled.

The amount of energy provided by a motor is measured in *horsepower*. A little less than a kilowatt, 746 watts (electrical power input) is required to produce one horsepower (mechanical power output).

SELECTION OF MOTORS

In the purchasing of a motor, the type of electrical power available, the type and size of load, and the conditions under which the motor will operate must all be taken into consideration. Local codes and regulations should be carefully followed in making each installation, and all parts of the electric system should be maintained in good condition.

Type of Electric Power Available. Electricity flows in either alternating current (a.c.) or direct current (d.c.) and is available in several voltages. Alternating current may be single-phase or three-phase and obtainable in several frequencies (cycles per second). Generally, electric power is single-phase, 60-cycle a.c. and is available at 120 and 240 volts. There is also a 208 volt. Power companies can tell you if service is 120/240 volt or 120/208 volt. The 240 volt motors should not be used on 208 or vice versa. The voltage ratings of single-phase motors are 115 and 230 volts which are slightly lower than the line voltages of 120 and 240 volts to allow for the voltage drop.

Starting (Inrush) Convert. When a motor starts there is an initial surge of power required to get the motor up to speed. This starting current may be many times the size of the motor's normal operating requirement, and it is inadvisable to add to it. For example, a power-driven grinding wheel, used for sharpening tools, should reach full operating speed before actual grinding is started.

Motors used on water pumps, milking machines, and refrigerators are constantly starting up. For this reason, they are built differently from motors which are run continuously.

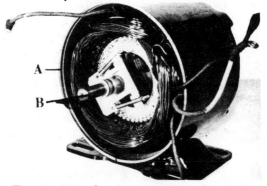

Fig. 1.   Typical motor stator (A) and rotor (B).

Operating Conditions. Since motors are used under all kinds of conditions, it is often necessary to choose the proper type of motor enclosure. Motors should be protected from such unfavorable factors as dust, dirt, excessive moisture, inflammable objects, explosive gases, and rodents.

STRUCTURE OF MOTORS

All motors consist of a rotating part, called the rotor, which revolves freely within a stationary part, called the stator (Fig. 1).

Rotor. The rotor consists of a slotted core made of thin sections of special soft steel. It is carefully balanced on a central shaft that has a ground bearing surface at each end of the core. This shaft extends beyond the bearing surface at one or both ends to provide for pulleys or other means of attachment to the device it drives. Rotors may be the *squirrel-cage type* (Fig. 2) or the *wound-rotor type* (Fig. 3).

Fig. 2. Squirrel-cage type of motor.

Fig. 3. Brush-lifting type of wound rotor.

The squirrel-cage rotor derives its name from the fact that it resembles the cage sometimes used to exercise pet squirrels. The slots of the rotor contain bare copper, brass, or aluminum bars which are short-circuited together at each end by the end rings. Most squirrel-cage rotors also have some type of cooling fan.

The single-phase wound rotor, used in a repulsion-start induction motor, has coils of insulated copper wire wound in the rotor slots. It also has a commutator made of copper segments that are insulated from the rotor shaft and from each other with mica or a similar substance. The ends of the rotor coils are soldered to individual commutator segments. When the motor brushes contact these segments, they complete the circuit, thus permitting currents to flow through the coils in the rotor in the proper sequence for starting. These rotors have a cooling fan and a centrifugal device for short-circuiting all the commutator segments when the rotor comes up to speed. The wound rotor shown in Fig. 3 also has a brush ring and a device for lifting the brushes away from the commutator at the same time that the segments are short-circuited.

Fig. 4.

Stator. The electrical part of the stator consists of a special laminated-steel slotted core. Insulated copper wire is wound in the slots to form one or more pairs of definite magnetic poles (Fig. 4). In so-called "constant-speed" motors, their speed is determined by the frequency of the power supply and the number of poles. With ordinary 60-cycle current, the full-load running speed of a 2-pole motor is about 3450 r.p.m.; a 4-pole motor, 1725 r.p.m.; and a 6-pole motor, 1140 r.p.m.

Motors also have a *frame, end shields,* and *through bolts* or *cap screws* (Fig. 5). The motor is mounted and supported in a frame and the end shields house the bearings. Usually either the starting switch or the brushes and the terminal box where the attachment is made to the line are enclosed in one of the shields. The through bolts or cap screws hold the motor together.

TYPES OF MOTORS

The three types of single-phase, a.c. motors generally used are the split-phase, capacitor, and repulsion-start induction motors. Two additional types of motors are the universal, which may run on either a.c. or d.c., and the three-phase induction motors. The first three types differ mainly in the manner in which they start and come up to running speed. If they have the same horsepower rating, there is no practical difference in the amount of work they

Fig. 5.

Fig. 6.

can do or in the current they require after they have reached operating speed.

Split-Phase Motor. The typical split-phase motor consists of a squirrel-cage rotor and a stator with two different sets of windings (Fig. 6), the main or running winding, and the auxiliary or starting winding. The running winding has a greater number of turns of larger diameter wire than the starting winding. Generally the running winding is wound in the stator slots first. The motor shown in Fig. 6 has four distinct poles. Both windings have the same number of poles. The poles of the starting winding are spaced halfway between the poles of the running winding.

When a motor is started, the starting switch is closed, allowing the current to flow through both windings. The rotor commences to turn, and when it reaches about three-fourths full speed a centrifugal device opens the starting switch. This disconnects the starting winding from the circuit, and the motor continues to operate on the running winding only. When the motor stops, the starting switch is closed, and both windings in the circuit are ready for restarting.

The split-phase motor is the simplest in construction and

Fig. 7. The split-phase motor is reversed by switching the starting winding leads.

least expensive. It has a relatively low starting torque (ability to start a load) and requires a high starting current. This limits its use to loads that are easy to start. Due to the large starting current required, split-phase motors are rarely made in sizes larger than ⅓ h.p. They are usually made for one voltage, either 115 or 230, and cannot readily be changed from one to the other.

Direction of rotation is determined by the direction the current flows through the starting winding in relation to the direction it flows through the running winding. Switching the starting-winding leads (Fig. 7) or the running winding leads will cause the motor to start and run in the opposite direction. A motor is described as rotating in a clockwise (c.w.) or a counterclockwise (c.c.w.) direction when viewed facing the end opposite the shaft extension. Usually this is the end where the motor lead connections are made.

Capacitor Motor. The two types of capacitor motors used are (1) the capacitor-start and (2) the capacitor-start-capacitor-run (two-value capacitor).

Both the capacitor-start motor and the split-phase motor have squirrel-cage rotors with two separate windings in the

Fig. 8. The capacitor-start motor has a capacitor or condenser, often mounted on top of the motor.

stator and a starting and running winding. The capacitor-start motor also has a capacitor (condenser) placed in series with the starting winding. The smaller sizes of capacitor-start motors are usually identified by the tube-shaped container on the top of the motor which holds the capacitor (Fig. 8). Some motors have the capacitor mounted inside one of the end shields or in the motor base.

Frequently in the larger horsepower ratings a capacitor-start-capacitor-run motor is used. This motor is similar in external appearance to the capacitor-start motor. It has a starting winding and a running winding in the stator, with a capacitor connected in series with the starting winding, and a squirrel-cage rotor. The starting switch in this motor does not remove the starting winding from the circuit as in the capacitor-start type, but serves only to disconnect the starting capacitor when the motor comes up to speed.

A capacitor-start motor is an improved split-phase type. The improvements in the starting circuit, made possible by the addition of the capacitor, give this motor a greater starting and accelerating torque for the same starting current, usually at least twice as great as a split-phase motor of the same horsepower rating. Capacitor-start motors are widely used for such appliances as water pumps, industrial refrigerators, air compressors, and other equipment which start and stop frequently and have relatively high starting torque requirements. Recent developments in capacitors make these motors practical in sizes up to 5 h.p. or 7½ h.p. or larger.

Capacitor motors smaller than ⅓ h.p. are usually wound for 115 volts. Many rated ⅓ h.p. and larger motors can be connected to either 115 volts or 230 volts by changing the lead wires. This information is indicated on the nameplate of the motor. Use the 230-volt connection when 230-volt service is available. The direction of rotation is reversed in the same manner as in the split-phase motor. The nameplate or instruction tag attached to the motor denotes whether it is a capacitor-start or a capacitor-start-capacitor-run motor.

Repulsion-Start Induction Motor. The repulsion-start

Fig. 9. The repulsion-start induction motor has a wound rotor and only a running winding in the stator.

Fig. 10.

induction motor has one winding in the stator which acts as a running winding. Instead of a squirrel-cage rotor, it has a wound rotor with commutator and brushes (Fig. 9). There is no direct connection between the line current and the brushes or rotor windings. The brushes serve to complete the circuit in certain rotor coils. This creates strong magnetic forces within the rotor which react with those of the stator, causing the motor to start. When the motor approaches full running speed, a centrifugal device within the rotor short-circuits all the commutator bars together, and the rotor operates at full speed like the squirrel-cage type.

Some replusion-start induction motors, called brush-riding motors, operate with the brushes in contact with the commutator at all times. They usually have an axial-type commutator (Fig. 10). In other types, called brush-lifting motors, a centrifugal device pushes the brushes away from the commutator as the rated speed is approached. These motors have radial-type commutators (Fig. 10).

Capacitor-start and repulsion-start induction motors have the same high-starting torque. They can be used interchangeably under normal voltage conditions. Since the repulsion-start induction motor has greater starting torque per ampere of current, it is less likely to aggravate a low-voltage condition or to cause trouble by a voltage drop. Both motors are ruggedly built and will give good service for steady or intermittent use.

Most repulsion-start induction motors, even in the fractional horsepower sizes, can be operated on either 120-volt or 240-volt current. The stator winding is usually divided into halves, and four leads are brought into the terminal

Fig. 11.

box. These halves are connected in parallel for 120-volt and in series for 240-volt operation (Fig. 11).

Direction of rotation is determined by the position of the brushes in relation to the centers of the stator coils. The motor can be reversed by shifting the brushes to a different position. Some motors have a brush-shifting lever that extends outside the motor (Fig. 12). In other motors it may

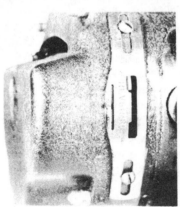

Fig. 12.

be necessary to remove a plate on the end shield, loosen a set-screw, and move a pointer (Fig. 13).

Universal Motor. The universal motor derives its name from the fact that it will operate on either direct or alternating current of the correct voltage. It has a wound rotor and brushes (Fig. 14). The line current in this motor flows through the brushes to the rotor as well as to the stator in such a way that the two windings are in series with each other. Both windings are constantly in the circuit during operation.

Fig. 13.

Universal motors do not operate at constant speed like induction motors. They operate like a gas engine with the throttle wide open. Since they run as fast as their load will permit, they are usually permanently and directly connected to some device or appliance. Universal motors have high starting torques and high starting currents. They are widely used on portable electric drills and on household appliances such as vacuum cleaners, food mixers, and sewing machines.

Three-Phase Motor. Up to the present time, the availability of three-phase power under certain local conditions has been limited. Three-phase electricity consists of three distinct currents that require three or more primary (highline) wires, two or three transformers, and three or more

Fig. 14.

secondary wires to the motor. The utilization voltage is generally 208 volts.

Three-phase motors are very simple in construction and

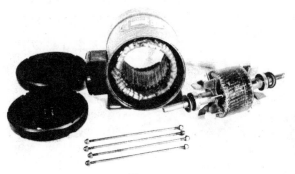

Fig. 15.

relatively low in initial cost. They have three-phase wind-ings in the stator and usually have a squirrel-cage type of rotor (Fig. 15).

The three-phase current produces a rotating magnetic field in the stator that enables the rotor to start and run without any special starting device. Squirrel-cage three-phase motors are free from trouble, for they have no brushes, starting switch, or short-circuiting device.

Direction of rotation of a three-phase motor is determined by the way the three line wires are connected to the motor. Interchanging the connections of any two line leads will cause the motor to rotate in the opposite direction; reversing is therefore a very simple process. Three-phase motors can-not be used on a single-phase line.

MOTOR ENCLOSURES

The type of frame protection or enclosure should be considered when choosing a motor.

The four main types of motor enclosures are open, splash-proof, totally-enclosed, and explosion-proof.

Open motors (Fig. 16) are used indoors where the motor is kept dry and the atmosphere is normally clean. Openings for ventilation are of drip-proof design to prevent objects or liquids from falling into the vital parts of the motor.

Splash-proof motors may be used indoors or sometimes outdoors in mild climates. Their construction will protect the vital parts of the motors even when it is necessary to wash the equipment with a hose. Splash-proof construction is seldom used for motors ¾ h.p. or smaller.

Totally-enclosed construction is designed to protect the motor from dirt and grit in the atmosphere as well as from moisture. These motors are recommended for use under extremely dirty conditions. They do not have ventilating openings. An internal and external frame is sometimes provided with a fan to carry the heat away from the surface of the frame.

Fig. 16. Open motors have openings in the end shields to permit ventilation.

Explosion-proof motors are available in two types. One is designed to withstand an inside explosion of gas or vapor without igniting the gas or vapor surrounding the motor. It is widely used around gasoline and similar vapors. The other is a dust-explosion-proof motor which is designed and built so as not to cause ignition or explosion of a hazardous dust concentration on or around the motor. This type is used in industrial plants where dangerous explosive dusts may be present in hazardous quantities.

BEARINGS

The type of bearings is another factor to consider when selecting a motor. In some sizes and types of motors there

Fig. 17.

may be a choice between sleeve bearings and ball bearings.

Sleeve bearings are steel-backed and babbitt-lined, and some are made of bronze or a similar alloy. Sleeve-bearing motors are oil-lubricated and are generally designed to operate only in a horizontal position.

Motors equipped with *ball bearings* may be operated in a vertical position. Some ball-bearing motors are designed for relubrication at infrequent intervals, while others have prelubricated sealed bearings without provision for re-lubrication.

SIZE OF MOTOR

Determining the size of motor to use for a given load is a frequent problem. Electric motors can withstand momentary overloads. To insure long and satisfactory

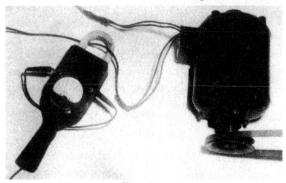

Fig. 18.

service, however, motors should not be subjected to frequent or continuous overloading. The equipment manufacturer should be consulted as to the proper size of motor to use with a given device.

To determine whether a motor is being overloaded, measure the current it consumes in doing its job. Use either a conventional a.c. ammeter of the proper capacity (Fig. 17) or a special clip-on ammeter (Fig. 18). Place the ammeter in series with the circuit when the motor is operating the load in question and compare the ammeter reading with the nameplate rating. Line voltage should also be checked and should not vary more than about 10 per cent from the rated voltage of the motor.

INSTALLATION

The steps for installing a motor include: connecting the motor to the load, mounting the motor, determining the proper size of wire, protecting and controlling the motor, and providing for safety. Installation should conform to provisions set forth in the National Electrical Code or local rules and regulations.

Connection to the Load. Motors are connected to the load by a system of belts and pulleys or by a direct drive.

Belts and Pulleys. V-belts have replaced flat belts for connecting motors to their driven loads. The advantages of the V-belt drive over the flat belt are: V-belts permit a lighter, easier, and more compact assembly. V-pulleys are narrower and will operate satisfactorily at close centers without idlers or belt tighteners. Also, the wedging action of V-belts provides a good grip between the belt and the pulley, thus reducing bearing wear and increasing belt life.

Standard V-belts are available in a variety of lengths and cross-section sizes. The cross-section sizes are designated by the letters A, B, C, D, and E. Types A and B cover most of the requirements for electric-motor drives from the fractional sizes up to and including 7½ h.p. Type B belts should not be used for pulleys smaller than 5½" in diameter.

The number and type of V-belts required for a given drive depend on the size of the motor pulley and the speed and horsepower of the motor (Table 14).

TABLE 14

NUMBER AND TYPE OF V-BELTS RECOMMENDED
FOR 1750 R.P.M. MOTORS

Diameter of Motor Pulley (in Inches)*	Horsepower							
	½	¾	1	1½	2	3	5	7½
	V-Belts Recommended							
2	1-A	2-A	......	......	......	......	......	......
2½	1-A	1-A	......	......	......	......	......	......
3	1-A	1-A	1-A	2-A	2-A	3-A	5-A	8-A
3½	1-A	1-A	1-A	2-A	2-A	3-A	4-A	7-A
4	1-A	1-A	1-A	1-A	2-A	2-A	3-A	5-A
4½	1-A	1-A	1-A	1-A	1-A	2-A	3-A	5-A
5	1-A	1-A	1-A	1-A	1-A	2-A	3-A	4-A
5½	1-A	1-A	1-A	1-A	1-A	1-B	2-B	3-B
6	1-A	1-A	1-A	1-A	1-A	1-B	2-B	2-B
7	1-A	1-A	1-A	1-A	1-A	1-B	2-B†	2-B
8	1-A	1-A	1-A	1-A	1-A	1-B†	1-B	2-B

* Pulleys less than 3″ in diameter should not be used for motors 1 h.p.
† Type A may be used.

Belt length should be determined by measuring around the pulleys with a tape or string while the motor is mounted in place. Measurement should be taken at the point of pitch diameter on the pulleys. Pitch diameter is ⅜″ less than the outside diameter if the pulley is used with a type A belt, and ½″ less with a type B belt. When mounting the motor on a machine or appliance, plan the distance between the shaft centers to be no more than three times the total of the pulley diameters, nor less than the diameter of the larger pulley.

Most V-belt drives are known as V-V drive and use a grooved pulley on both the motor and the driven machine as shown in Fig. 19. When a drive known as V-flat is used, in which the motor has a V-pulley but the driven machine has a flat pulley, a V-belt of the type shown in Fig. 20 is provided. This drive is satisfactory when the flat pulley is large, as the area of contact supplies enough friction to prevent slippage.

Fig. 19.

Sometimes the driving and driven pulleys cannot be arranged in parallel and must be placed at right angles to each other. This type of drive is known as a quarter-turn drive. It is satisfactory if the speed ratio between the pulleys is not greater than 2½ to 1 and the distance between centers is about 6 to 6½ times the diameter of the larger pulley.

On some machines a variable-speed drive is necessary and may be achieved in two ways. Multiple-step pulleys may be used on the motor, on the driven machine, or on both as shown in Fig. 21. If used on both, the same belt can be used for all speeds without changing the motor position. Pulleys that have adjustable space between the groove walls can also be used. This allows the V-belt to ride higher or lower in the groove and, in effect, changes the pitch diameter of the pulley (Fig. 22).

Fig. 20.

Direct Drive. When both the driven machine and the motor have similar operating speeds and satisfactory mounting conditions, a

Fig. 21.

direct, end-to-end shaft coupling between the motor and the load (Fig. 23) should be used. Direct connections are used in driving rotary pumps, blowers, fans, and numerous other machines. Direct drive requires careful alignment and the use of a coupling device that is partly flexible. If properly mounted, direct-driven machines entail a minimum amount of wear on motor and shaft bearings.

Methods of Mounting Fixed Motors. Sleeve-bearing motors, unless specially designed, should always be used in a position with the shaft level. Nevertheless, motors can be mounted on the floor, on a side wall, or on the ceiling, by rotating the end shields to keep the oil holes and reservoirs in an upright position (Fig. 24). Ball-bearing motors can be mounted in any position, including a vertical one.

Motors should be mounted with some provision for tightening and loosening the belt. Only a reasonable tension is necessary with a V-belt drive. When the motor is in operation, the tight side of the belt should form a straight line from pulley to pulley. Running belts tighter than necessary to pre-

Fig. 22.

vent slippage causes extra wear on the belts and motor bearings. Belts should never be forced over pulleys. Loosen the motor mounting so that belts can be slipped on easily. Care should be taken not to draw a motor down tight on an irregular surface as this will tend to place a strain on it and will throw the bearings out of line.

Portable Motors. A portable ¼ to ½ h.p. motor is inexpensive, can be used almost anywhere without special wiring, and can be plugged into a regular 115-volt outlet. Equipment needed to change a small motor into a portable one is shown in Fig. 25. No. 10 insulated wire is twisted

Fig. 23. Many devices are connected to motors by direct drive.

together to make a carrying handle. Short pieces of pipe are used to make the motor rails. The motor is equipped with a hard service cord of ample size and a multiple-step V-pulley.

A small portable motor can be attached to its load and held in position with pipe straps (Fig. 26) or with wooden cleats (Fig. 27).

Large motors are too heavy to carry and should be mounted on a cart or dolly. A two-wheel cart is recommended. Motor control and protective devices can be mounted on the cart in a convenient location. To lessen the number of heavy-duty power outlets that will be re-

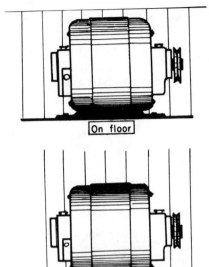

On floor

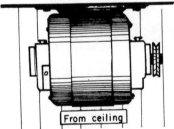

On wall

From ceiling

Fig. 24. Motors may be mounted on the floor, wall, or ceiling by rotating the end shields to keep oil holes upright.

quired, a 40-foot or 50-foot extension cord connected to the motor is practical.

Motor Protection. To keep motors from burning out, it is necessary to use wires of the correct sizes and to provide protection against overloads.

Correct Wire Sizes. Power is produced in a motor by voltage and current. If the circuit conductors (wires) are too small, there is a voltage drop at the motor terminals. This drop causes an increase in the current which, in turn, raises the heating effect of the motor. Under such conditions the motor will overheat and burn out. Table 15 shows the minimum size of wire to use in connecting motors according to the distance from the center of distribution.

Overload Protection. Motors may also burn out if not protected by some device that prevents them from being overloaded. The ordinary household fuse does not offer this protection. Such a fuse, if large enough to carry the starting current, will be too large to take care of the motor while it is running. The starting current of a motor is always several times greater than its running current.

Fig. 25.

Fig. 26.

Fig. 27.

Some household appliances now have motors with *built-in* overload protection which may be reset automatically or by hand. In the automatic reset type, the motor stops when it becomes overheated and starts again after it has cooled. In the hand reset type, it is necessary to push a button to reset or start the motor.

TABLE 15

WIRE SIZES FOR SINGLE-PHASE MOTORS

Horsepower	Distance in Feet from Motor to Center of Distribution																	
	20	30	40	50	60	70	80	90	100	120	140	160	180	200	240	280	320	360
	Wire Sizes (Gauge)																	
115 volt:																		
¼	14	14	14	14	14	14	14	14	14	12	12	10	10	10	10	8	8	8
½	14	14	14	14	12	12	12	12		10	10	10	8	8	8	6	6	6
¾	14	14	14	12	12	10	10	10	10	8.	8	8	6	6	6	4	4	4
1	14	14	12	10	10	10	8	8	8	8	6	6	6	4	4	4	2	2
230 volt:																		
¼	14	14	14	14	14	14	14	14	14	14	14	14	14	14	14	14	14	14
½	14	14	14	14	14	14	14	14	14	14	14	14	14	14	14	12	12	12
¾	14	14	14	14	14	14	14	14	14	14	14	14	12	12	12	10	10	10
1	14	14	14	14	14	14	14	14	14	14	12	12	12	10	10	10	8	8
2	14	14	14	14	14	14	12	12	12	10	10	10	10	8	8	8	6	6
3	14	14	14	14	12	12	10	10	10	10	8	8	8	6	6	6	4	4
5	14	14	12	12	10	10	10	8	8	8	6	6	6	4	4	4	2	2
7½	12	12	12	10	10	8	8	8	8	6	6	6	4	4	4	2	2	2

Motors that do not have this protection can be protected by using a *time-lag fuse* or a *thermal overload relay*. Thermal overload relays may be hand-operated or automatic. The hand-operated type is used for small motors, but does not protect them against low voltage, which may also cause them to burn out. The automatic type is used for larger motors and protects them against both overloading and low voltage. Both devices should have the proper ampere rating for the motor used, and the manufacturer's directions for installing should be followed.

MAINTENANCE

Under ordinary conditions a motor will operate for a long period of time without requiring a thorough cleaning. General-purpose single-phase motors have either a starting switch or brushes that operate only while the motor is starting. However, dirt or corrosion on the starting switch or brushes may make it impossible for the motor to start or may cause overheating. Even if the motor does operate, excessive dirt will cause the moving parts to wear rapidly. A periodic inspection should be made to determine whether the motor requires cleaning.

Methods of Cleaning. When it becomes necessary to disassemble and thoroughly clean an electric motor, first wipe the outside to remove all dirt and grease. Care should be taken in disassembling to avoid damage. Before taking the

motor apart, mark the exact position of the end shields on the motor frame with a sharp center punch or file as shown in Figs. 28 and 29. This will help maintain the true bearing alignment when reassembling the motor. Remove the nuts and through bolts and cap screws which hold the end shields in place and carefully remove the rotor with its end shield. If the motor has brushes, it is advisable to remove them first to avoid breakage. The end shield opposite the shaft extension has the motor lead wires attached to it. Be careful not to tear these wires loose from the motor windings.

Fig. 28.

Fig. 29.

Use compressed air at low pressure, if available, or a vacuum cleaner to remove dust and loose dirt from the inside of the motor. A soft brush may also be used to clean out loose dirt. To remove grease and oil, apply a safe cleaning solvent with a small paint brush and wipe clean with a cloth. Do not use excessive amounts of cleaning fluid directly on the windings as it may damage the insulation.

If the motor has sleeve bearings, remove the yarn or oil wick and clean the oil well. Replace the yarn or oil wick. In some types of ball-bearing motors, bearings can be cleaned and relubricated while the motor is disassembled. Be sure the cleaning fluid does not enter the bearings. After all parts of the motor have been cleaned, place them on a clean surface and wipe dry with a clean cloth. If a quantity of cleaning fluid has been applied, use an electric heater, a heat lamp, or a large light bulb to dry out the windings.

When the motor is clean and dry, reassemble it carefully. Be sure the motor leads are pulled out of the way of the rotor fan or other moving parts, as they might be caught and torn loose when the motor starts. Tighten the through bolts or cap screws gradually and evenly, making sure that the end shields fit tightly all the way around and that the motor shaft turns freely.

Lubrication. Proper lubrication of an electric motor is essential. The correct amount of lubricant will remain in the bearings to reduce friction heat and wear. Do not overlubricate as excess oil or grease will spread to other parts of the motor, causing the motor to plug with dirt and the insulation to break down. Manufacturers' directions should be followed closely.

For sleeve-bearing motors use a good grade of SAE 10 or 20 oil. Lighter or heavier oil may be used if temperatures are extremely low or high. There is a wide variation in the oil storage capacity of various types of sleeve-bearing motors. One type uses an oil well below the bearing with a wick to carry the oil up to the shaft (Fig. 30). Twice a year, the oil well should be unscrewed, the old oil cleaned out, and the well refilled about two-thirds full with new oil. Another system uses a yarn-packed bearing to which a few drops of oil should be added every few months. If there is a drain plug at the bottom, accumulated oil can be drained off occasionally (Fig. 31). A third type has a ring-oiled bearing. Oil is carried from an oil reservoir below the bearing onto the shaft by a loose ring that turns while the motor runs. Keep the oil level up to the filler hole by checking periodically. Every two or three years drain off the old oil, flush out the reservoir, and add new oil.

Lubrication is less important in a ball-bearing than in a sleeve-bearing motor. Ball bearings carry the load by direct contact. Sleeve bearings carry the load on an oil film. A ball-bearing motor could be operated dry, but it

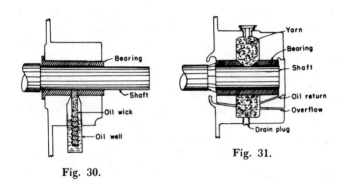

Fig. 30.

Fig. 31.

needs lubrication to protect it from dirt, corrosion, friction heat, and other adverse factors that are always present.

Ball bearings that have been prelubricated and sealed by the manufacturer should not be disturbed. Unsealed ball bearings can be relubricated either by disassembling the motor or by oiling through the lubrication openings. Disassembled bearings must be wiped clean of old grease with a soft cloth and repacked half to two-thirds full of electric-motor ball-bearing grease recommended by the motor manufacturer.

If the bearing has lubrication openings, remove both the filler and the drain plugs. If the old grease has hardened, run the motor to warm it up and add light oil until the grease softens and runs out. Stop the motor and add new grease until it has forced out any remaining old grease or the new grease starts to appear at the drain opening. Run the motor again with both holes open and let the motor force out the excess grease. Before the plugs are replaced, remove some grease with a rod or wire to allow for expansion when the motor is operating.

Commutator and Brush Maintenance. In commutator motors it is important to take proper care of the commutator and brushes for satisfactory service and long life. Sluggish starting and excess sparking at brushes are indications that trouble has developed.

Badly worn brushes must be replaced. Secure the proper brushes from the manufacturer. New brushes must be fitted to the contour of an axial type of commutator. Wrap fine sandpaper around the commutator and place the new brush in its holder. To grind the brush to the proper contour, hold it against the commutator with one hand and turn the rotor back and forth with the other hand. After the motor has been run for a short time, examine the brushes to ascertain whether the fitting has been correctly seated. A well-seated brush will appear shiny all over the contact surface.

Sometimes brushes will fail to make positive contact with the commutator because of a gummy accumulation of oil and dirt in the holders which causes them to stick or because of weak or broken springs. Remedy these conditions by cleaning or replacing the springs.

Trouble may also be caused by a dirty or worn commutator. In some types of motors, the commutator can be cleaned with a clean, lint-free cloth or fine sandpaper without taking the motor apart. As emery dust is a conductor of electricity and may cause short circuits, emery cloth should never be used. If the commutator is rough, pitted, or worn, take the motor to a service station to resurface the commutator and undercut the mica insulation between the commutator bars; this service requires special equipment.

STORAGE

To prepare a motor for storage proceed as follows: Wipe the outside of the motor with a cloth to remove dirt and grease; check bearings for lubrication and add fresh oil or grease if required; cover the shaft extension with a coating of grease to prevent rust; wrap the motor with heavy paper to keep dust and dirt from accumulating.

TROUBLE SHOOTING

Sudden failure of a motor to operate may be due to a number of causes, some of which are: failure of power supply; excessive load; frozen or worn motor bearings; operation of built-in thermal protection; failure of the motor-starting mechanism; failure of motor windings. Diagnosing the trouble is the first step toward getting the machine back into operation.

Failure of power supply is the first thing to check. Use a test lamp or voltmeter to make sure that proper voltage is available up to the motor terminals. Check motor control and protective devices.

Excessive load conditions in the driven device can be checked by removing the belt (if belt driven) and turning the device by hand. If the driven machine is at fault, the motor will run normally with the belt off.

If the motor will not run idle, shut off the current and try turning the shaft by hand. If it does not turn freely, the trouble may be a *frozen or worn bearing*. Lubrication may remedy a frozen bearing, but it may be necessary to take the motor apart to free a bearing that is frozen or stuck. If the motor shaft has any up and down play, bearings may be worn to the extent that the rotor is dragging, particularly when belt tension is applied. Do not confuse this with end play, as a slight amount of end play is necessary. Worn bearings must be replaced by a motor service station.

Motors with *built-in thermal* protection have an identification on the nameplate or elsewhere on the motor. If the motor has a reset button, press the button to see whether it starts the motor. If it is the automatic reset type, wait until the motor cools, and then turn it on to see if it will run. A tripped thermal protector usually indicates trouble.

Failure of the motor-starting mechanism is usually indicated by a hum when the motor is turned on, although the rotor will run normally when given a spin by hand. If the motor is a split-phase or capacitor-start type, the starting circuit may be open at some place. Failure of the starting switch to make contact may be due to a number of reasons. The contacts may be dirty, burned, or pitted. A wire may be burned in two. Failure of the centrifugal device to close the switch may be due to too much end play in the rotor shaft. With a capacitor-start motor, the capacitor may be burned out.

Failure of motor windings may be due to a burned-out winding, resulting in a characteristic odor. Charred insulation can often be seen through the openings in an open-type motor. At times, however, and most frequently in the starting winding, a single strand of wire quickly burns in two.

CHAPTER 8

Planning Your Home Lighting

Good lighting—properly placed—not only makes a home comfortable and easy to live in, but enhances furnishings, textures, and colors as well.

If you are building a new home or remodeling an older one, the lighting should be a part of the structural and interior plan. Or you can simply start where you are with the lighting you have and improve it. Much can be done at a surprisingly low cost—once you understand and apply the principles of good lighting.

HOW YOU SEE

To apply the principles of good lighting in your home and to safeguard the eyesight of your family, you need to understand *how you see*. Seeing is not done by the eyes alone—the nerves, muscles, and the brain are also involved.

Prolonged visual work in too little light or in glare can be as tiring as working hard all day. The resulting exertion does not damage your eyes, but it can cause eyestrain and nervous fatigue. (Glare is direct or reflected light that is visually disturbing or that interferes with seeing.)

Consider the eye as a living camera that transforms light energy into sight. Light rays—from the sun, from a lamp, or reflected from any object—pass through the cornea and lens of the eye and are focused on the retina. The retina continually transfers pictures to the optic nerve and on to the brain. The final step in seeing is performed in the brain.

During your waking hours this rapid, complex process constantly gives back the shape, color, size, and motion of the things you see. More than three-fourths of sensory impressions come through the eyes.

Four factors determine visibility (the ease and accuracy with which you see). They are as follows.

1. *Size.* The larger the object, the easier it is to see.

2. *Contrast.* To be seen, an object must contrast to some extent with its background—a dark wood carving against a white wall can be easily seen, but a white figurine against a white wall is difficult to see.

3. *Time.* It takes time to see clearly. For this reason, rapidly moving blades of an electric fan blur. There is not enough time to see individual blades.

4. *Brightness.* This is the amount of light the eye actually sees. It may be direct or reflected from objects and surfaces. There is no visibility in darkness and little in dim light. As light increases, surrounding surfaces become brighter and reflect more light to the eyes. You see more efficiently and effortlessly.

Of the four factors, *brightness* is the one over which you have control. Usually you cannot change the size or form of an object or slow the speed with which it moves. You may not be able to put the object against another background that gives more contrast. *But you can add more and better* light to make seeing easier.

WHERE YOU NEED LIGHT

Does the lighting in your house make seeing easy wherever eyes do their work? In many homes, there are 50 or more seeing jobs that need specific lighting.

An easy way to spot the places in your home where light is needed is to make a room-by-room check. Jot down the activities that take place in each room, remembering each member of the family as you do so. You may be surprised at the length of your list.

The family living room, *for example,* is the setting for a wide range of activities. Here family members may entertain friends and relatives, read, study, telephone, watch television, write letters, or engage in other work or play that requires different kinds of lighting.

Once you are aware of the places where specific lighting is needed, you can plan the lighting for the activities that take place there. Good lighting, you will find, greatly increases the usable space within your home.

Compare the seeing jobs in your home with the following suggested checklist.

CHECKLIST OF ACTIVITIES

Living Areas (living room, dining room, family or recreation room):

Reading or studying (prolonged)
Reading (casual, intermittent)
Viewing television
Visiting and conversation
Playing games (adult, children)
Reading music
Setting table
Dining

Work Areas (kitchen, laundry, workroom, home office):

Reading recipes and measuring ingredients
Reading labels and following directions
Inspecting and sorting foods
Reading dials and checking foods as they cook
Washing dishes and cleaning equipment
Sorting and pretreating laundry
Ironing
Sewing (reading directions, cutting, fitting, machine and hand stitching)
Repairing small appliances
Working on hobbies (art work, collections, model building, photography)

Housekeeping (dusting, vacuuming, waxing, washing walls, woodwork, and other surfaces)

Deskwork (recordkeeping, studying, typing, reading telephone directory)

Private Areas (bedrooms, bathrooms):

Personal grooming (bathing, shaving, manicuring, shampooing and arranging hair, applying cosmetics)

Assembling clothes from closet or storage unit

Dressing

Caring for infants, small children, or the sick

Reading medicine labels or taking temperatures

Reading in bed or at bedside

Halls, Stairways, Entrances:

Moving between rooms

Using steps and stairs

Reading house numbers

Identifying callers

Safe access to house or garage

HOW MUCH LIGHT IS NEEDED

For comfortable, easy seeing throughout your home, you need to find out how much light is needed in various areas.

The output from fixtures, portable lamps, and built-in units in any room or area should provide general lighting, and also give the right kind and amount of lighting for specific activities.

How to Check Your Present Lighting. Lighting authorities of the Illuminating Engineering Society have determined the minimum amounts of light needed to do certain tasks. These levels of illumination are measured in foot-candles. (*See* Table 16.)

You can check your present lighting with a light meter. Sometimes your electric company or county extension agent can help you locate a light meter. Compare your foot-candle readings with the minimum levels recommended. Then you can increase light if needed.

TABLE 16

MINIMUM LEVELS OF ILLUMINATION
(Recommended by Illuminating Engineering Society)

Specific visual task	Amount of light on task [1] (foot-candles) [2]
Reading and writing:	
Handwriting, indistinct print, or poor copies	70
Books, magazines, newspapers	30
Music scores, advanced	70
Music scores, simple	30
Studying at desk	70
Recreation:	
Playing cards, table games, billiards	30
Table tennis	20
Grooming:	
Shaving, combing hair, applying makeup	50
Kitchen work:	
At sink	70
At range	50
At work counters	50
Laundering jobs:	
At washer	50
At ironing board	50
At ironer	50
Sewing:	
Dark fabrics (fine detail, low contrast)	200
Prolonged periods (light-to-medium fabrics)	100
Occasional (light-colored fabrics)	50
Occasional (coarse thread, large stitches, high contrast of thread to fabric)	30
Handicraft:	
Close work (reading diagrams and blueprints, fine finishing)	100
Cabinetmaking, planing, sanding, glueing	50
Measuring, sawing, assembling, repairing	50

TABLE 16—*Continued*

General lighting	Average light throughout area [3] (foot-candles) [2]
Any area involving a visual task_____	30
For safety in passage areas_____	10
Areas used mostly for relaxation, recreation, and conversation_____	10

[1] Average of light measured over the task area.
[2] A foot-candle is the amount of light falling on a surface 1 foot away from a standard candle (ordinary 1½-inch diameter candle).
[3] Average of light measured on a horizontal plane 30 inches above the floor.

HOW TO GET GOOD QUALITY LIGHTING

Try to improve the quality or comfort of light in your home as well as the amount of light.

In an effectively lighted room, light is well distributed and free from glare, bright spots, and deep shadows. Large surfaces—walls, ceilings, and floors—should have favorable reflectances.

Light affects color. A lovely room by day can be even lovelier by night. Before you choose wall and fabric colors, try them under the same combination of lighting you will be using at home. Remember, too, that dimming of incandescent bulbs changes the color of their light—makes it yellower.

The light from some fluorescent tubes appears blue-white when combined with the yellowish light from incandescent lamp bulbs. To minimize color distortion, get deluxe fluorescent tubes in warm or cool white for both wall and ceiling lighting units.

You will find that light from deluxe warm white tubes (marked WWX) blends well with light from incandescent bulbs and enhances complexions, foods, fabrics, and paints. Deluxe cool white tubes (marked CWX) create a cool atmosphere and are effective in rooms with blue or green color schemes.

Favorable Light Reflectances. When planning the lighting of any interior, consider color and finish of walls, ceilings, wood floors or floor coverings, and large drapery areas. These large surfaces reflect and redistribute light within a room. Their lightness or darkness greatly affects the mood of a room.

White surfaces reflect the greatest amount of available light. Light tints of colors reflect light next best. Somber color tones absorb much of the light that falls upon them and reflect little light. (*See* Table 17.)

If a large room gets plenty of daylight, you probably can use fairly strong color. Light tints on walls make a small room seem larger.

TABLE 17
REFLECTANCE RANGE

Color	Approximate percent reflection
Whites:	
Dull or flat white	75–90
Light tints:	
Cream or eggshell	79
Ivory	75
Pale pink and pale yellow	75–80
Light green, light blue, light orchid	70–75
Soft pink and light peach	69
Light beige or pale gray	70
Medium tones:	
Apricot	56–62
Pink	64
Tan, yellow-gold	55
Light grays	35–50
Medium turquoise	44
Medium light blue	42
Yellow-green	45
Old gold and pumpkin	34
Rose	29
Deep tones:	
Cocoa brown and mauve	24
Medium green and medium blue	21
Medium gray	20
Unsuitably dark colors:	
Dark brown and dark gray	10–15
Olive green	12
Dark blue, blue-green	5–10
Forest green	7
Natural wood tones:	
Birch and beech	35–50
Light maple	25–35
Light oak	25–35
Dark oak and cherry	10–15
Black walnut and mahogany	5–15

Whatever the room size, try to keep colors within the 35 to 60 per cent reflectance range. Ceilings should have reflectance values of 60 to 90 per cent, and floors at least 15 to 35 per cent. Matte finishes (flat or low gloss surfaces) on walls and ceilings diffuse light and reduce reflections of light sources. Glossy, highly polished or glazed surfaces produce reflected glare.

LIGHT SOURCES

Incandescent bulbs and fluorescent tubes—in portable and wall lamps, in ceiling and wall fixtures, and in built-in lighting units—are the usual sources of electric light in homes.

But bulbs and tubes do not insure good lighting by themselves. You must *select* the *right bulb or tube* for the *purpose* you have in mind. Then the bulb or tube has to be placed in an *appropriately designed lamp or fixture.* And finally, the lamp or fixture must be *correctly placed in the room.*

The following sections in this chapter will help you gain a working knowledge of the bulbs and tubes available and how to use them.

Incandescent Bulbs. The incandescent bulb (principal home light source for approximately 75 years) comes in a wide assortment of shapes, colors, sizes and wattages. (*See* Tables 18 and 19.)

Types of bulbs. General household bulbs, the most commonly used type, range from 15 to 300 watts. They are available in three finishes: inside frost, inside white (silica coated), and clear.

Inside frost is the older bulb finish still in general use. Use bulbs of this type in well shielded fixtures.

Bulbs with *inside white* finish (a milky-white coating) are preferred for many home uses. They produce diffused, soft light and help reduce bright spots in thin shielding materials.

Decoratively shaped *clear* bulbs add sparkle to chandeliers or dimmer controlled simulated candles.

Three-way bulbs have two filaments and require three-way sockets. Each filament can be operated separately or in

TABLE 18

<small>GUIDE FOR SELECTING INCANDESCENT BULBS</small>

Activity	Minimum recommended wattage [1]
Reading, writing, sewing:	
Occasional periods	150.
Prolonged periods	200 or 300.
Grooming:	
Bathroom mirror:	
1 fixture each side of mirror	1–75 or 2–40's.
1 cup-type fixture over mirror	100.
1 fixture over mirror	150.
Bathroom ceiling fixture	150.
Vanity table lamps, in pairs (person seated)	100 each.
Dresser lamps, in pairs (person standing)	150 each.
Kitchen work:	
Ceiling fixture (2 or more in a large area)	150 or 200.
Fixture over sink	150.
Fixture for eating area (separate from workspace)	150.
Shopwork:	
Fixture for workbench (2 or more for long bench)	150.

[1] White bulbs preferred.

combination. Make sure that a three-way bulb is tightened in the socket so both contacts in the screw-in base are touching firmly.

The three lighting levels offered by these bulbs are particularly nice in portable lamps and pulldown fixtures. You can turn the lamp high for reading and sewing, on medium for televiewing, conversation, or entertaining, and on low for a night light or a soft, subdued atmosphere.

Dimmer switches are now available on some lamps. They make it possible to light from very low to the maximum output of the bulb.

Tinted bulbs create decorative effects both indoors and out. Silica coatings inside these bulbs produce delicate tints of colored light—pink, aqua, yellow, blue, and green. Home uses of these bulbs are best limited to lighting plants, flowers, or art objects. You will need to buy tinted bulbs of higher wattage because they give less light than white bulbs.

Silver-bowl bulbs are standard household bulbs with a silver coating applied to the outside of the rounded end. They are used base up, and direct light upward onto the ceiling or into a reflector. You can get them in 60-, 100-, 150-, and 200-watt sizes. They are generally used with reflectors

TABLE 19

SIZES AND USES FOR 3-WAY BULBS

Socket and wattage	Description	Where to use
Medium: 30/70/100	Inside frost or white.	Dressing table or dresser lamps, decorative lamps, small pin up lamps.
50/100/150	Inside frost or white.	End table or small floor and swing-arm lamps.
50/100/150	White or indirect bulb with "built-in" diffusing bowl (R–40).	End table lamps and floor lamps with large, wide harps.
50/200/250	White or frosted bulb.	End table or small floor and swing-arm lamps, study lamps with diffusing bowls.
Mogul (large): 50/100/150	Inside frost	Small floor and swing-arm lamps and torcheres.
100/200/300	White or frosted bulb.	Table and floor lamps, torcheres.

in basements, garages, or other work areas. Fixtures for silver-bowl bulbs are widely available.

Reflector bulbs are available with silver coatings either on the inside or outside of the bulbs. The *spotlight bulbs* direct light in a narrow beam and generally accent objects. The *floodlight bulbs* spread light over a larger area, and are suitable for floodlighting horizontal or vertical surfaces. Typical floodlight sizes include 30, 50, 75, and 150 watts.

Heat-resistant bulbs, called PAR bulbs because of their parabolic shape, are used *outdoors.* They are resistant to rain and snow. Common sizes are 75 watts, 150 watts, and up.

Bulbs in decorative shapes are designed to replace bare bulbs in older fixtures and sockets. Some shapes and sizes are made for traditional fixtures (chandeliers and wall sconces); others combine contemporary styling and function. Bulb shapes include globe, flame, cone, mushroom, and tubular.

Some of these bulbs are made of diffusing type glass and are tinted to produce colored lighting effects. Clear bulbs may be needed to produce sparkle in crystal chandeliers. When selected to harmonize with fixtures and room decor, these decorative bulbs may offer a pleasing, low cost solution to a lighting problem.

Colored floodlight bulbs are available for indoor or outdoor use. The tints—particularly pink and blue-white—create nice effects on planters or flowers and are acceptable for lighting people and furnishings. Strong colors—blue, green, and red—are best reserved for holiday and party decorations.

Fluorescent Tubes. Most households use fluorescent lighting in some form. Although know-how is needed to select and use this light source correctly, it does offer advantages in home lighting. For guidance in selecting fluorescent tubes and fixtures *see* Tables 20 and 21.

Fluorescent tubes must be used in fixtures that contain the necessary electrical accessories. The basic structure of a fluorescent fixture is shown in Fig. 1. The fluorescent phosphor coating on the inside of the tube is activated by electric energy passing through the tube, and light is given

TABLE 20

SELECTION GUIDE FOR FLUORESCENT TUBES

[All are T12 (1½ inch diameter) tubes]

Use	Wattage and color [1]
Reading, writing, sewing:	
Occasional	1 40w or 2 20w, WWX or CWX.
Prolonged	2 40w or 2 30w, WWX or CWX.
Wall lighting (valances, brackets, cornices):	
Small living area (8-foot minimum)	2 40w, WWX or CWX.
Large living area (16-foot minimum)	4 40w, WWX or CWX.
Grooming:	
Bathroom mirror:	
One fixture each side of mirror	2 20w or 2 30w, WWX.
One fixture over mirror	1 40w, WWX or CWX.
Bathroom ceiling fixture	1 40w, WWX.
Luminous ceiling	For 2-foot squares, 4 20 w, WWX or CWX
	3-foot squares, 4 30w, WWX or CWX
	4-foot squares, 4 40w, WWX or CWX
	6-foot squares, 6 to 8 40w, WWX or CWX.
Kitchen work:	
Ceiling fixture	2 40w or 2 30w, WWX.
Over sink	2 40w or 2 30w, WWX or CWX.
Counter top lighting	20w or 40w to fill length, WWX.
Dining area (separate from kitchen)	15 or 20 watts for each 30 inches of longest dimension of room area, WWX.
Home workshop	2 40w, CW, CWX, or WWX.

[1] W W X = warm white deluxe; C W X = cool white deluxe; C W = cool white.

off. The starter in standard starter type fixtures permits preheating of the electrodes in the ends of the tube to make it easier to start. The ballast limits the current to keep the tube functioning properly. The channel holds ballast and wiring and spaces the lampholders.

White fluorescent tubes are labeled "standard" and "deluxe." The whiteness of a standard tube is indicated by letters, WW for warm white; CW for cool white. The addition of an "X" to these letters indicates a deluxe tube.

A deluxe warm white (WWX) tube gives a flattering light, can be used with incandescent light, and does not

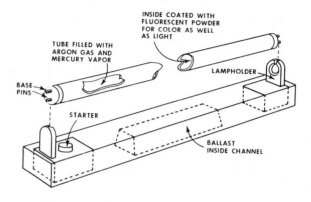

Fig. 1. Basic structure of fluorescent fixture.

distort colors any more than incandescent light does. A deluxe cool white (CWX) tube simulates daylight and goes nicely with cool color schemes of blue and green. Deluxe tubes are the only fluorescent tubes recommended for home use. They are worth waiting for if your dealer has to order them for you.

TABLE 21

Type of Tube for Ballast Used

[Fluorescent tube wattages vary with length of tube and are not interchangeable]

Length of tube (inches)	Watt-age	Diameter of tube	Ballast marking [1]
	Watts	*Inches*	
15_____	14	1½ (T12)___	Preheat start.
18_____	15	1 (T8)_____	Trigger start or preheat.
		1½ (T12)___	Trigger start or preheat.
24_____	20	1½ (T12)___	Preheat or rapid start.
36_____	30	1 (T8)_____	Rapid start.
		1½ (T12)___	Rapid start or dimming.
48_____	40	1½ (T12)___	Rapid start or dimming.

[1] Only preheat ballasts require starters and use standard-type tubes.

CHOOSING AND USING PORTABLE LAMPS

Most living rooms need at least five portable lamps, and most bedrooms need three.

Well-designed floor and table lamps may be difficult to find, but keep looking. Look for lamps that combine function and beauty. Pay close attention to what is under the shade. Choose lamps that make seeing comfortable and, at the same time, harmonize with your furnishings, color schemes, and with other lamps and accessories.

Generally speaking, the *design* of a lamp should be akin to the style and decoration of the room; its *scale* in accord with the furniture it appears on or over; and the *amount of light* it radiates suitable for the purpose intended.

Dimmer switch controlled lamps give greater flexibility than three-way lamps that use three-way sockets and require three-way bulbs.

The small *high intensity lamps* now on the market are not designed for study, reading, or general work. They can, however, provide a concentrated area of high-level light for special tasks, such as sewing, crafts, or fine-detail work. They should always be used in combination with good general lighting.

If you select a lamp for style or color alone, do not expect to use it for close work. It is a *decorative lamp*. Use it for the following purposes:

to give limited general lighting;
to brighten a corner, foyer, or hallway; or
to display an object of art or an accessory.

Any portable lamp, regardless of size, should be sturdily built. See that the power cord is well protected from sharp edges where it enters the lamp base and the vertical pipe.

Table Lamps. Before you shop for table lamps, jot down the heights of tables on which lamps will be placed, and the height of any chair or sofa seat on which a person using the lamp will be seated. Take these figures with you as you shop. You may also want to consider the eye level height of persons using a lamp, particularly if an individual is unusually tall or short.

As you shop, keep the following rule in mind. *Table height plus lamp base height (to the lower edge of shade) should equal the eye height of the person using the lamp.*

Eye height depends on the height of the chair or sofa seat and the eye level of the person seated. Generally, eye height is 40 to 42 inches above the floor. For comfortable seeing, the bottom of the lampshade should also be 40 to 42 inches above the floor. For the proper placement of table lamps, see Figs. 2 and 3.

Figure 4 shows how to place a lamp on a table for reading in bed. Good positioning of vanity lamps for a seated person is shown in Fig. 5. A lighting arrangement that works well in a bathroom is shown in Fig. 6. Three shielded fluorescent fixtures provide the right lighting for grooming at a bathroom mirror. In a small bathroom, the overhead fixture, as mounted in the illustration, also gives general lighting. If a bathroom has a ceiling fixture, the overhead tube can be mounted directly above the mirror.

When selecting and using table lamps, other points to consider are as follows.

Shade dimensions. Shades on lamps for reading, sewing, or studying should be 16 inches wide or more at the bottom; 9 inches wide at the top; and at least 10 inches deep.

Minimum bulb wattages. Bulbs in single-socket lamps for reading, sewing, or studying should be 150 watts; multiple-socket lamps should be at least three 60-watt bulbs.

Sockets. Center of the light source (bulb) should always be located in the lower third of the shade.

Diffusers. Diffusers soften and spread light.

Floor Lamps. In choosing a floor lamp, keep in mind exactly where it is to be located in your home. Floor lamps should harmonize with furnishings, and be carefully scaled to space. Choose lamps sized and constructed for proper placement without interfering with house traffic.

Small floor lamps (standard, swing arm, or bridge type)

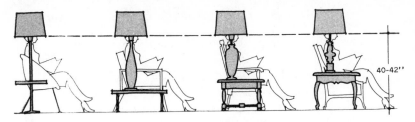

TYPICAL HEIGHTS OF LAMPS AND TABLES FOR SHADE AT EYE LEVEL

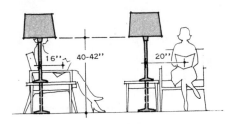

PLACEMENT DIMENSIONS FOR SHADE
AT EYE LEVEL

Fig. 2.

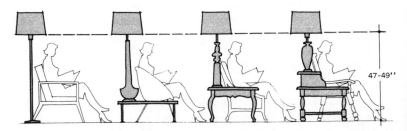

TYPICAL HEIGHTS OF LAMPS AND TABLES FOR SHADE ABOVE EYE LEVEL

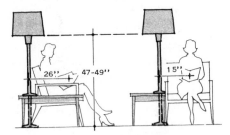

PLACEMENT DIMENSIONS FOR SHADE
ABOVE EYE LEVEL

Fig. 3.

Fig. 4. Correct position for study lamp.

Fig. 5. Dressing table lamps with translucent shades.

may be 43 to 47 inches from the floor to the bottom of the shade. Large lamps (standard or swing-arm) measure 47 to 49 inches from the floor to the bottom of the shade.

For reading, a floor lamp with a fixed or swing arm is correctly placed when the light comes from behind the shoulder of the reader, near the rear of the chair—either at the right or the left—but never from directly behind the chair. (*See* Figs. 2 and 3.)

If a floor lamp is used for prolonged reading or sewing, it should have a bulb wattage of 200 or 300 watts; the minimum bulb wattage that may be used for reading is 150 watts.

Figure 7 shows the correct placement of a floor lamp for hand sewing or needlework. A double swing-arm lamp concentrates light on hand sewing and needlework. Sewing requires twice as much light as casual reading. To position lamp, measure 15 inches to the left of center of sewing, and place center of shade 12 inches back from this point. A good lighting arrangement for machine sewing is shown in Fig. 8. For machine sewing, mount a wall lamp 14 inches above the working surface, 12 inches to the left of the needle, and 7 inches from the wall. The bottom of the shade should be at eye level.

Desk and Study Lamps. A well-lighted desk or study center makes it easy to concentrate. Good quality light falls on the task at hand. Eyestrain and physical tension are reduced to a minimum because the worker has enough light and little or no glare.

For continued study or deskwork, be sure your lighting arrangement gives an average of 70 foot-candles over the work area. A lamp, correctly placed and fitted with a 200-watt bulb, will give this amount of light. If you do not have a diffuser on the lamp, using a white bulb helps to reduce glare.

The study lamp shown in Fig. 9 is located correctly for a right-handed person. The center of the lampshade should be about 15 inches to the left of the work center, and the

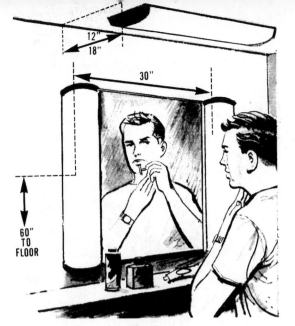

Fig. 6.

Fig. 7.

Fig. 8. Use of wall lamp for machine sewing.

Fig. 9. Use of double swing-arm lamp for hand sewing and needlework.

lamp about 12 inches back of the front edge of table. Place the lamp to the right for a left-handed person.

To serve the growing student population in the United States, several lamp manufacturers are now making and selling specially designed, higher-wattage study lamps. Tags attached to the lamps state, "This is a Better Light Better Sight Bureau Study Lamp." These improved lamps reduce glare to a minimum. They also give nearly twice as much light as other study lamps and spread light over a wider area. For these lamps, 200-watt bulbs are recommended.

When a study lamp is placed as shown in Fig. 9, it gives excellent lighting for intensive and prolonged desk work.

A swing arm on a floor lamp makes it possible to place light in good position for study or desk work. Such a floor lamp should measure 47 to 49 inches from the floor to the bottom of the shade. It is usually fitted with a white glass diffusion bowl and a mogul socket for a 100/200/300-watt three-way bulb or a medium socket for a 50/100/150 or 50/200/250 three-way bulb or a single 150-watt bulb.

If you plan to study under a wall bracket or "pin-up" type lamp, choose a bracket with a swing arm so you can move the center of the shade forward. Then you can position the light to your best advantage.

Other suggestions to improve light in a study are:

Choose a desk with a light-colored, nonglossy finish.

Use a light-colored blotter or desk pad.

Paint walls in neutral or light colors or select a plain wallpaper or one with a small, quiet design.

Make sure that the desk does not face a window.

Shades and Bases. Shades made of translucent materials are usually chosen for portable lamps.

Desk lamps use slightly translucent shades to avoid uncomfortable brightness in the eyes of the person using the lamp. Shades for dressing table or dresser lamps should be highly translucent because light reaching the face must come through the shade. Shade materials suitable for vanity lamps are too translucent for reading, study, and decorative lamps.

For effective light reflectance, the inside of the shade needs to be white or near-white. Good color choices for the outside of the lampshade are neutral or pale tints, off-white, beige, and light gray. Try to avoid excessive contrast between the color of the shade and adjacent walls.

The shape of a lampshade also affects lighting. Straight or nearly straight lines are preferable to extreme curves. Look for a shade that gives a wide spread of downward light as well as some upward light.

Choose lamps tall enough for the dressing table so the center of the shade is 15 inches above the table, approximately at eye level. Be sure to use high wattage bulbs.

The base of a portable lamp needs to be heavy enough to support the lamp firmly and keep it from upsetting easily. The design of the base should be appropriate to its function; it should please but not distract. Grotesquely shaped bases are seldom in good taste.

Materials in lamp bases often relate to certain furniture styles or periods. Some popular furniture styles and appropriate base materials include: *Early American*—pewter, brass, stoneware, copper, wrought iron, pottery, and wood; *18th Century or traditional*—silver, porcelain, china, cloisonne, crystal, and marble; *contemporary*—metals, glass, wood, cork, and ceramics.

Diffusers and Shields. Diffusers are bowl- or disc-shaped devices that surround the lamp bulb under the shade. They scatter and redirect light, soften shadows, and reduce reflected glare.

Effective diffusion materials in order of preference include blown milky glass, enameled glass, flashed opal, and plastics.

Undershade diffusers are now being offered by manufacturers for use in study and reading lamps. One is a highly reflective, inverted metal cone. Other new diffusers are bowl-shaped, prismatic reflectors.

Shields are also used to prevent glare. Perforated metal shields or plastic louvered shields, placed above the bulb, keep direct glare from reaching the eyes of the passerby. The mushroom shaped (R-40 type) 150-watt bulb made of white diffusing glass needs no diffuser or shield. It serves well for casual reading.

Figure 10 shows the various types of diffusers. Fig. 10, A shows a *molded prismatic diffuser*. It has a louvered top shield that diffuses light and prevents direct glare from the bulb. Study lamps with such diffusers give an exceptionally good spread of light over desk areas. *Prismatic refractor* at base of bulb (Fig. 10, B) helps distribute light and reduce direct and reflected glare. Top diffusing shield also restricts glare. *Certified Lamp Makers* (CLM) *glass diffuser* shields bulb from top viewing (Fig. 10, C) and gives an efficient spread of light below. *White glass bowl-shaped diffuser* (Fig. 10, D) reduces direct glare from bulb and provides both up and down light. Large size and whiteness of *R-40 white indirect bulb in wide harp* helps diffuse light (Fig. 10, E). *Plastic or white glass diffusing disc* at the bottom spreads light and reduces glare from the bulbs; *perforated* disc at top shields the bulb from top viewing and redirects light downward (Fig. 10, F).

HOW TO IMPROVE YOUR PRESENT LIGHTING

1. When you redecorate, finish walls in light pastel colors and ceilings in white or near white or a pale tint. Flat or low gloss paint on walls and ceilings helps diffuse light and makes lighting more comfortable. Use sheer curtains or draperies in light or pastel tints.

2. Add portable lamps for better balance of room lighting.

3. Install structural lighting (valance and cornice) in living areas where there is only one ceiling light or none. Eight to 20 feet of wall lighting will add a feeling of spaciousness to an average sized room and make the lighting more flexible.

4. Replace present bulbs with those of higher wattage, but do not exceed the rated wattage of the fixture. A minimum of 150 watts is needed in many single-socket lamps. For better control of lighting, use three-way bulbs or dimmer switches. If you want a higher wattage fluorescent

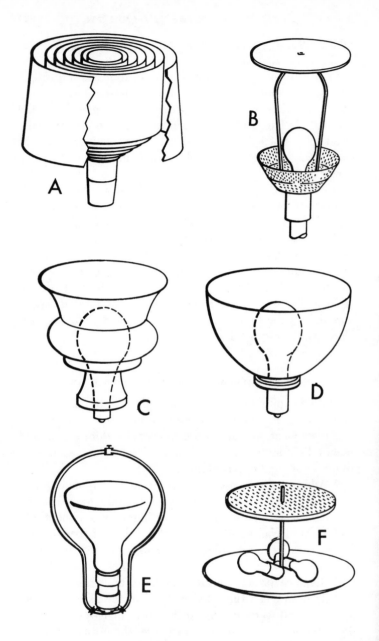

Fig. 10. Types of diffusers.

TABLE 22

PROS AND CONS OF LIGHT SOURCES

Incandescent Bulbs	*Fluorescent Tubes*
Can be concentrated over a limited area or spread over a wide area.	Provide more diffused lighting—a line of light, not a spot.
Initial cost less than fluorescent tubes.	Higher initial cost, but greater light efficiency— three to four times as much light per watt of electricity.
Designed to operate at high temperature.	Cool operating temperature. Generally about one-fifth as hot as incandescent bulbs.
Have average life of 750 to 1,000 hours.	Operate seven to ten times longer than incandescent bulbs.
Wattages range from 15 to 300 watts.	Wattages for home use range from 14 watts (15 inches long) to 40 watts (48 inches long).
Amount of light can be increased or decreased by changing to bulbs of different wattage because most bulbs have same size base.	Cannot be replaced by higher or lower wattage tubes.
Require no ballast or starter.	Require ballasts, and in some cases, starters.
Do not interfere with radio reception.	May cause noise interference with radio reception within 10 feet of the tube location.
Suitable for use in less expensive fixtures.	Adaptable to and commonly used in custom-designed installations and in surface-mounted and recessed fixtures.
Available in colors to enliven decor and accessories. Colored bulbs are 25 to 50 percent less efficient than white bulbs.	Available in many colors (plus deluxe cool white, CWX, and deluxe warm white), WWX, at much higher light output than colored incandescents.
Gain flexibility by use of three-way bulbs and multiple-switch controls or dimmer controls designed for incandescent bulbs.	Gain flexibility by use of dimming ballasts combined with dimmer controls designed for fluorescent tubes.

unit, the fixture must be changed.

5. For efficiency, use one large bulb rather than several small ones. A 100-watt bulb gives as much light as six 25-watt bulbs, but only uses about two-thirds as much current.

6. Replace outmoded bare bulb fixtures with well-shielded ones.

7. Cover all bare bulbs or tubes in a ceiling fixture with a shade or diffuser. Some of these diffusers clip to the bulb. Others hang from small chains attached to the husk of the fixture. Large diffusers, sometimes called adapters, may have supporting frames that are screwed on the sockets of single-bulb fixtures. An inexpensive way to avoid the glare of bare bulbs in a ceiling fixture is to replace these bulbs with silver bowl bulbs or decorative mushroom shaped bulbs.

8. Keep all light sources operating efficiently by replacing blackened bulbs and tubes promptly.

9. Put in additional convenience outlets if needed for correct lamp placement. Added outlets prevent the use of dangerous "cube" plugs and extension cords. Surface wiring strips may be attached along baseboards or counter tops. These strips may be more economical than adding built-in convenience outlets. *Be sure* any surface wiring system you choose is of the correct size and carries the Underwriters' Laboratories seal of approval. Ask your electric power supplier or electrical contractor for the correct size.

10. Lift the lampshade on a portable lamp with a riser if the bulb is too high. When a bulb is too high it restricts the downward circle of light and shines into the eyes of persons standing near. Risers come in multiples of one-half inch and can be screwed to the top of the harp to lift the shade the amount needed.

11. Replace the lampshade with a deeper shade if bulb is too low in lamp and bulb shows beneath lower edge of shade. Or, if you prefer, use a shorter harp or a different diffusing bowl.

12. If a lamp base is too short, set the base on wood, marble, ceramic, or metal blocks to raise the lamp to proper height. For ease in handling, cement the block to the base.

13. Get replacement shades for table lamps if present shades do not meet specifications. Choose shades made of translucent materials with white linings and open tops. For shade dimensions see Figs. 2 and 3.

14. Invert and rewire the socket of old style bridge lamps. Then add 6-inch diffusing bowls and larger, wider shades that give softer, better lighting.

LIGHTING FIXTURES

Lighting fixtures usually provide the general lighting in a home. When they are well chosen, they also add decorative tone and a pleasant atmosphere.

The basic principles of lighting—quantity, quality, color, and reflectance of light—should be considered in selecting fixtures. You will find that individual fixtures may be combined with structural lighting for pleasing effects.

The manufacturers' wattage rating and the size of the fixture must be large enough to accommodate the largest wattage bulb needed to light the area. Often, you need more than one fixture. *For example,* a large rectangular kitchen may need two 48-inch two-tube fixtures placed end to end.

Checking Fixtures. Check fixtures carefully before buying them, keeping in mind the following points.

Incandescent bulbs should be no closer than ¼ inch to enclosing globes or diffusing shields.

Top or side ventilation is desirable in a fixture to keep temperatures low and to extend bulb life.

Inside surfaces of shades should be of polished material or furnished with white enamel.

Shape and dimension of a fixture should help direct light efficiently and uniformly over the area to be lighted.

Plain or textured glass or plastic is preferred for enclosures and shades.

Choosing Fixtures. When choosing fixtures, see section on Light Sources; also see Tables 18, 19, and 21; and section on Shades and Bases in this chapter.

Fixtures shown in Figs. 11, 12, 13, 14, 15, and 16 have design features that function well in the house areas designated. *They are intended as a guide only.* They do not include many kinds of fixtures on the market. You may prefer, *for example,* to light your dining room with a chandelier instead of the fixture types shown. You can get chandeliers with or without sparkle-type bulbs that can be controlled by dimmer switches.

Kitchen. *Closed globe unit* (Fig. 11, A). Minimum diameter of bowl is 14 inches, and white glass gives good diffusion of light. *Shielded fixture* (Fig. 11, B), has three or four sockets, 14 to 17 inch diameter. Shallow wide bowl is desirable. *Fluorescent fixture with diffusing shield* (Fig. 11, C) has two or four tubes as needed in a 48-inch unit. For a large kitchen, two 2-tube fixtures can be placed end to end.

Dining Room. *Lantern style pulldown* (Fig. 12, A) unit has a three-way socket, and a diffusing globe, and takes a 50/100/150-watt bulb. *Ventilated ceiling fixture* (Fig. 12, B) has a bent glass diffuser, 14-inch minimum diameter. Interior reflecting surfaces should be white or polished. *Pulldown fixture* (Fig. 12, C) with a ventilated unit has three-way single socket or three sockets and white glass diffuser.

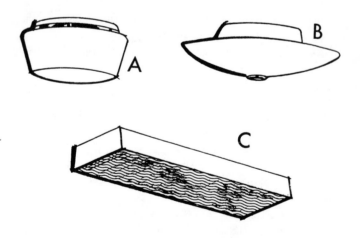

Fig. 11. Types of kitchen fixtures.

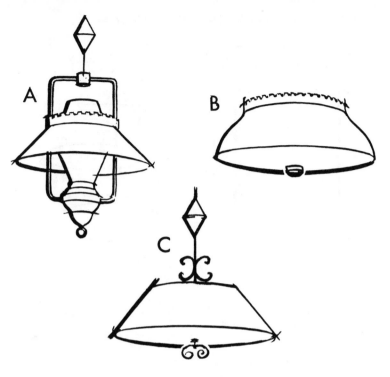

Fig. 12. Types of dining room fixtures.

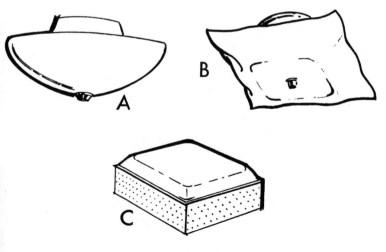

Fig. 13. Types of bedroom fixtures.

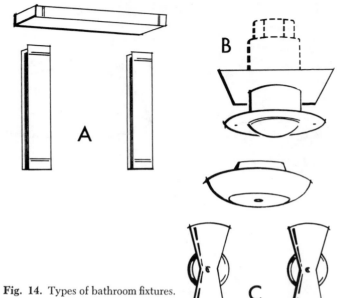

Fig. 14. Types of bathroom fixtures.

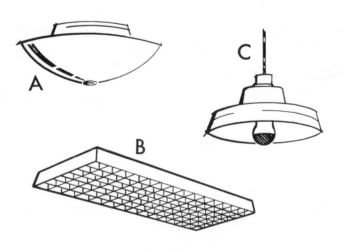

Fig. 15. Types of utility room fixtures.

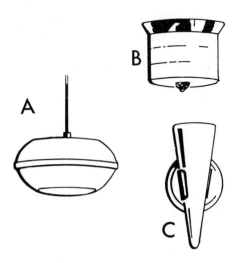

Fig. 16. Types of hallway fixtures.

Bedroom. *Surface mounted ceiling fixture* (Fig. 13, A) 12-inch minimum diameter, single socket or three sockets. Shallow wide diffuser is desirable. *Ceiling fixture* (Fig. 13, B) is similar to the fixture shown in Fig. 13, A. It is 12 or 14 inches wide, surface mounted, with plain or textured glass or plastic diffuser. *Ventilated ceiling fixture* (Fig. 13, C) has one or two sockets, diffusing shade to extend below trim to give side lighting. Unit is surface mounted on ceiling.

Bathroom. *Side and overhead fluorescent fixtures* (Fig. 14, A)—a pair of 24-inch long fixtures are spaced 30 or more inches apart at mirror sides. Use fixture above mirror if no ceiling light in room. *Vapor proof ceiling fixture* (Fig. 14, B)—a good type of fixture for a shower stall. Use a 60-watt bulb. *Be sure* that the switch is located outside of the shower. *Side and overhead incandescent units* (Fig. 14, C)—one- or two-socket fixtures at mirror sides are centered 60 inches above floor. *Note* overhead fixture. Bulbs are well shielded to reduce glare.

Utility Room. *Surface mounted ceiling fixture* (Fig. 15, A)—minimum diameter of 12 inches is desirable. Unit may have one or two sockets. *Shielded fluorescent fixture* (Fig. 15, B) is a two- or four-tube fixture, and can be centered in ceiling or mounted over work area. *Reflector and reflector bowl bulb unit* (Fig. 15, C) has a 12- or 14-inch minimum diameter. Used to reduce glare and to spread light.

Hallway. *Hanging bowl fixture* (Fig. 16, A) is a good choice for lighting a high-ceilinged hall or stairway. It is eight inches in diameter. *Closed globe fixture* (Fig. 16, B) is mounted on ceiling. Choose a white glass globe for diffusion of light. *Wall bracket fixture* (Fig. 16, C) may be used to supplement general lighting. It can be mounted on wall near a mirror.

STRUCTURAL LIGHTING

Valance. A lighted valance makes a room appear more spacious and dramatizes colors and textures. Valance boards are open at top and bottom, and are usually mounted above draperies. They provide up light that spreads across ceiling for general lighting and down light that accents draperies. Allow from 10 to 12 inches between the ceiling and top edge of the valance. For efficient structural lighting, follow dimensions exactly (Fig. 17). Valance lighting fixtures can be wired for entrance switching, and dimmer controlled fixtures may be installed to increase or decrease the level of the lighting.

Cornice. *Cornice lighting*, generally mounted at junction of wall and ceiling, is closed at top and extends full length of wall. A lighted cornice may be used with or without draperies. Cornices direct light downward to enliven wall textures, murals, scenic wallpaper, picture arrangements, art objects, and draperies. Lighted cornices are effective in low-ceilinged rooms where they give an illusion of height. Exteriors of faceboards may be painted, wallpapered, or covered with fabric. Since lighted cornices do not supply upward light, it is well to use open top lamps in the same room. (*See* Fig. 18.)

Soffit. In the *soffit* shown in Fig. 19 two 30- or 40-watt fluorescent tubes (deluxe warm white) are recessed in soffit above a kitchen sink. The fixture may be shielded with frosted glass or plastic to make it similar to a recessed unit.

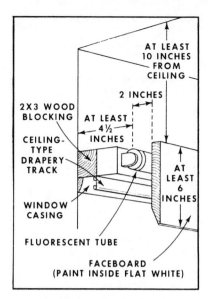

Fig. 17. Lighted valance.

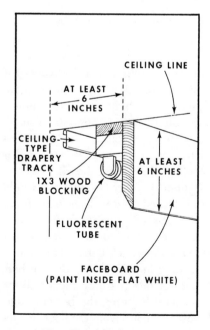

Fig. 18. Lighted cornice.

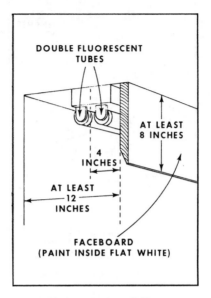

Fig. 19. Lighted soffit.

The entire interior of soffit should be painted flat white to reflect the light downward. Such an installation gives comfortable, diffused down light for easy seeing at sinks or along dressing counters in bathrooms. You can also use incandescent lighting in a soffit. Two 75-watt inside frosted bulbs, spaced 15 inches apart, can be mounted on ceiling or front edge of soffit behind a faceboard.

Bracket. A *lighted bracket* is similar to a lighted valance except that it is located on a wall instead of over windows. A bracket above a sofa provides both general and local lighting. Brackets can be mounted at suitable levels for lighting work counters, snack bars, pictures and wall hangings, and for reading in bed. For reading in bed, place bottom edge of bracket faceboard 30 inches above the top of mattress. For general lighting, the bottom edge of bracket should be at least 65 inches from the floor; for local lighting, bottom edge of bracket should be about 55 inches from the floor (*See* Fig. 20).

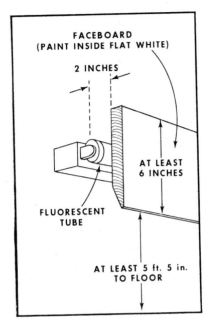

FACEBOARD
(PAINT INSIDE FLAT WHITE)

2 INCHES

AT LEAST
6 INCHES

FLUORESCENT
TUBE

AT LEAST 5 ft. 5 in.
TO FLOOR

Fig. 20. Lighted bracket.

USE OF DIMMERS

Add convenience, safety, and flexibility to your home lighting by using dimmer controls on fixtures in bedroom, bathrooms, halls, and living rooms. Gradations of light—from full bright to very dim—are possible simply by turning a knob.

A low level of lighting is helpful in the care of small children, sick persons, and others who need assistance during the night.

You can make dramatic changes in the mood of a room by softening lights with a dimmer switch. Lights can be lowered when listening to music or enjoying a fire on the hearth.

Dimmers for incandescent bulbs are simple, compact, and can be mounted in walls in much the same way as off-on switches. Be sure that the wattage capacity of the dimmer control is equal to or more than the total wattage to be controlled.

Dimmer controlled fluorescent fixtures must be pre-planned with your power supplier or electric contractor before installation. They are considerably more expensive than incandescent dimmer units. The control combines with a special built-in ballast, and can operate one or more specially designed fluorescent fixtures as a unit.

LIGHTING MAINTENANCE

Home lighting equipment needs regular care and cleaning to keep it operating efficiently. A collection of dirt and dust on bulbs, tubes, diffusion bowls, lampshades, and fixtures can cause a substantial loss in light output.

It is a good idea to clean all lighting equipment at least four to six times a year. Bowl-type portable lamps should be cleaned monthly.

Here are some suggestions for taking care of lamps and electrical parts.

1. Wash glass and plastic diffusers and shields in a detergent solution, rinse in clear warm water, and dry.

2. Wipe bulbs and tubes with a damp, soapy cloth, and dry well.

3. Dust wood and metal lamp bases with a soft cloth and apply a thin coat of wax. Glass, pottery, marble, chrome, and onyx bases can be washed with a damp soapy cloth, dried, and waxed.

4. Lampshades may be cleaned by a vacuum cleaner with a soft brush attachment, or dry-cleaned. Silk or rayon shades that are hand sewn to the frame, with no glued trimmings, may be washed in mild, lukewarm suds, and rinsed in clear water. Dry shades quickly to prevent rusting of frames.

5. Wipe parchment shades with a dry cloth.

6. Remove plastic wrappings from lampshades before using. Wrappings create glare and may warp the frame and wrinkle the shade fabric. Some are fire hazards.

7. Replace all darkened bulbs. A darkened bulb can reduce light output 25 to 50 per cent, but uses almost the same amount of current as a new bulb operating at correct wattage. Darkened bulbs may be used in closets or hallways where less light is needed.

8. Replace fluorescent tubes that flicker and any tubes that have darkened ends. A long delay in starting indicates that a new starter is probably needed. If a humming sound develops in a fluorescent fixture, the ballast may need to be remounted or replaced.

PROPER WIRING FOR GOOD LIGHTING

Good quality, adequately protected wiring makes it possible to light your home well. It also provides for convenient use of electrical equipment.

Safety is assured by careful inspection and follow-up maintenance. *Be sure* that all wiring complies with the National Electrical Code, and meets local and area requirements. Each fixture, control, or electrical part should carry the label of the Underwriters' Laboratories (UL).

Efficient wiring has outlets located so that no point along the floor line in any usable wall space is more than six feet from an outlet. Wire size is large enough to prevent excessive voltage drop that results in poor lighting (a five per cent voltage loss produces a 17 per cent loss of light from an incandescent bulb).

Install enough circuits to provide electricity where you want it without overloading any one circuit. Locate switch controls at all principal doorways. Modern wiring systems generally use standard switches, but low-voltage relay switching for multiple point light and equipment control is on the increase. Special controls—dimmers, timers, and photocell units—all have a place in the effective performance of equipment and lighting.

A service entrance geared to present and future family needs is essential. A 100-ampere service entrance, the minimum code requirement, provides for modern living in a small house. If you have a large house or expect to add large electrical appliances, consider installing a 150- or 200-ampere service entrance. The 200-ampere service provides for electric space heating and other possible applications.

TELEVIEWING

Viewing television in a darkened room is extremely tiring to the eyes because of the sharp contrast between the bright screen and unlighted surroundings. To avoid eyestrain and fatigue, provide a low to moderate level of lighting throughout the viewing area.

Wall lighting from valances and brackets creates a delightful background for watching television. When you use these types of lighting, position your TV set to the side or in front of the lighted walls.

Another way to offset the brightness of the screen and make viewing comfortable is to place one or two portable lamps behind or at the sides of the set. This helps prevent reflections on the TV screen. If the lamps have three-way controls, turn them on the low settings.

CHAPTER 9

Electrical Projects

The electrical projects and improvements in this chapter are simple enough to be accomplished by the homeowner. Information about the required tools, materials, and basic procedures may be found in this and in Chapters 2, 3, and 6.

BASIC INSTRUCTIONS FOR MAKING AND WIRING LAMPS

Many old glass, china, and metal vases and kerosene lamps in many homes today are valuable not only because of the fact that they may be wired for much less than the price of a new lamp but also because of their beauty, their age, and the uniqueness of their designs. Novelty lamps can be made from jugs, vases, rolling pins, old coffee pots, and similar materials. While these lamps do not have the beauty of the old lamps, they add a note of color and serve as interesting conversation pieces. Although the various types of lamps may differ in their wiring, all of them require some of the following basic steps:

Cord. Lamp cord is available in different colors and may be chosen to match the wall, the lamp, or the furniture as desired. Much of the lamp cord on the market is covered with rayon fabric ranging in quality from a very loose weave that ravels easily to a close weave that will give longer wear. Some types of cord, such as those covered with rubber or with a plastic material, will stand harder wear than the rayon-covered cord, but care should be taken to avoid using wire that is too heavy and unattractive.

Convenience Outlet Plugs. Most plugs used for lamps are made of bakelite and are dark brown, ivory, or white. Care should be taken to select a plug which may be pulled easily from the outlet without pulling on the cord.

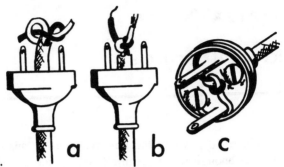

Fig. 1.

Attachment of a Convenience Outlet Plug to a Cord. To attach the plug to the cord, proceed as follows: Pull the cord through the plug, remove the fabric covering from the end of the cord, separate the wires, and tie a knot (Fig. 1, A). Bind the end of the fabric covering with a piece of thread. Remove ½″ of rubber insulation (Fig. 1, B). Twist the strands of wire tight. Connect the wire around the screw in the direction in which the screw tightens (Fig. 1, C). Be sure that insulation reaches the screw terminal, but does not extend under it.

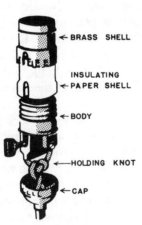

← BRASS SHELL

INSULATING
← PAPER SHELL

← BODY

← HOLDING KNOT

← CAP

Fig. 2.

Attachment of a Lamp Socket to a Cord. To attach the socket to the cord, proceed as follows: Release the brass shell (Fig. 2) by pressing it at the point marked "press" near the switch. (Some sockets come apart by twisting.) Then remove the cap and body. Pass the cord through the cap. Tie a knot if desired (not necessary in the case of a lamp since there is little strain at this point). Remove the insulation and attach the wire. Reassemble by reinserting the brass shell into the cap.

OLD GLASS FONT LAMPS

The following equipment is needed for wiring glass font lamps:

> Glass font lamp with metal collar.
> 1 convenience outlet plug with neck.
> No. 1 and No. 2 adapter, depending upon lamp size.
> 1 harp and finial (optional depending upon type of shade selected).
> 1 lamp socket.
> 8′ of lamp cord.

The harp (Fig. 3, A) is attached between the socket and the adapter. The harp (Fig. 3, B) screws onto the top of the socket. A finial (Fig. 4) is screwed onto the top of the harp after the shade is in place. This holds the shade to the harp.

To wire the lamp, attach the plug (Fig. 1). Pass the cord through the adapter and cap of the socket. Screw

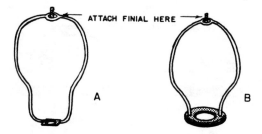

ATTACH FINIAL HERE

A   B

Fig. 3. Two types of harps.

the socket cap to the adapter. Be sure the paper insulation material on the cap is not lost. When using a type A harp, place it on the top of the adapter before screwing on the cap (Fig. 5). Attach the cord to the socket (Fig. 6). Screw the adapter into the lamp. Turn the lamp, instead of the adapter and socket, to avoid twisting the cord.

CHINA KEROSENE LAMPS

Many china lamps can be wired in the same way as glass font lamps. Use a silk shade with a harp. If a glass dome or globe shade is to be used, the wiring will be a little different. Often the old burner is used to make an

Fig. 4. Two types of finials.

adapter (Fig. 7). The holder for the ball (globe) shade (Fig. 8) fits on this type of homemade adapter (Fig. 7).

A converter (Fig. 9) may be used for china lamps. The

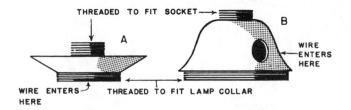

Fig. 5. Two types of adapters.

socket is in the converter and is screwed onto the adapter. With this method, extra parts must be purchased, but it is not necessary to insert a switch in the base of the lamp or in the cord, as may be the case if an adapter made from the old burner is used.

In wiring a type A lamp (Fig. 10), a hole must be bored near the base to let the cord pass up through the lamp, but be careful not to break the lamp. There are four methods of wiring this type of lamp: (1) Follow the directions for wiring glass font lamps and use a silk shade. (2) Use the converter (Fig. 9) or (3) use the adapter (Fig. 5). (4) If a dome-shaped glass shade is used, make an adapter from the original burner (Fig. 7) and fit a tripod (Fig. 11) on

Fig. 6. Adapter.

it to hold the shade. A switch should be set in the cord (Figs. 14 and 15). A tripod (Fig. 12) can be used with the adapter (Fig. 5, B). It will be necessary to use a

Fig. 7. (a) Original burner. (b) Pull wick holder out, cut away top part and file smooth. (c) Set socket in and solder. Use brass.

chimney holder (Fig. 13) which clips onto the socket.

To insert a switch in the cord, proceed as follows: Split the cord at a point (about 12″ to 18″ from the base of the

Fig. 8. Ball shade holder. Use with adapter made from burner.

Fig. 9. Converter, adapter, shade holder, and chimney holder.

lamp) where the switch is to be inserted. Cut only one wire, remove the insulation (Fig. 14), and twist the wires tight. Attach the wires under the screws as in the case of the socket and plug (Fig. 15).

In wiring a type B lamp (Fig. 10), carry the cord down through the lamp, since there is usually a hole through the center. A converter can be made from the old burner (Fig. 7), and a switch can be inserted in the base. To insert a switch in the base of the lamp, proceed as follows: Bore a hole in the metal base with a drill (if no suitable holes are already in it). Attach the switch to the cord at a point which will come inside the lamp near the base, split the cord, remove the insulation from one wire (Fig. 16), and attach the wires of the switch to the exposed

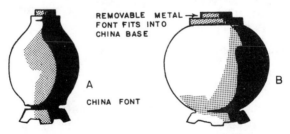

Fig. 10.

Fig. 11. Tripod used with oil burner converter.

wires of the lamp cord by twisting them firmly. Wrap each set of wires separately with friction tape, making sure there are no exposed wires (Fig. 17). Remove the holding nut from the switch and insert the switch in the hole in the base of the lamp from the bottom so that the switch comes out at the top. Replace the nut to hold the switch in place (Fig. 18).

METAL LAMPS

Before wiring clean the lamp thoroughly with steel wool, soap, or a liquid metal cleaner. Then solder socket into the center tube. Saw about an inch off the top of the tube so that the light bulb fits lower in the lamp. This will vary

Fig. 12. Tripod shade holder. Fits between socket and adapter.

Fig. 13. Chimney holder to fit socket.

with different types of lamps. Bore the hole with a hand drill and set the switch in the base (Figs. 17 and 18), leaving enough wire to reach the socket. Attach the wire to the socket of the lamp and attach the other end of the wire to the convenience outlet plug. Use a tripod to hold the dome-type glass shade.

MODERNIZATION OF LAMPS

Most of the lamps in homes have a low wattage and are equipped with inside frosted bulbs in narrow harps

Fig. 14.

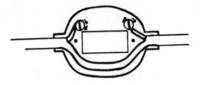

Fig. 15.

Fig. 16.

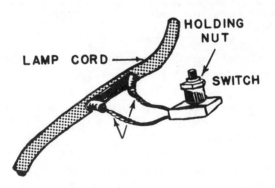

Fig. 17. Twisted wires wrapped separately with friction tape.

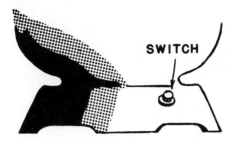

Fig. 18. Lamp base.

Fig. 19.

or in two-socket assemblies. Such lamps, if acceptable in size and shape, can be modernized to provide more efficient lighting (Fig. 19).

Height of Table Lamps. The height of a table lamp and the size and shape of its shade are important factors in good lighting. For the most satisfactory results a table lamp base should be at least 12″ high. With a 9″ harp,

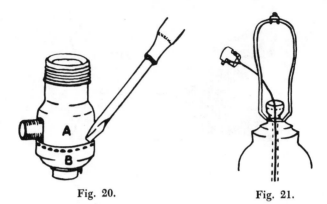

Fig. 20.   Fig. 21.

the minimum over-all height should be 21″. If a new shade seems desirable, use a shade with a bottom diameter that is approximately two-thirds of the over-all height of the lamp.

One-Socket Table or Floor Lamp with a Narrow Harp. The materials and tools required are a 150-watt white indirect-lite lamp, a wide harp, a lamp cord, and a screwdriver.

To modernize a one-socket table or floor lamp, proceed as follows: Remove the upper half of the socket cover

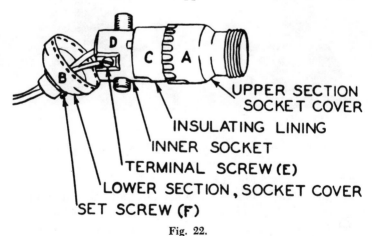

UPPER SECTION SOCKET COVER
INSULATING LINING
INNER SOCKET
TERMINAL SCREW (E)
LOWER SECTION, SOCKET COVER
SET SCREW (F)

Fig. 22.

from the lower section (Fig. 20, A, B). Pull about 6″ of the cord through one lamp base for easy workability (Fig. 21). Loosen the terminal screws to detach the wires from the inner socket (Fig. 22, D, E). Loosen the set screw (Fig. 22, F). Unscrew and remove the lower section of the socket cover and the narrow harp from the threaded pipe of the lamp base. Thread the cord through the new

wide harp and the lower section (B) of the socket cover. Screw the lower socket cover (B) tight onto the threaded pipe to hold the new harp securely in place. Tighten the set screw (F). If additional thickness is needed to give a rigid joint, washers of the appropriate size can be inserted above or below the new harp. Rewire the socket by looping one wire end around each screw terminal (E) and tightening the screws. Pull the slack cord through the bottom of the lamp base seating the inner socket (D) into the lower section of the socket cover (B). Insert insulating lining (C) around the socket, taking care that it covers both terminals (E). Snap the top section cover (A) into the lower section socket (B). Install a white indirect-lite lamp in the socket and place a shade on the threaded pin of the harp. Secure the shade with a finial. If the finial-

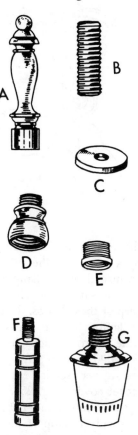

Fig. 23. **A**, finial; **B**, nipple; **C**, washer; **D**, nozzle; **E**, bushing reducer; **F**, shade riser; **G**, bottle adapter.

opening is too large for the threaded pin of the harp, reduce the finial opening with the bushing (Fig. 24). (Fig. 23 shows lamp and lampshade accessories.)

Alternative Method of Modernizing One-Socket Lamps. In some tall lamps of the table, bridge, and wall types, the

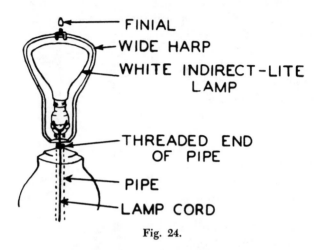

Fig. 24.

shades clip onto the lamp bulb instead of being supported by a harp. Further, the construction of the conventional up-turned bridge lamp and goose-neck lamp does not provide a satisfactory base for anchoring the under-socket harp of these lamps. It is possible to modernize these types of lamps by the use of a wide harp which has a threaded base to screw onto the upper section of the socket cover (Fig. 25). But this installation does not give so steady a support for the lamp shade as the under-socket harp, since the entire weight of the lamp shade is centralized on the socket instead of the lamp base.

Fig. 25.

The 4-prong clamp-on (clamped to the top of the bulb to hold the shade) is not recommended because direct contact between the metal and the hot glass may cause a sudden failure of the bulb. Also, the weight of the shade is placed on the bulb instead of on the base of the floor lamp or table lamp.

A bridge lamp with an up-turned socket, or a bridge lamp with a socket that can be put in a vertical position can be modernized with a screw-on wide harp, a white indirect-lite bulb, and a new shade with a bottom diameter approximately 11″.

To modernize a gooseneck lamp, remove the old metal reflector, straighten the gooseneck shaft, use a screw-on wide harp, a white indirect-lite bulb, and a new shade approximately 14″ across the bottom.

Two-Socket Table Lamps. In general, two-socket table lamps can be classified as one of the two types of construction. The lamp shown in Fig. 26 is constructed so that the pipe (housing the cord) protrudes above the lamp base about ⅜″ and is threaded. A threaded sleeve conceals the joint of the pipe and the socket stem. In the second type of lamp, the pipe and the socket stem are one continuous length of pipe from the bottom of the lamp to the socket assembly.

Materials and *tools required* are: a socket, a wide harp, a new electric cord (optional) 6′ to 8′, a 150-watt white indirect-lite, bushing and washers, a small screwdriver, pliers, and a sharp knife.

To modernize a two-socket table lamp, proceed as follows: For the first type of lamp, unscrew the two-socket assembly, the socket stem, and the threaded sleeve from the pipe and cut the electric cord. Remove the cord if a new one is to be installed. Insert the new cord through the pipe. For the second type of lamp, unscrew the two-socket assembly from the pipe, and cut and remove the electric cord. Mark the pipe for cut-off ⅜″ above the top of the lamp base. There are two alternate methods of procedure. The first method is to remove the pipe from the lamp base and measure the length from the bottom end to the cut-off mark.

The ⅛" threaded pipe is available at most local electric shops. Fit the new pipe with a washer and nut from the old pipe and insert it in the lamp base. Insert the cord through the new pipe. The second method is to remove the pipe from the base, cut off as marked, and thread the pipe end. If the necessary equipment for threading pipe

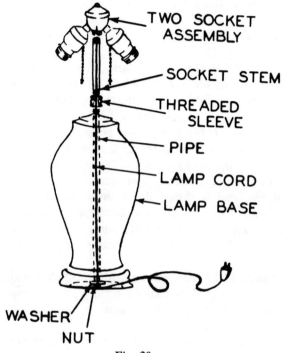

TWO SOCKET ASSEMBLY

SOCKET STEM

THREADED SLEEVE

PIPE

LAMP CORD

LAMP BASE

WASHER

NUT

Fig. 26.

is not available, have it done at a plumbing or electrical repair shop. Reinsert the pipe in the lamp base, and insert the cord through the pipe.

Split the cord back 2" from the end and peel ¾" of insulation from each copper wire. Disassemble the new socket (Figs. 19 and 21). Thread the cord through the wide harp and through the lower section of the socket cover (B). Screw socket cover base (B) down onto the threaded end of the pipe to anchor the harp securely in position. Washers of appropriate size can be used above and below the harp if necessary to obtain a rigid joint. Tighten the set screw (Fig. 22, F). Twist the copper strands, loop one wire around each terminal screw (E), and tighten the screws. Pull the slack cord through the lamp base seating the inner socket (D) into the lower section of the socket cover (B). Insert the insulating lining (C) around the socket, taking care that it covers both terminals. Snap the top section cover (A) into the lower section socket cover (B). Install a white indirect-lite in the socket and place a shade on the threaded pin of the harp. Secure the shade with a finial. If the finial opening is too large for the threaded pin of the harp, reduce the finial opening with a bushing. If a new cord has been installed, attach the plug to the cord end.

Two-Socket Floor Lamps. The two-socket floor lamp is most often constructed with the pipe extending through the lamp shaft to the two-socket assembly. There is no threaded joint ⅜″ above the top of the lamp shaft in this type of lamp. The threaded joint can be cut in the home workshop if the necessary equipment is available or at a local metal or lamp repair shop. The pipe, which is generally ¼″ iron pipe size, must be cut off, threaded, and fitted with a nipple or nozzle onto which the new socket can be screwed.

Floor lamps should range in over-all height from 55″ to 59″. Decide on the over-all height, then subtract 9″ from the height of the harp; the difference will be the new height. Cut the floor lamp shaft accordingly, but first check this height in relation to the design and decorative features of the lamp. It may be desirable, though not generally necessary, to adjust the established height.

Materials and tools required are: a socket, a wide harp, a white indirect-lite lamp, a new cord (optional) 8′ to 10′, ⅛″ brass nipple or ¼″ F x ⅛″ M nozzle, washers (of appropriate size) and bushing if required, a pipe saw, a screwdriver, pliers, pipe-threading equipment, and a sharp knife.

To modernize two-socket floor lamps, proceed as follows:

Unscrew the two-socket assembly from the pipe. Cut the electric cord and remove it from the lamp base. Remove the pipe through the bottom end of the lamp base. If the lamp is to be shortened, cut the shaft to the desired height (Fig. 27). There are at least two methods for providing an ⅛″ threaded pipe on which the socket can be screwed. Cut the pipe so that it protrudes ¼″ above the top of the shaft. Thread the outside of the pipe and fit it with a

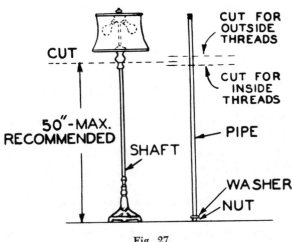

Fig. 27.

¼″ F x ⅛″ M nozzle (Fig. 28, A); or cut the pipe so that the top end sinks ⅛″ below the top of the shaft. Thread the inside of the pipe to receive a ⅛″ brass nipple (Fig. 28, B). A washer of appropriate size may be used to cover the cut edge of the lamp shaft. Insert the pipe through the lamp shaft. Thread the cord through the pipe, and re-assemble.

INSTALLATION OF A FEED-THROUGH SWITCH

If a feed-through switch is used, an appliance can be disconnected without disconnecting at either end of the cord (Fig. 29, A). When installing a switch of this type on an appliance cord, proceed as follows: Remove the screws or clamp that hold the two halves of the switch together. Remove the outer insulation of the appliance cord at the point where the switch is to be connected.

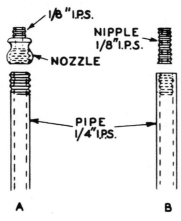

Fig. 28.

Separate the two wires and make necessary inside loops where the switch is to be installed (Fig. 29, B). Cut one of these wires and make necessary connections around the terminal screws in the switch. Place the uncut wire in the groove provided and reassemble the switch.

INSTALLATION OF ADDITIONAL BASE RECEPTACLES

Determine the circuit to be tapped and make a rough

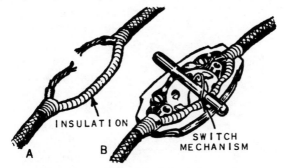

Fig. 29. Installing feed-through switch.

sketch of the location and subsequent line of wiring, indicating the points in walls, floors, basements, studs, or

joists where holes for wire or for cable should be drilled. The Underwriters' Code Requirements covering materials and procedures must be followed. Draw a line on the baseboard at the location where the receptacle will be installed (Fig. 30, A). Place the center of the wall box on this line and draw an outline on the baseboard. Using a ¾" bit and brace, drill four holes, one in each corner of the outline (Fig. 30, B). Cut from one hole to the other with a key saw or a compass saw. Cut to the drawn lines and smooth all sides of the opening with a ¾" or 1" chisel. Insert the wall box into the opening to make sure it is the right size and, if necessary, enlarge the opening. Drill holes to the connecting or tapping point for the cable and pull the cable through (as described in Chapter 3). Insert the box in the baseboard opening and

Fig. 30. Installing base receptacle. mark the outline of the box supports on the baseboard (Fig. 30, C). Then cut out the wood the thickness of the box supports. Remove knockouts from the receptacle box where wires or cables enter. Pull the wires into the box and mount the box with flat head wood screws (as described in Chapter 3). Remove insulation from the wires and make connections to the terminals on the receptacle. Fasten the receptacle to the box with screws recommended by the manufacturer. Be sure that the current is cut off at the location of the added base receptacle, and continue with the splicing and connecting.

GARDEN LIGHTING

For a family whose garden happens to be in the planning stage, one of the best procedures is to extend underground UF cable into the garden (Fig. 31). Then, some weather-

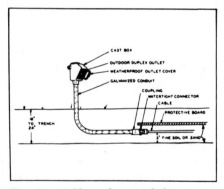

Fig. 31. Cable underground for permanently installed garden lighting system.

proof convenience outlets can be installed at the very outset. These outlets come in handy for the use of hedge clippers and other electrical garden equipment or for an outdoor radio. Homeowners who wish a permanent connection can always add a cable which should be buried in a trench 18″ to 36″ deep (as described in Chapter 3).

Connections for Garden Lighting. For temporary or seasonal lighting, it is possible to plug in on existing circuits in the house or garage. But avoid overloading any one household circuit (an average circuit carries only about 1700 watts) and protect extension cords, wherever they leave the house or garage, against damage from closing

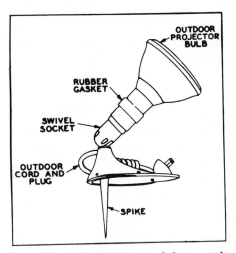

Fig. 32.   Stick-it-in-the-ground fixture with
150-watt outdoor-projector type
of bulb.

doors and windows. Insert a wood panel the width of the window about 2″ or 3″ high and bring the cord out through a hole cut in the board.

Safety Measures. Because water and electricity do not mix, observe the following precautions when installing or maintaining garden lighting:

Weatherproof cords and *moisture-proof connections* are a must for temporary and permanent installations alike.

Work in a dry garden when installing garden-lighting fixtures, whether experimentally or permanently. Any kind of moisture must be guarded against.

Turn off the electricity before installing the lighting equipment and before making any replacement, even if it be only a light bulb.

Bulbs and fixtures should be so placed that the light is confined to the garden. Keep the light source out of sight by hiding bulbs in shrubbery or by using a shielding reflector that conceals the light bulb. Blinding lights spoil what should be a restful garden scene. (A practical fixture and light are shown in Fig. 32.) White lights usually look best on flower beds; green and yellow lights are complementary to fountains, pools, and trellises. All fixtures used should be portable, decorative by day as well as by night, and equipped with weatherproof cord and plug, all ready to be installed (Fig. 32). In addition to fixtures that highlight favored spots, there are others available that make the pathways and grounds safe for nighttime walking or for special outdoor activities. Light bulbs of various sizes can be used, depending on what is to be illuminated and where it is located. A Christmas-tree bulb may be adequate for use behind a small leaf shield. A 150-watt projector-type bulb made of heavy glass is excellent for general garden use. It can be installed on the roof line of the house or under the eaves, or it can be mounted in an adjustable bracket at ground level.

SINGLE BELL AND PUSH-BUTTON CIRCUIT

For bell installation, use double-cotton-covered annunciator wire, rubber-covered fixture wire, or rubber-covered No. 18 B&S-gauge signal wire. In damp locations use a damp-proof or rubber-covered wire. In existing homes run the signal wires either behind or at the top of the baseboards, under mouldings, or in any other inconspicuous place. (See Chapter 3 for instructions on securing wires, staples, and nails.) All wires should be run on straight horizontal or vertical lines with right-angle turns throughout (Fig. 33).

The parts needed in bell installation are the bell battery wire, the section wire, and the button battery wire. The bell battery wire connects the bell or bells directly to the battery or a 6-18 volt transformer without interruption. The section wire runs to the individual bell or bells. The button battery wire, which connects the push button to the battery, is not run directly to the battery, but is connected to the

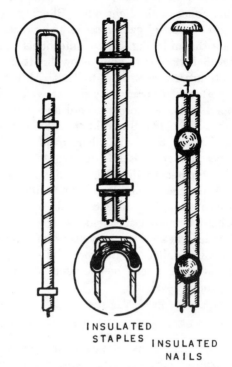

INSULATED
STAPLES   INSULATED
NAILS

Fig. 33.   Method of signal wiring.

dead side of the push button or other circuit-breaking device. Figure 34 shows a wiring diagram of a single bell or buzzer operated from a dry cell. It is a simple type of bell-ringing circuit and operates as follows:

The current starts out of the positive pole of the cell and travels along the button battery wire as far as the push button. The circuit is closed by pushing the button and bringing the two contact springs together; the current travels along the section wire through the bell and returns to the negative pole of the cell through the bell battery wire.

To install a single bell or buzzer (Fig. 34) the following is required: a mounted push button, a buzzer or bell, several dry batteries connected in series, No. 18 annunciator wire, and insulated staples. Drill the hole for passage of the wire and temporarily install the push button. Set the batteries in an accessible location. Run the wiring as indicated in the diagram. Allow enough wire to form pigtails at both the dry cell and the bell. When fastening wire under screws or binding posts, bend the wire around the screw or post in the direction that the screw or nut turns. When using metal push buttons, bring the wire through the grooves in the flange to avoid cutting through the insulation when the button is screwed tight. Test the circuit after installation has been completed. If the bell or buzzer does not sound when the button is pressed, check the wiring and all terminal connections and make necessary corrections.

Single Bell Operating From An A.C. Circuit. A bell-

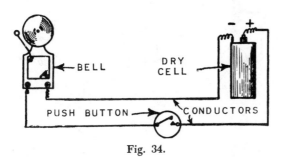

Fig. 34.

ringing transformer is used to step the voltage down to 6 volts to 18 volts from an a.c. lighting circuit. A high-reactance transformer will deliver a maximum current of only a few amps. even when short-circuited (Fig. 35). All local rules and regulations applying to light and power wires should be followed. The primary side of the transformer must be connected with at least a No. 14 B&S-gauge wire. Install the secondary side from the transformer to the push button and bell, using the same procedures as for battery-powered bells.

Single Bell Controlled By Three Single Push Buttons. Figure 36 shows the wiring of a bell so that it can be rung from three points. Methods of connecting this system and running the wire are the same as for the installation of a single-button system. In this type of system one terminal of each push button is connected to the button battery wire and the remaining terminal is connected to the section wire.

FLUORESCENT LAMPS

The auxiliary equipment necessary for installing fluorescent lamps are ballasts, starters, and lampholders. *Note* that fluorescent lamps will not operate on d.c. without a d.c. adapter.

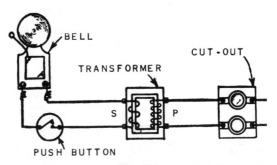

Fig. 35.

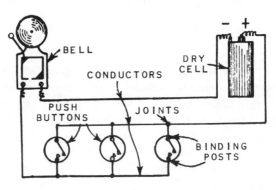

Fig. 36.

Ballasts. A ballast for fluorescent lamps preheats the electrodes to make available a large supply of free electrons, provides a surge of relatively high potential to start the arc between the electrodes, and prevents the arc current from increasing beyond the limit set for each size of lamp. The ballast commonly used is a current limiting device which consists either of a reactor alone or a reactor combined with a step-up transformer enclosed in a metal case. Figure 37 shows a schematic diagram of a single lamp ballast and its circuit. Figure 38 shows the two-lamp ballast and its circuit. It is characteristic of fluorescent lamps (which are actually a type of arc lamp) that some method must be

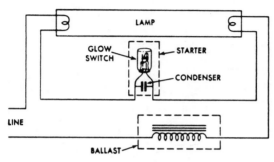

Fig. 37.

provided for limiting the current drawn by the discharge. Without the ballast as a limiting device, the current would rise to a value that would destroy the lamp. A starting compensator is included with all two-lamp ballasts except for 90-watt lamps. One is required for each two-lamp ballast and it is connected in the starting circuit of the lead lamp (Fig. 38). It functions only while the lamp is starting and is automatically disconnected from the circuit when the starting switch opens. Starting compensators are generally built into the ballast.

Starters. The glow type of starting switch is generally used (Fig. 37). It is enclosed in a small glass bulb and consists of two electrodes (one of which is made from a bimetallic strip) in an inert gas such as neon or argon (Fig. 39). These electrodes are separated under normal conditions, but when closed they form part of a series circuit through the lamp electrodes and the reactance. When voltage is applied, a small current flows as a result

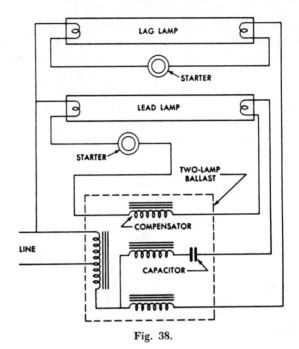

Fig. 38.

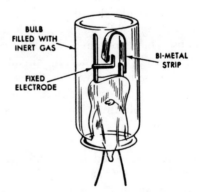

Fig. 39. Glow type of starter switch used almost universally for pre-heat circuits for fluorescent lamps. One required for each lamp.

of the glow discharge between two electrodes of the switch. The consequent heating of the electrodes expands the bi-metallic element, causing the electrodes to touch. This short-circuiting of the switch stops the glow discharge but allows a substantial flow of current to preheat the lamp electrodes. There is enough residual heat in the switch to keep it closed for a short time while the electrodes are

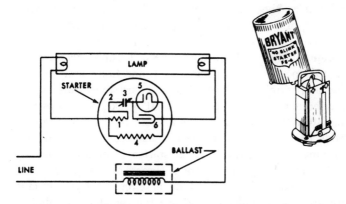

Fig. 40. "No-Blink" starter eliminates blinking of worn-out lamps.

being preheated. After the glow disappears, the bimetal cools, the switch opens, and the resulting high-voltage surge starts normal lamp operation. If the lamp arc fails to strike, the cycle is repeated. The switch does not again glow (if the lamp arc is established) since it is so designed that the remaining available electrical potential is insufficient to cause a breakdown between its electrodes. Thus, it consumes no power and, if the lamp is turned out is available for immediate restarting.

The conventional glow switch type of starter consists of a glow switch plus a condenser, both enclosed in a small cylindrical container with contacts which may be easily inserted in a bayonet type of socket. This socket may be an integral part of the standard lampholder (to which it is attached by a single screw) or it may be merely connected to the lampholder electrically. Usually the starter is mounted separately so that it may be readily replaced without removing the lamp itself. The switch provides the preheating of the lamp electrodes and the starting surge; the condenser suppresses radio interference. Since the glow switch is designed to operate between critical voltage limits, the proper start must be used for each lamp to insure satisfactory starting. The "No-Blink" starter (Fig. 40) contains the regular glow switch and, in addition, a bimetal element which automatically cuts the lamp out of the circuit when it reaches the end of life. This eliminates annoying blinking of a worn-out lamp, protects the ballast, prolongs starter life, and reduces the cost of maintenance.

Lampholders and Starter Sockets. Lampholders are employed to provide the electric connections and to hold the lamp firmly in place. The standard twist-turn type of lampholder is shown in Fig. 41. It is used for lamps with miniature, medium, and mogul bipin bases. For the electrode preheat type of operation, starters are needed. For "instant start" operation, the starter is omitted and the ballast and lampholders are designed and applied in the manner required for this type of circuit. Starter circuits are lower in initial cost and are somewhat more efficient,

Fig. 41. Standard twist-turn lampholder for fluorescent lamps with medium bipin bases. Base pins are inserted in groove and lamp is locked in position with a 90° twist turn.

and the voltages involved are low, but the starterless types have the advantages of prompt starting and a simple electrical circuit without moving parts. A typical starter and separate starter socket are shown in Fig. 42. Some available starter sockets are mounted on the base of one of the two lampholders. Figure 43 shows the push contact type of fluorescent lampholder used for relatively simple lamp replacement. These lampholders are available for two or three lamps having a medium bipin base.

Fig. 42. Fluorescent lamp starter and starter socket.

The instant start, slimline, and rapid start lamps do not require starters for operation. Sufficiently high voltage is placed across the instart start and slimline lamps to cause them to start immediately. The rapid start lamps start within 1 sec. and at a much lower voltage because of the continuous preheating current flowing through the electrode coils.

Installation and Maintenance of Fluorescent Lamps. Wiring procedures for fluorescent lighting are the same as those outlined for incandescent lighting fixtures, but the voltage requirements are slightly different. Keep the voltage at the fixture within the range required. Low voltage is as undesirable as high voltage. In the case of filament lamps, low voltage reduces light output and efficiency but

Fig. 43. Push contact fluorescent lampholder for medium bipin base.

prolongs lamp life. With fluorescent lamps, however, low voltage or high voltage will reduce both lamp life and lumen maintenance; low voltage may cause instability in the arc and starting difficulty.

Fluorescent lamps and auxiliaries are designed to operate well together over specific voltage ranges. The lamps with "low" voltage ballast equipment are designed for operation on circuit voltages of 110 volts to 125 volts, inclusive; in some cases they may operate satisfactorily on circuits as low as 105 volts or as high as 130 volts. Lamps with ballasts designed for high-voltage service should operate satisfactorily on circuits using 220 volts to 250 volts, with

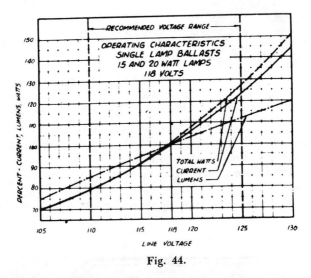

Fig. 44.

possible satisfactory operation on voltages as low as 210 volts or as high as 260 volts. The characteristics of lamps and ballasts for a range of voltages are shown in Figs. 44 and 45. On voltages above the specified range, the operating current becomes excessive and may not only overheat the ballast but may cause premature end blackening and early lamp failure. Voltages below the specified range may lower the preheat current to a point where the electrodes fail to emit their proper quota of electrons. Such a condition may cause the lamps to flash frequently during the starting cycle. This results in a more rapid wasting away of the emission material, thereby shortening lamp life.

When a fluorescent lamp is first lighted, its output is abnormally high. During the first 100 hours of burning, the output drops about 10 per cent. The deterioration during the remaining life of the lamp is less rapid. The depreciation in light output is due chiefly to a gradual deterioration of the phosphor powders and a blackening of the inside of the tube. If the lamp is started less frequently, its life span will increase. Its lumen maintenance is indicated by the dotted extension of the curve. All fluorescent lamps blacken

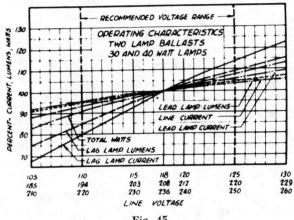

Fig. 45.

gradually and continuously as they burn. A slight blackening at the ends of the lamp is ordinarily a normal sign of age in service and is the result of bombardment of the electrodes by the arc stream which causes a deposit of mercury and electrode material on the bulb. This darkening is more noticeable in lamps in which the electrode is close to the glass wall. In the last few hours of lamp life a somewhat dense deposit will develop at the end of the lamp where the electrode is deactivated. This darkening will be especially noticeable if the lamp is allowed to flash on and off before it is replaced.

Fixtures of a decorative type in which the ends of the lamp are exposed to view are not recommended because the normal lamp blackening may be objectionable and lamps may need to be replaced before they have given full life. Shields, which conceal this blackening but reduce light output somewhat, may be used where appearance is more important than efficiency.

Radio Interference. Most radio interference from fluorescent lamps is eliminated by a small condenser mounted in the starter switch container (Fig. 37). If special measures prove to be necessary, radio interference filters which are commercially available will give excellent results when properly installed. The simplest of these is a three-section delta-connected capacitor which is grounded to the metal fixture and connected across the supply lines as they enter the fixture. In most cases, these filters will reduce the interference caused by fluorescent lamps to a level that is not detected by the ear. A filter should be installed on each fixture and as close to the lamps as possible.

Installation of Bed Lighting. The entire cost of installing a modern fluorescent lighting system in a bedroom should not exceed ten dollars. This system includes a new ceiling fixture, materials for building a fluorescent bed-lighting unit, a wall switch control for the new fixture, and a new convenience outlet.

A bed-lighting unit can be an attractive item as well as an efficient lighting circuit.

Materials. For construction of a bed-lighting unit, the following materials are required:

> 1 piece white pine ½″ x 5¾″ x 30″ (valance front).
> 2 pieces white pine 1″ x 5¾″ x 5¾″ (valance ends).
> 1 piece white pine 1″ x 4″ x 28⅜″ (mounting board).
> 1 piece half-round molding 8′0″.
> 4½″ corner irons with ¾″ flat head screws.
> 1 box of brads (for fastening half-round).
> 1 channel unit with auxiliaries for 20-watt fluorescent lamp.
> 1 through-switch.
> 1 electric cord.
> 1 plug.
> 1 frosted glass panel.
> Plastic wood.
> Paint.

Construction. To construct this lighting fixture, proceed

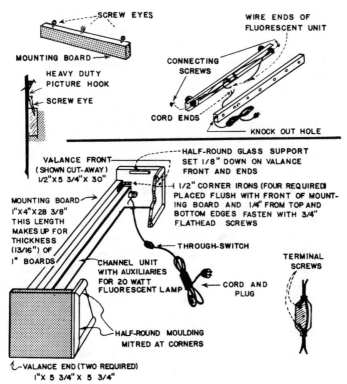

SCREW EYES

WIRE ENDS OF
FLUORESCENT UNIT

MOUNTING BOARD

HEAVY DUTY
PICTURE HOOK

SCREW EYE

CONNECTING
SCREWS

CORD ENDS

KNOCK OUT HOLE

HALF-ROUND GLASS SUPPORT
SET 1/8" DOWN ON VALANCE
FRONT AND ENDS

VALANCE FRONT
(SHOWN CUT-AWAY)
1/2"X 5 3/4"X 30"

MOUNTING BOARD
1"X 4"X 28 3/8"
THIS LENGTH
MAKES UP FOR
THICKNESS
(13/16")
1" BOARDS

1 1/2" CORNER IRONS (FOUR REQUIRED)
PLACED FLUSH WITH FRONT OF MOUNT-
ING BOARD AND 1/4" FROM TOP AND
BOTTOM EDGES FASTEN WITH 3/4"
FLATHEAD SCREWS

THROUGH-SWITCH

CHANNEL UNIT
WITH AUXILIARIES
FOR 20 WATT
FLUORESCENT LAMP

TERMINAL
SCREWS

CORD AND
PLUG

HALF-ROUND MOULDING
MITRED AT CORNERS

VALANCE END (TWO REQUIRED)
1"X 5 3/4" X 5 3/4"

Fig. 46.

as follows: Construct a valance (Fig. 46). Check the wall where the valance is to be mounted to determine where the 2 x 4 upright studs in the house framing are located. These are generally 16″ apart and can sometimes be detected by tapping along the wall. Attach the mounting board to the wall, nailing it securely through the studs. The bottom of the mounting board should be 31″ above the top of the bed. If the valance is not to be mounted permanently on the wall, three screw eyes can be attached to the top edge of the mounting board. After the unit has been completed, it can be hung on heavy-duty picture hooks. Screw eyes for hanging the picture hooks should be inserted at a slight angle. (Note sturdy results obtained from method shown in Fig. 46.)

Detach the top cover of the channel of the fluorescent fixture by removing screws that are generally located at the ends of the unit. Knock out indented circle at one end of the channel, and insert the cord end through the hole. Connect one wire of the cord end to one loose wire end of the fluorescent unit. Connect the other wire of the cord end to the remaining loose wire end of the fluorescent unit. Make the splices carefully, taping them securely or using a solderless connector. Attach the plug to the cord end (Fig. 46). Separate the through-switch into two halves. Split the two wires of the cord apart for 2″ at the place where the switch is to be attached. Cut one wire and connect each end to a terminal screw. The other wire runs through the switch without a break (Fig. 46).

Center the back section of the wiring channel on the mounting board and attach it with screws. Screw the top section of the wiring channel to the back section. Attach the valance to the mounting board with the top edge of corner irons ¼″ below the top edge of the mounting board After the valance unit has been attached to the wall, make an exact measurement to determine the size of frosted glass to be purchased. Install the frosted glass, insert a 20-watt T-12 fluorescent lamp, and plug it into the nearest convenient outlet.

Valances and Cornices for Fluorescent Lighting. Ready-to-install valances as well as plastic and glass channels for custom-designed installations using standard fluorescent or slimline tubes are available. Individually designed valance or cornice boards can be made out of wood about ¾″ thick and continuous in length. They should conform and harmonize with architectural and decorative elements in the room. The outside face of each board may be painted, wallpapered, covered with fabric, upholstered, or left in a natural wood finish. For the best reflection of light, the inside face and the wiring channel should be painted white.

Valances. Over windows and French doors, extend the valance beyond each side of the frame so that the draperies will only partially cover the wall and will allow a greater amount of natural light to enter the room. A more pleasing

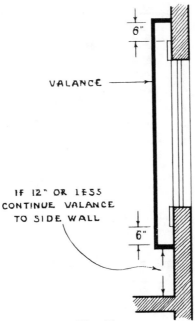

6"

VALANCE

IF 12" OR LESS
CONTINUE VALANCE
TO SIDE WALL

6"

Fig. 47.

effect is often gained if the valance extends from wall to wall and draperies extend to the corner (Fig. 47). See Fig.

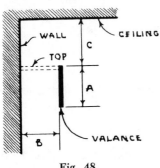

WALL

CEILING

C

TOP

A

VALANCE

B

Fig. 48.

48 for depth, return, and distance from the ceiling. The depth (A) should be a minimum of 6"; as the valance length and ceiling height increase, the depth may be increased to 10" to 12". The return (B) should be a minimum of 6". A return of 7" to 8" improves light distribution. Where the valance extends from wall to wall, place the board 8" to 10" from the wall. For a light ceiling and a light wall a minimum distance of 10" from the ceiling (C) is required; for a light ceiling and a dark wall,

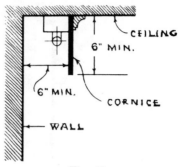

Fig. 49.

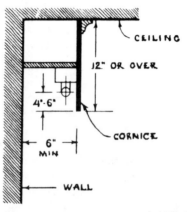

Fig. 50.

a minimum distance of 12″ is required. If the distances are less than the minimum, close the top of the valance.

Cornices. If the distance (Fig. 48, C) from the ceiling is less than the minimum, either a closed valance top or a cornice lighting system should be considered (Fig. 49). If a cornice 12″ deep or deeper is desired, place the light sources as shown in Fig. 50.

Types of Tubes. Use a combination of fluorescent lengths to fill the required space. Whenever possible, use the same diameter of tubes: ¾″ (T-6), 1″ (T-8), or 1-1½″ (T-12).

The following lists of sizes and diameters can be used as a guide:

Preheat (Slight starting delay)   (*Length given includes sockets.*)

14 watt T-12	15" length	25 watt	T-12	33" length
15 watt T-8 or T-12	18" length	30 watt	T-8	36" length
20 watt T-12	24" length	40 watt	T-12	48" length

Slimline (Instant start)   (*Length given includes sockets.*)

120, 200 or 300° T-6	42" length	425°	T-12	48" length
120, 200 or 300° T-6	64" length	425°	T-12	72" length
120, 200 or 300° T-8	72" length	425°	T-12	96" length
120, 200 or 300° T-8	96" length			

° Ma—milliamperes or current value. Slimline tubes in any one valance should all be operated at the same current value.

Placement of Channel and Light Source. The channel and light source may be on the wall or on the valance board.

On Wall. If a fluorescent channel is used with ballast enclosed, mount the channel on a wood strip or wood blocks that have previously been attached securely to the wall.

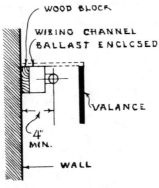

Fig. 51.

For greater strength the screws attaching the wood strips or blocks should be anchored through plaster and lath onto the house studs. The wood strips or blocks should be thick enough so that the center of the light source will be 4" from the wall (Fig. 51).

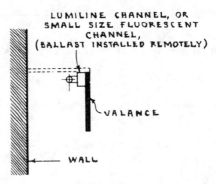

Fig. 52.

On Valance Board. When ballasts are not in the channel but are installed elsewhere, a smaller channel of less weight may be used and mounted directly on the inside face of the valance. The light sources should be positioned as high as possible in the valance (Fig. 52). Keep the lamp sockets back to back, to avoid shadows between lamps. To avoid dark areas at each end of a valance, the line of tubes should extend to within 3″ to 4″ of the ends of the valance board.

CHAPTER 10

Electrical Planning for New and Modernized Homes

A good *home wiring system* is a necessity for modern electrical needs. Even though a full supply of electricity is made available by the power supplier, the extent to which it can be used in any home is determined by its interior wiring. Starting from the service entrance, where electricity enters, it is the responsibility of the home's wiring system to deliver a full supply of electricity to every outlet. In a properly planned system each circuit and outlet is designed to serve a specific purpose and switch controls are located with both convenience and safety in mind.

Appliances are popular aids to homemaking. They are constantly being improved to provide higher and higher standards of performance and convenience. The resulting increases in wattage are reflected in increasing demands upon the home wiring system.

The *number of appliances* in homes is increasing. Work saving, timesaving, wife saving appliances, like the automatic washer, electric clothes dryer, dishwasher, and food waste disposer, are no longer considered luxuries. The average family also looks forward to the convenience and economy of home freezers and electric water heaters, and the comfort and relaxation of air conditioning and television, if these appliances are not already installed. All appliances require proper wiring facilities for satisfactory operation.

Air conditioning is rapidly being accepted for residential use in areas where temperatures exceed comfort levels. Many of the new units are designed to provide some heating as well. Proper wiring for these installations will aid in avoiding difficulties such as service interruptions due to overloaded circuits and, at the same time, will help to assure better operating efficiencies from each unit.

Home lighting is being revolutionized. Portable lamps of modern design use bulbs of high wattage, with lighting fitted to the seeing task. The newer tubular light sources are bringing built-in lighting, for both decorative and utilitarian uses, within the economic reach of most families, and these applications are multiplying rapidly.

The *tangle of extension cords* in countless homes; the frequent need to replace fuses and reset circuit breakers; unlighted stairways and entrances; the necessity of groping in the dark to locate light switches—and dozens of similar conditions—all result from the fact that home wiring practice has not kept pace with the progress of electrical utilization.

When load outstrips the capacities of the wiring system, operating efficiency suffers. When voltage drop becomes excessive in the circuit between the service panel and the appliance or light source it results in poor lighting and inefficient or totally ineffective appliance operation. *For example,* a five per cent voltage loss produces a ten per cent loss of heat in any heating appliance or a 17 per cent loss of light from an incandescent lamp.

GENERAL REQUIREMENTS

The basic requirement is compliance with the safety regulations. Next, service entrance equipment of ample size, sufficient circuits of adequate capacity, and sufficient convenience and lighting outlets, properly placed, with switches for convenient control are a must.

Compliance with Safety-Regulations. A fundamental prerequisite of any adequate wiring installation is conformity to the safety regulations applicable to the dwelling. The approved American Standard for electrical safety is the National Electrical Code. Usually, local safety regulations in the form of a municipal ordinance or state law are based upon the National Electrical Code.

Inspection service, to determine conformance with safety regulations, is available in most communities. Where available, a certificate of inspection should be obtained. In the absence of inspection service, an affidavit should be obtained from the electrical contractor attesting to conformity with the applicable safety regulations.

Location of Lighting Outlets. Proper illumination is an essential element of modern electrical living. The amount and type of illumination required should be fitted to the various seeing tasks carried on in the home. There must also be planned lighting for recreation and entertainment. In many instances, best results are achieved by a blend of lighting from both fixed and portable luminaries. Good home lighting, therefore, requires thoughtful planning and careful selection of fixtures, portable lamps, and other lighting equipment to be used.

Where lighting outlets are mentioned in these standards, their types and locations should conform with the lighting fixtures or equipment to be used. Unless a specified location is stated, lighting outlets may be located anywhere within the area under consideration to produce the desired lighting effects.

Location of Convenience Outlets. Convenience outlets should be located preferably near the ends of wall space, rather than near the center, thus reducing the likelihood of being concealed behind large pieces of furniture. Unless otherwise specified, outlets should be located approximately 12 inches above the floor line.

Location of Wall Switches. Wall switches should normally be located at the latch side of doors or at the traffic side of arches and within the room or area to which the control is applicable. Some exceptions to this practice are the control of exterior lights from indoors; the control of stairway lights from adjoining areas, when stairs are closed off by doors at head or foot; and the control of lights from the access space adjoining infrequently used areas, such as storage areas. Wall switches are normally mounted at a height of approximately 48 inches above the floor line.

Multiple-Switch Control. All spaces for which wall switch controls are required, and which have more than one entrance, should be equipped with multiple-switch control of each principal entrance. If this requirement would result in the placing of switches controlling the same light within ten feet of each other, one of the switch locations may be eliminated.

Where rooms may be lighted from more than one source, as when both general and supplementary illumination are provided from fixed sources, multiple switching is required for one set of controls only. Usually this is applied to the means of general illumination.

Principal entrances are those commonly used for entry to or exit from a room when going from a lighted to an unlighted condition, or the reverse. For instance, a door from a living room to a porch is a principal entrance to the porch. However, it would not necessarily be considered a principal entrance to the living room unless the front entrance to the house is through the porch.

Number of Lighting or Convenience Outlets. Where required on a basis of linear or square-foot measure in these standards, the number of outlets should be determined by dividing the total linear or square footage by the required distance or area; the number so determined to be increased by one if a major fraction remains.

> *Example:*   Required: one outlet for each 150 square feet
> Total square feet of area: 390
> 390 divided by 150 = 2.6
> 3 outlets are required

Dual Purpose Rooms. Where a room is intended to serve more than one function, such as a combination living-dining room or a kitchen-laundry, convenience and special-purpose outlet provisions of these standards are separately applicable to the respective areas. Lighting outlet provisions may be combined in any manner that will assure general overall illumination as well as local illumination of work surfaces. In considering locations for wall switches, the area is considered as a single room.

Dual Functions of Outlets. In any instance where an outlet is located so as to satisfy two different provisions of the standard of adequacy, only one outlet need be installed at that location. In such instances, particular attention should be paid to any required wall switch control, as additional switching may be necessary. *For example,* a lighting outlet in an upstairs hall may be located at the head of the stairway, thus satisfying both a hall lighting outlet and a stairway lighting outlet provision of the standards with a single outlet. Stairway provisions will necessitate multiple-switch control of this lighting outlet at head and foot of the stairway. In addition, the multiple-switch control rule, previously described, when applied to the upstairs hall, may require a third point of control elsewhere in the hall because if its length.

Interpretation of Requirements. The standards in this chapter are necessarily general in nature. They apply to most situations encountered in normal house construction and are to be considered as minimum standards of adequacy for such construction. Situations will arise from time to time, because of unusual design or unusual construction methods, when it will be impossible to satisfy the wording of a particular provision of the standards. The following rules will serve as a guide for meeting such situations.

1. The wiring installation should be fitted to the structure. If compliance with a particular provision would require alteration of doors, windows, or structural members, alternate provisions should be made. *For example,* some types of building construction may make it desirable to alter the recommended outlet or switch heights presented in this chapter, or to resort to surface-type wiring construction.

2. Each provision of the adequacy standard is intended to provide for one or more specific usages of electricity. If

such usage is appropriate to the particular home under study, and cannot be provided in accordance with the standard, an alternate provision should be made. If the construction peculiarity eliminates the need for the particular electrical facility, no alternate provision is needed.

3. If certain functions are omitted from the plans, such as a workshop, a laundry, and so on, electrical wiring serving these functions may be similarly omitted.

4. If certain facilities are indicated as future additions to the plans, such as a basement recreation room, initial wiring should be so arranged that none of it need be replaced or moved when the ultimate plan is realized. Final wiring for the future addition may be left to a later date if desired.

5. Standards given in this chapter may be applied to dwelling units of multi-family dwellings. At the present time there are no nationally recognized standards for common use spaces in such buildings, and good judgment as to lighting and other needs will have to be applied. Particular attention should be paid to limiting voltage drop in feeders to individual apartments to assure proper operation of appliances.

OUTLETS

Exterior Entrances—Lighting Provisions. These should include one or more lighting outlets, as architecture dictates, wall switch controlled, at front and trade entrances. Where a single wall outlet is desired, location on the latch side of the door is preferable.

It is recommended that lighting outlets, wall switch controlled, be installed at other entrances. The principal lighting requirements at entrances are the illumination of steps leading to the entrance and of faces of people at the door. Outlets in addition to those at the door are often desirable for post lights to illuminate terraced or broken flights of steps or long approach walks. These outlets should be wall switch controlled inside the house entrance.

Convenience Outlets. A weatherproof outlet (Fig. 1), preferably near the front entrance, should be located at least 18 inches above grade. It is recommended that this outlet be controlled by a wall switch inside the entrance for convenient operation of outdoor decorative lighting. Additional outlets along the exterior of the house are recommended to serve decorative garden treatments and for the use of appliances or electric garden tools, such as lawn mowers and hedge trimmers. Such outlets should also be wall switch controlled.

LIVING ROOMS

Lighting Provisions. Some means of general illumination is essential. This lighting may be provided by ceiling or wall fixtures, by lighting in coves, valances, or cornices, or by portable lamps (Fig. 2). Provide lighting outlets, wall switch controlled, in locations appropriate to the lighting method selected.

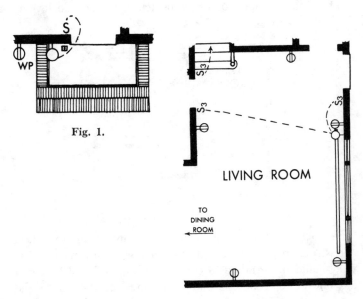

Fig. 1.

LIVING ROOM

TO
DINING
ROOM

Fig. 2.

These lighting provisions also apply to sun rooms, enclosed porches, television rooms, libraries, dens, and similar areas.

The installation of outlets for decorative lighting accent

is recommended, such as picture illumination and bookcase lighting.

Convenience Outlets. Convenience outlets should be placed so that no point along the floor line in any usable wall space is more than six feet from an outlet in that space. Where the installation of windows extending to the floor prevents meeting this requirement by the use of ordinary convenience outlets, equivalent facilities should be installed using other appropriate means.

If, in lieu of fixed lighting, general illumination is provided from portable lamps, two convenience outlets or one plug position in two or more split receptacle convenience outlets should be wall switch controlled (*See* Fig. 3).

In the case of switch-controlled convenience outlets, it is recommended that split receptacle outlets be used in order not to limit the location of radios, television sets, clocks, and

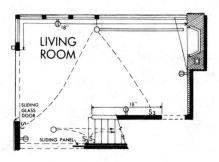

Fig. 3.

the like. Install one convenience outlet in a mantel shelf, if construction permits. In addition, a single convenience outlet should be installed in combination with the wall switch at one or more of the switch locations for the use of the vacuum cleaner or other portable appliances. Outlets for the use of clocks, radios, decorative lighting, and so on in bookcases and other suitable locations are recommended.

Special Purpose Outlets. One outlet for a room air conditioner should be installed wherever a central air-conditioning system is not planned.

DINING AREAS

Lighting Provisions. Each dining room, or dining area combined with another room, or breakfast nook, should have at least one lighting outlet, wall switch controlled (Fig. 4). Such outlets are normally located over the probable location of the dining or breakfast table to provide direct illumination of the area.

Convenience Outlets. Convenience outlets should be placed so that no point along the floor line in any usable wall space is more than six feet from an outlet in that space. When dining or breakfast table is to be placed against a wall, one of these outlets should be placed at the table location, just above table height.

Where open counter space is to be built in, an outlet should be provided above counter height for the use of portable appliances.

Convenience outlets in dining areas should be of the split receptacle-type for connection to appliance circuits. (*See* section on Branch Circuits later in this chapter.)

BEDROOMS

Lighting Provisions. Good general illumination is particularly essential in the bedroom. This should be provided from a ceiling fixture or from lighting in valances, coves, or cornices. Provide outlets, wall switch controlled, in locations appropriate to the method selected.

Light fixtures over full-length mirrors, or a light source at the ceiling located in the bedroom and directly in front of the clothes closets, may serve as general illumination. Master-switch control in the master bedroom, as well as at other strategic points in the home, is suggested for selected interior and exterior lights.

Convenience Outlets. Outlets should be placed so that there is a convenience outlet on each side and within six feet of the center line of each probable individual bed location. Additional outlets should be placed so that no point along the floor line in any other usable wall space is more than six feet from an outlet in that space. (*See* Figs. 5 and 6.)

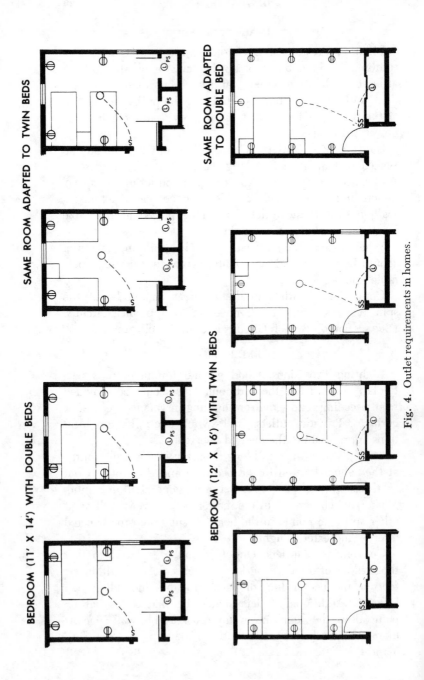

SAME ROOM ADAPTED TO TWIN BEDS

SAME ROOM ADAPTED TO DOUBLE BED

BEDROOM (11' X 14') WITH DOUBLE BEDS

BEDROOM (12' X 16') WITH TWIN BEDS

Fig. 4. Outlet requirements in homes.

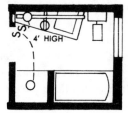

Fig. 5.

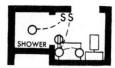

Fig. 6.

It is recommended that convenience outlets be placed only three to four feet from the center line of the probable bed locations. The popularity of bedside radios and clocks, bed lamps, and electric bed covers makes increased plug-in positions at bed locations essential. Triplex or quadruplex convenience outlets are used at these locations. Also, at one of the switch locations, a receptacle outlet should be provided for the use of a vacuum cleaner, floor polisher, or other portable appliances.

Special Purpose Outlets. The installation of one heavy-duty, special-purpose outlet in each bedroom for the connection of room air conditioners is recommended. Such outlets may also be used for operating portable space heaters during cool weather in climates where a small amount of local heat is sufficient.

Figures 5 and 6 show the application of these standards to both double- and twin-bed arrangements and also their application where more than one probable bed location is available within the room.

BATHROOMS AND LAVATORIES

Lighting Provisions. Illumination of both sides of the face when at the mirror is essential. There are several methods that may be employed to achieve good lighting at this location and in the rest of the room. Lighting outlets should be installed to provide for the method selected, bearing in mind that a single concentrated light source, either on the ceiling or the side wall, is not acceptable. All lighting outlets should be wall switch controlled. (*See* Figs. 7 and 8.)

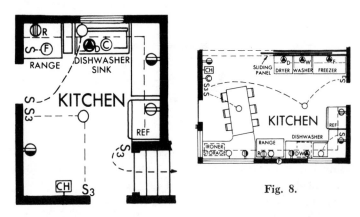

Fig. 7.

Fig. 8.

A ceiling outlet located in line with the front edge of the basin will provide improved lighting at the mirror, general room lighting, and safety lighting for combination shower and tub. When more than one mirror location is planned, equal consideration should be given to the lighting in each case.

It is recommended that a switch controlled night light be installed. Where an enclosed shower stall is planned, an outlet for a vapor-proof luminaire should be installed, controlled by a wall switch outside the stall.

Convenience Outlets. One outlet near the mirror, three to five feet above the floor, should be installed. Also, an outlet should be installed at each separate mirror or vanity space, and also at any space that might accommodate an electric towel dryer, electric razor, and the like. A receptacle which is a part of a bathroom lighting fixture should not be considered as satisfying this requirement unless it is rated at 15 amperes and wired with a least 15-ampere rated wires.

Special Purpose Outlets. Each bathroom should be equipped with an outlet for a built-in type space heater, and an outlet for a built-in ventilating fan, wall switch controlled.

KITCHEN

Lighting Provisions. Provide outlets for general illumination and for lighting at the sink. These lighting outlets should be wall switch controlled. Lighting design should provide for illumination of the work areas, sink, range, counters, and tables. Undercabinet lighting fixtures within easy reach may have local-switch control. Consideration should also be given to outlets to provide inside lighting of cabinets. (*See* Figs. 9 and 10.)

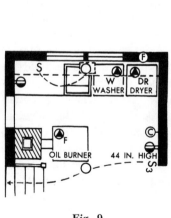

Fig. 9.

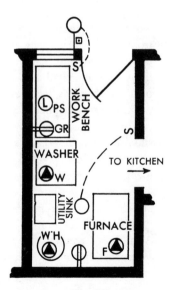

Fig. 10.

Convenience Outlets. One outlet for the refrigerator and one outlet for each four linear feet of work-surface frontage, with at least one outlet to serve each work surface, should be provided. Work surface outlets should be located approximately 44 inches above floor line. If you are installing a planning desk, one outlet should be located to serve this area. Table space should have one outlet, preferably just above table level.

An outlet is recommended at any wall space that may be used for ironing or for an electric roaster.

Convenience outlets in the kitchen, other than for the refrigerator, should be of the split-receptacle type for connection to appliance circuits. (*See* section on Appliance Circuits later in this chapter.)

LAUNDRY AND LAUNDRY AREAS

Lighting Provisions. For *complete laundries,* lighting outlets should be installed to provide proper illumination of work areas, such as laundry tubs, sorting table, washing, ironing, and drying centers. At least one lighting outlet in the room should be wall switch controlled. For *laundry trays* in unfinished basements, one ceiling outlet centered over the trays is sufficient. All laundry lighting should be wall switch controlled. (See Fig. 11.)

Convenience Outlet. At least one convenience outlet should be installed. In some instances, one of the special-purpose outlets, properly located, may satisfy this requirement. The convenience outlet is intended for such purposes as laundry hot plate, sewing machine, and the like. These outlets in the laundry area should be of the split-receptacle type for connection to appliance circuits. (*See* section on Appliance Circuits later in this chapter.)

Special Purpose Outlets. One outlet should be used for each of the following appliances: automatic washer; iron; or clothes dryer. The installation of outlets for a ventilating fan and a clock are highly desirable. If an electric water heater is to be installed, the requirements may be obtained from your local utility dealer.

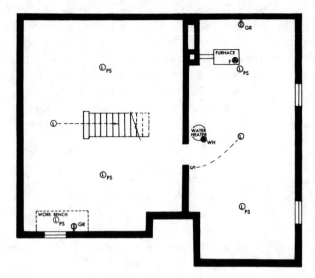

Fig. 11.

CLOSETS

Lighting Provisions. One outlet should be installed for each closet. Where shelving or other conditions make the installation of lights within a closet ineffective, lighting outlets should be so located in the adjoining space as to provide light within the closet. The installation of wall switches near the closet door, or door-type switches, is recommended.

HALLS

Lighting Provisions. Lighting outlets, wall switch controlled, should be installed for proper illumination of the entire area. Particular attention should be paid to irregularly shaped areas. These provisions also apply to passage halls, vestibules, entries, foyers, and similar areas. A switch controlled night light should be installed in any hall giving access to bedrooms.

Convenience Outlets. One outlet should be placed for each 15 linear feet of hallway, measured along center line. Each hall over 25 square feet in floor area should have at least one outlet. In reception halls and foyers, convenience outlets should be placed so that no point along the floor line in any usable wall space is more than ten feet from an outlet in that space. A convenience receptacle should be provided at one of the switch outlets for connection of vacuum cleaner, floor polisher, and the like.

STAIRWAYS

Lighting Provisions. Wall or ceiling outlets should be installed to provide adequate illumination of each flight of stairs. Outlets should have multiple-switch control at the head and foot of the stairway, so arranged that full illumination may be turned on from either floor, but that lights in halls furnishing access to bedrooms may be extinguished without interfering with ground-floor usage. These provisions are intended to apply to any stairway at both ends of which are finished rooms. (For stairways to unfinished basements or attics, *see* sections on Basements and Attics later in this chapter.) Whenever possible, switches should be grouped together and never located so close to steps that a fall might result from a misstep while reaching for a switch.

Convenience Outlets. At intermediate landings of a large area, an outlet should be provided for decorative lamp, night light, vacuum cleaner, and the like.

RECREATION ROOM, PLAY ROOM, OR TV ROOM

Lighting Provisions. Some means of general illumination is essential. It may be provided by ceiling or wall fixtures, or by lighting in coves, valances, or cornices. Lighting outlets, wall switch controlled, should be provided in locations appropriate to the lighting method selected. The lighting method used in the recreation room, play room, or TV room should be selected by the type of major activities for which the room is planned. (*See* Fig. 12.)

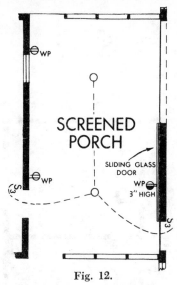

SCREENED
PORCH

SLIDING GLASS
DOOR

WP
3" HIGH

Fig. 12.

Convenience Outlets. Convenience outlets should be placed so that no point along the floor line in any usable wall space is more than six feet from an outlet in that space. One convenience outlet should be installed flush in the mantel shelf, where construction permits. Outlets for the use of a clock, radio, television, ventilating fan, motion picture projector, and the like should be located in relation to their intended use.

UTILITY ROOM OR SPACE

Lighting Provisions. Lighting outlets should be placed to illuminate furnace area and work bench, if planned. At least one lighting outlet should be wall switch controlled. (*See* Fig. 13.)

Convenience Outlets. One convenience outlet should be provided, preferably near the furnace location or near any planned work bench location.

Special Purpose Outlet. One outlet should be installed for electrical equipment used in connection with furnace operation.

BASEMENT

Lighting Provisions. Lighting outlets should be placed to illuminate designated work areas or equipment locations,

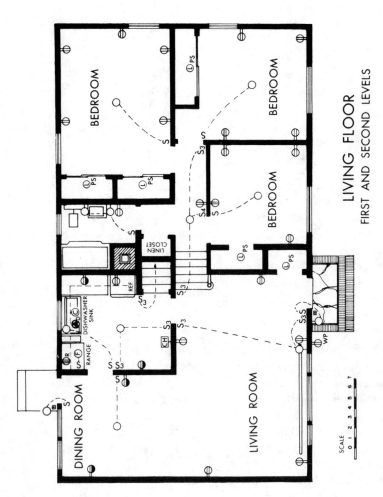

Fig. 13. Outlet requirements in homes.

LIVING FLOOR

FIRST AND SECOND LEVELS

BEDROOM

BEDROOM

BEDROOM

LINEN CLOSET

DINING ROOM

LIVING ROOM

DISHWASHER

SINK

RANGE

REF

WP

CH

SCALE

0 1 2 3 4 5 6 7

such as at furnace, pump, work bench, and the like. Additional outlets should be installed near the foot of the stairway, in each enclosed space, and to open spaces so that each 150 square feet of open space is adequately served by a light in that area (*See* Fig. 14).

In *unfinished basements* the light at the foot of the stairs should be wall switch controlled near the head of the stairs. Other lights may be pull-chain controlled.

In basements with *finished rooms,* with garage space, or with other direct access to outdoors, the stairway lighting provisions apply.

In basements which will be infrequently visited a pilot light should be installed in conjunction with the switch at the head of the stairs.

Convenience Outlets. At least two convenience outlets should be provided. If a work bench is planned, one outlet should be placed at this location. Basement convenience outlets are useful near furnace, at play area, for basement laundries, dark rooms, hobby areas, and for appliances, such as dehumidifier, portable space heater, and the like.

Special Purpose Outlet. One outlet for electrical equipment used in connection with furnace operation; and an outlet for a food freezer should be provided.

ATTIC

Lighting Provisions. For an accessible attic one outlet is needed for general illumination, wall switch controlled from foot of stairs. When no permanent stairs are installed, this lighting outlet may be pull-chain controlled, if located over the access door. Where an unfinished attic is planned for later development into rooms, the attic-lighting outlet should be switch controlled at top and bottom of stairs. One outlet for each enclosed space should be installed. These provisions apply to unfinished attics; for attics with finished rooms or spaces, see appropriate room classifications for required outlets. A pilot light in conjunction with the switch controlling the attic light should be provided.

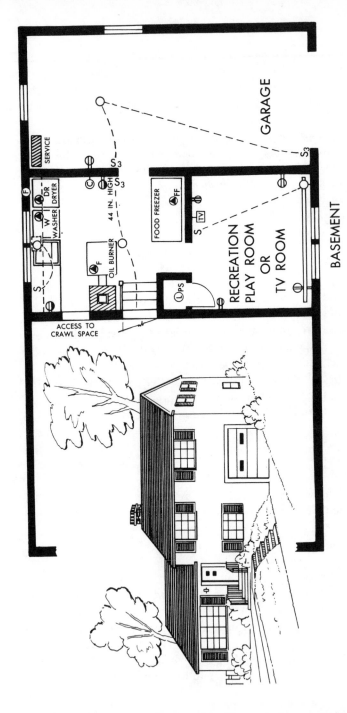

Fig. 14. Outlet requirements in three-level house.

Convenience Outlets. One outlet for general use should be provided. If open stairway leads to future attic rooms, provide a junction box with direct connection to the distribution panel for future extension to convenience outlets and lights when the rooms are finished. A convenience outlet in the attic provides additional illumination in dark corners, and also for the use of a vacuum cleaner and its accessories in cleaning.

Special Purpose Outlet. The installation of an outlet, multiple-switch controlled from desirable points throughout the house, is recommended in connection with the use of a summer cooling fan.

PORCHES

Lighting Provisions. Porches, breezeways, or other similar roofed areas of more than 75 square feet in floor area should have a lighting outlet, wall switch controlled. Large or irregularly shaped areas may require two or more lighting outlets. Multiple-switch control should be installed at entrances when the porch is used as a passage between the house and the garage.

Convenience Outlets. For each 15 feet of wall bordering porch or breezeway use one convenience outlet. Weatherproof if porch is exposed to moisture. All such outlets should be controlled by a wall switch inside the door. The split-receptacle convenience outlet shown in Fig. 15 is intended to be connected to a three-wire appliance branch circuit. See relationship of screened porch to kitchen shown in Fig. 19.

TERRACES AND PATIOS

Lighting Provisions. The installation of an outlet on the house wall or on a post centrally located in the area provides fixed general illumination. These outlets should be wall-switch controlled just inside the house door opening onto the area.

Convenience Outlets. Use one weatherproof outlet located at least 18 inches above grade line for each 15 linear feet of the house wall bordering terrace or patio. These outlets should be wall switch controlled from inside the house.

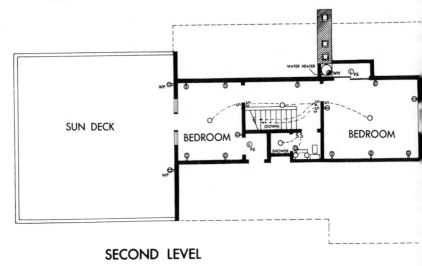

SECOND LEVEL

Fig. 15.

GARAGE OR CARPORT

Lighting Provisions. For one- or two-car storage area, use at least one ceiling outlet, wall switch controlled (Fig. 16). If the garage has no covered access from the house, one exterior outlet, multiple-switch controlled, should be provided. If the garage is to be used for other purposes in addition to storing the car, such as to include work bench, closets, laundry, attached porch, and the like, regulation rules should be followed. An exterior outlet, wall switch controlled, should be installed for all garages. Additional interior outlets are sometimes used, even if no specific additional use is planned for the garage. For long driveways, post lighting should be used. These lights should be wall switch controlled from the house.

Convenience Outlets. Use at least one outlet for one- or two-car storage area.

Special Purpose Outlets. When a food freezer, work bench, or an automatic door opener is planned for installation in the garage, outlets appropriate to these uses should be installed.

EXTERIOR GROUNDS

Lighting Provisions. Lighting outlets for floodlights are often used for illumination of grounds surrounding the house. These outlets may be installed on the exterior of the house or the garage, or on posts or poles placed at suitable places. All these outlets should be switch controlled from within the house. Multiple- and master-switch control from strategic points may be used.

CIRCUITS

Branch Circuits—General-Purpose Circuits. General-purpose circuits should supply all lighting outlets throughout the house, and all convenience outlets except the convenience outlets in the kitchen, dining room or dining areas of other rooms, breakfast room or nook, and laundry or laundry area. These general-purpose circuits should be pro-

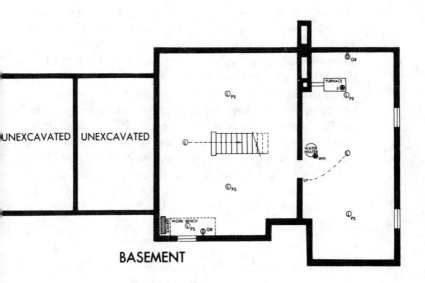

Fig. 16.

vided on the basis of one 20-ampere circuit for not more than each 500 square feet, or one 15-ampere circuit for not more than each 375 square feet of floor area. Outlets supplied by these circuits should be divided equally among the circuits.

These requirements for general-purpose branch circuits take into consideration the provision in the current edition of the National Electric Code for either 20- or 15-ampere general-purpose circuits. The floor area designations are in keeping with present-day usage of such circuits. It is recommended that separate branch circuits be provided for lighting and for convenience outlets in living rooms and bedrooms, and that the branch circuits serving convenience outlets in these rooms be of the three-wire type, equipped with split-wired receptacles.

Appliance Circuits. At least one three-wire, 115/230 volt, 20-ampere branch circuit, equipped with split-wired receptacles for all convenience outlets in the kitchen, dining room or dining areas of other rooms, and breakfast room or nook should be provided. This circuit should also be extended to the laundry or laundry area to serve any convenience outlets not otherwise required to be served by individual equipment circuits.

The use of three-wire circuits for supplying convenience outlets in the locations mentioned is an economical means for dividing the load and offers practical operating advantages. Such circuits provide greater capacity at individual outlet locations and lessen voltage drop in the circuit. They also provide greater flexibility in the use of appliances. For maximum effectiveness in utilization the upper half of all receptacles should be connected to the same side of the circuit.

Individual-Equipment Circuits. Circuits should be provided for the equipment shown in Table 23.

TABLE 23

INDIVIDUAL EQUIPMENT CIRCUITS

Item	Conductor Capacity
Range (Up to 21-kw rating)	50A-3W-115/230V
Combination Washer-Dryer or	40A-3W-115/230V
Automatic Washer	20A-2W-115V
Electric Clothes Dryer	30A-3W-115/230V
Fuel-Fired Heating Equipment (if installed)	15A or 20A-2W-115V
Dishwasher and Waste Disposer (if necessary plumbing is installed)	20A-3W-115/230V
Water Heater (if installed)	Consult Local Utility

Spare circuit equipment should be provided for at least two future 20-ampere, 2-wire, 115-volt circuits in addition to those initially installed. If the branch circuit or distribution panel is installed in a finished wall, raceways should be extended from the panel cabinet to the nearest accessible unfinished space for future use. Consideration should also be given to the provision of circuits for the commonly used household appliances and items of equipment shown in Table 24. This table does not list all the equipment available. In some instances, one of the circuits recommended may serve two devices which are not liable to be used at the same time, such as an attic fan and a bathroom heater. The majority of appliances for residential use are made for 110 to 120 volts. There is, however, a growing tendency to make fixed appliances for use on 220-240 volt circuits. In those cases where a choice exists, it is recommended that the higher voltage be preferred. Consideration should also be given to the provision of circuits for patios or *outdoor* living rooms and for exterior decorative or flood lighting.

A number of different systems of electrical house heating are now being used in some areas. These systems are presently a matter of individual engineering design for the house in question and local climatic conditions.

TABLE 24

CIRCUITS FOR HOUSEHOLD APPLIANCES

Item	Conductor Capacity
Attic Fan	20A-2W-115V (Switched)
Room Air Conditioners or	20A-2W-230V or 20A-3W-115/230V
Central Air-Conditioning Unit	40A-2W-230V
Food Freezer	20A-2W-115 or 230V
Water Pump (where used)	20A-2W-115 or 230V
Bathroom Heater	20A-2W-115 or 230V
Work Shop or Bench	20A-3W-115/230V

Feeder Circuits. It is strongly recommended that consideration be given to the installation of branch-circuit protective equipment, served by appropriate-size feeders, at locations throughout the house, rather than at a single location.

Service Entrance. All services should be three-wire. For homes up to 3,000 square feet in floor area, the size of the service entrance conductors should not be less than No. 2 AWG, type R (or No. 3, type RH) and the rating of the service equipment should not be less than 100 amperes. These capacities are sufficient to provide for lighting, for portable appliances, and for the equipment for which individual-equipment circuits are required, as listed in Table 23. For larger homes, or for homes in which more than one additional unit selected from the table is installed, a larger service is required.

Experience indicates that the most serious obstacle to the use of items of electrical equipment and appliances, in addition to those initially provided for, has been inadequate service conductors and related equipment. To provide for probable increase in electrical use, it is recommended that service conductors be of No. O AWG and that service equipment be of sufficient rating to match the carrying capacity of these wires. Because of the many major uses of electricity in the kitchen requiring individual equipment circuits, it is recommended that the electric service equipment be located near or on a kitchen wall to minimize installation and wiring costs. Such locations will usually also be found to be convenient to laundries, therefore minimizing circuit runs to laundry appliances as well.

SPECIAL REQUIREMENTS

Entrance Signals. *Entrance push buttons* should be installed at each commonly used entrance door and connected to the door chime, giving a distinctive signal for front and rear entrance. Electrical supply for entrance signals should be obtained from an adequate bell-ringing or chime transformer installed at the fuse box or circuit-breaker location. In the *smaller home* the door chime is often installed in the kitchen, providing it will be heard throughout the house. If not, the door chime should be installed at a more central location, usually the entrance hall. In the *larger home* extension signals are often necessary to insure being heard throughout the living quarters. *Entrance-signal conductors* should be no smaller than the equivalent of a No. 18 AWG-gauge copper wire.

Communication. For the residence with accommodations for a resident servant, additional signal and communication devices are recommended. These include a dining-room-to-kitchen signal operated from a push button attached to the underside of the dining table or a foot-operated floor tread placed under the table; flush annunciator in the kitchen with push-button stations in each bedroom, the living room, recreation room, porch, and the like; flush intercommunicating telephone installed in the kitchen, with corresponding telephone stations in master bedroom, recreation room, and at other desirable locations. Intercommunicating telephones should be operated from a power unit recommended and furnished by the manufacturer of the telephones, and connected to the a-c lighting service.

Home Fire Alarm. Automatic fire-alarm protection is recommended, with detectors in the basement oil burner or furnace area and the storage space as a minimum. Alarm bell and test button should be installed in the master bedroom or other desirable location in the sleeping area.

Operating supply should be from a separate bell-ringing transformer recommended and furnished by the manufacturer of the alarm equipment. Alarm system should operate from a separate and independent transformer, rather than from the entrance-signal or other transformer, because of the increased possibility of failure due to causes originating in the entrance-signal or other system.

Television. For the television antenna connection, provide a non-metallic outlet box of the *300-ohm* type, connected to an accessible location in the attic by a suitable transmission line (unshielded in a UHF service area). Sufficient slack of the transmission line should be provided in the attic to reach any reasonable location of the antenna. If the transmission line is shielded, the shield should be grounded. Where an unshielded transmission line is used, grounding should be in accordance with the National Electrical Code and the line kept away from pipes and other wires. A convenience outlet should be provided adjacent to the television outlet, which outlet will be in addition to the convenience outlets provided for other purposes in the room.

Radio. In their modern form, radio receivers used purely for the reception of broadcast communication seldom require antenna and ground connections. If FM reception is desired, it is recommended that provisions similar to those for television be included.

Telephone. It is recommended that at least one outlet be installed, served by a raceway extending to unfinished portions of the basement or attic. The outlet should be centrally located and readily accessible from the living room, dining room, and kitchen. A desirable location for a second outlet is the master bedroom. Where finished rooms are located on more than one floor, at least one telephone outlet should be provided on each such floor. It is suggested that the *local telephone company* be consulted for details of service connection prior to construction, particularly in regard to the installation of protector cabinets and raceways leading to finished basements, where contemplated.

CHAPTER 11

Interior Living Space

————————————•————————————
•
•

The importance of lighting as a design element can hardly be overemphasized. Well done, lighting will contribute greatly to accomplishing the goal of home interiors to provide security, convenience, and aesthetic appeal.

LIGHT AS AN ELEMENT OF DESIGN AND ENVIRONMENT

Light is an element of design which should be used not only for visual comfort but also to achieve predetermined emotional responses from the lighted environment. Light has certain characteristics that affect the mood and atmosphere of the space—influencing the emotional responses of people who occupy the space. The definition and character of space is greatly dependent on the distribution and pattern of illumination. Luminaires themselves have dimensional qualities that may be used to strengthen or minimize architectural line, form, color, pattern, and texture.

QUANTITY OF LIGHT

High levels of controlled general lighting are usually cheerful and stimulate people to alertness and activity. *Low levels* tend to create an atmosphere of relaxation, intimacy, and restfulness. Control of the illumination level is often desirable and sometimes necessary. Such control makes it possible to change the mood or tone of the room to suit its various uses and the requirements of the occasion.

QUALITY OF LIGHT

Lighting can be described as being *soft* or *hard*. Soft or diffused light minimizes shadows and provides a more relaxing and less visually compelling atmosphere. When used alone the effect can be lacking in interest, like an outdoor scene on an overcast day. The artful use of hard light can provide highlights and shadows that emphasize texture and add beauty to form, as a shaft of sunlight. An effect of brilliance or sparkle is obtained from small unshielded sources such as a bare lamp or a candle flame. Such sources are seldom used as the prime origin of illumination; they are generally decorative, and must be supplemented by other means of illumination. The glitter of crystal, jewels, and polished brass, and the luster and sheen of table settings and some types of surface materials, create a sense of life and gaiety. Occasionally it is desirable to have more than one lighting system in a given space, so that distinct changes in atmosphere and mood can be created.

COLOR

Color, one of the chief factors in the emotional effect of environment, is an additional dimension of lighting design. Color of light has an important effect on colors of objects, and therefore makes a major contribution to the atmosphere of an interior. Association of color with objects, experience, locale, or cultures is a powerful factor in the personal reaction to color. It is usually desirable that the color or tint of light enhance the color identification of the object or area. However, *colored light* used as a general light source should be low in saturation, since the use of saturated or special colors can destroy the accepted appearance of materials and people.

LUMINANCE RELATIONSHIPS

Seeing Zones. From the practical standpoint of achieving visual comfort, efficient performance of seeing tasks, and safety, the lighting of a home must fulfill two major requirements—adequate illumination and suitable luminance ratios in the visual field. The *visual field* of a person engaged in a seeing task is considered to consist of three major zones. First is the task itself; second, the area immediately surrounding the task; and third, the general surroundings. The luminances visible in all three zones are important; unsuitable relationships among them can cause distraction, fatigue, and even difficulty in seeing. For optimum visual comfort in the performance of such tasks as studying, reading, sewing, or any activity demanding close vision (especially if continued for some time) the following conditions should prevail. Luminances in the immediate surroundings should be no greater than, and no less than, one-third of the luminance of the task. Substantial areas in the general surroundings should not exceed five times the task luminance or be less than one-fifth of it. These limitations become more important as illumination levels increase and as the duration of tasks is lengthened. In locations where critical eye work is never done, substantially higher luminance ratios may be permissible.

Achieving Luminance Ratios. There are a number of ways in which luminances in the visual field may be brought into balance. In general, they involve the following.

1. Limitation of the luminance of luminaires.

2. For daytime activities, control of high luminance of window areas by blinds, shades, or draperies; use of high-reflectance colors on walls adjacent to fenestration.

3. For nighttime activities, use of light-colored covering for windows in the field of view.

4. Use of materials and finishes of favorable reflectance for large areas of room surfaces and furnishings. (*See* section on Light and Color, Reflectance, later in this chapter.)

5. Provision of additional lighting in the visual surroundings from other sources not specifically directed to the task.

Veiling Reflections. (1) An often-unrecognized factor which can reduce the ability to see a task is known as *veiling reflections*, the reflections of a light source in a task reducing the contrast between the critical detail and the background against which it must be discerned. Veiling reflections are most troublesome when the task involves a glossy surface (such as the paper used for reproduction of photographs as found in many magazines and books), but it is present to some degree in nearly every situation and can seldom be completely eliminated. Its effects can be minimized by control of source or luminaire luminance, and by placement of the source so that reflections are directed away from the eye. This type of reflection approximately follows the familiar law of the mirror—angle of reflection equals angle of incidence. Luminaires with strongly directional components and high luminance, such as downlights, are potential sources of difficulty. Large-area low-luminance luminaires, multiple sources, or any system in which much of the light reaching the task plane is reflected from walls or ceiling cause less veiling reflections because the light comes from many directions.

(2) When the reflections of light sources are not in the task, but in its surroundings, the factor is known as *reflected glare*. If these reflections are bright enough and large enough in area they may cause visual discomfort. These reflections can be reduced by controlling source of luminaire luminance, placement of source as previously described, and by using matte finishes on room and surrounding surfaces.

LIGHT AND COLOR

For an understanding of the relationship of colored surfaces and light, the designer or home owner must be aware that there are two aspects, other than physiological, to color recognition—*light source color* (the spectral characteristics of the light falling on an object) and *object color* (the selective reflectance characteristics of the object itself). The object, because of its inherent selective reflectance characteristics, reflects a certain portion of the particular light (illuminant color) which falls on the object, and the result is what we see as a *color*. If either the object color is changed (if a room is repainted) or the *illuminant color* is changed (from blue light bulbs to pink light bulbs), the resultant *color* which one sees will appear to change.

Light Source Color. Since all color comes from light, color is seen by an observer when spectral energy of a given light source is reflected from an object with selective reflection. A *white light,* such as daylight or sunlight, contains energy throughout the visible spectrum. *Incandescent* light sources radiate energy throughout the visible spectrum but with the greater proportion in the yellow to red end. When dimmed, there is noticeable shift to the red end of the spectrum. *Fluorescent* light sources have a different spectral energy distribution, usually lacking in the far red, but high in blue, green, and yellow where there are peaks from the spectral lines of the mercury arc used to excite the phosphors. However, with the introduction of appropriate phosphors, the spectral distribution can be readjusted to enhance skin tones and to flatter general decor. Though there is usually a loss of efficiency, these color-improved lamps are recommended for residential type spaces so that objects will appear in what seems to be their *natural* color.

Object Color. The color of an object can be strictly defined only if the color of the light source, the characteristics of the observer, and the conditions of observation are all specified. The "Munsell System," which is the most widely accepted method of accurately describing object color, assumes a normal observer, daylight illumination, and observation of the color samples against a gray to white background. In the Munsell System colors are specified in terms of three attributes—hue, value, and chroma.

Hue. Hue indicates whether a color is red (R), yellow (Y), green (G), blue (B), or purple (P). These with their five intermediates—yellow-red (YR), green-yellow (GY), blue-green (BG), purple-blue (PB), and red-purple (RP) —make up a hue circle. The Munsell hue circle (Fig. 1) shows 10 hues spaced at equal visual intervals. The number preceding the hue letters indicates the relationship of the hue with adjacent hues, five indicating the major hue. *For example,* among the reds, hues in the direction of 4R, 3R,

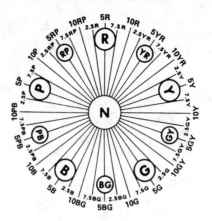

Fig. 1. "Munsell" hue circle.

2R, and 1R indicate that the color is becoming progressively more purple, while hues 6R, 7R, 8R, 9R, and 10R are progressing toward more of a yellow.

Value. Value indicates the degree of lightness or darkness of a color in relation to a scale of neutral grays. In the Munsell color designation, value is specified by the number immediately following the hue letters.

Chroma. Chroma indicates strength or saturation, the degree of departure from a neutral gray of the same value. Chroma is specified by the last number of the Munsell designation, following /. The designation 5YR6/12, *for example,* means that the color has a hue of 5 yellow-red, a value of 6, and a chroma of 12. By this system any color sample, lighted and viewed as specified, has its own numerical reference. To the design profession, the value of the system is that it provides an accurate and universally understood vocabulary of object color.

Reflectance. (1) The amount of light a surface reflects, expressed as a percentage of the amount of light it receives, is called its *reflectance.* This is the same variable as Munsell *value.* However, reflectance and value are not proportional to each other, because the Munsell value scale is empirically selected to have equal visual intervals, such as a color sample having a value of five appears to be halfway in lightness between pure black and pure white (0/ and

10/). Because of the nature of the human visual apparatus, a sample with a reflectance of 50 per cent as measured with a reflectometer or calculated from spectral reflectance characteristics and the spectral energy distribution of the light source, does not appear half-way in lightness between samples at 0 and 100 per cent reflectance. An understanding of the relationship between Munsell value and reflectance, shown in Table 25, is valuable in many lighting and decorating applications. A rough rule-of-thumb is that Munsell value is approximately the square root of reflectance.

(2) Light, high-reflectance colors for room surfaces and furnishings are important, and often essential, in achieving desirable luminance ratios in the visual field. Recommended reflections are shown in Table 26.

TABLE 25

RELATIONSHIP BETWEEN MUNSELL VALUE AND REFLECTANCE[*]

MUNSELL VALUE	REFLECTANCE
9/	79%
8.5/	68%
8/	59%
7/	43%
6/	30%
5/	20%
4/	12%
3/	7%
2/	3%

[*] FOR STANDARD SOURCE C AS SPECIFIED BY THE INTERNATIONAL COMMISSION ON ILLUMINATION, APPROXIMATING DAYLIGHT AT 6500° K

Color Rendition. (1) Differences in the spectral energy distributions of light sources (Fig. 1) are usually apparent not as differences in the color of the light itself, but as differences in the colors of objects. Object color appears more intense when the light source is rich in the wave-

TABLE 26

RECOMMENDED REFLECTANCES AND MUNSELL VALUES FOR HOMES

SURFACE	REFLECTANCE (%)	APPROX. MUNSELL VALUE
CEILING	60 TO 90	8 AND ABOVE
FABRIC TREATMENT ON LARGE WALL AREA	45 TO 85	7 TO 9.5
WALLS	*35 TO 60	6.5 TO 8
FLOOR	*15 TO 35	4 TO 6.5

*IN AREAS WHERE LIGHTING FOR SPECIFIC VISUAL TASKS TAKES PRECE-DENCE OVER LIGHTING FOR ENVIRONMENT, MINIMUM REFLECTANCES SHOULD BE 40% FOR WALLS, 25% FOR FLOORS.

lengths that are most highly reflected by the object, and appears grayer (less saturated, of lower chroma) when the source is low in energy in that spectral region. *Incandescent* light accentuates reds and oranges and slightly dulls, or grays, the blues. *Fluorescent* lamps, except the ones specially modified for red-rendition, tend to make blues and greens vivid, and to dull the reds. Although color-rendering differences among the light sources commonly used in the home are seldom striking, it is possible to change the entire color atmosphere and feeling of a room by changing from incandescent to fluorescent light, or even by changing from *cool* to *warm* fluorescent lamps. Illumination level also affects perceived object color; low levels tend to render colors more gray, higher levels to intensify them. For these reasons, selection of interior colors under the type of light source with which they will be used, and at approximately the same illumination level, is always a wise precaution. The use of strongly tinted light should be confined to decorative effects or accent lighting.

(2) Light is often modified by reflection between the time it leaves the source and the time it reaches the object. If daylight from a bright blue sky is reflected from the

green leaves of a tree into a room with a red rug and pink walls, the color of the light falling on an object in the room is not that of daylight, but the result of the daylight as modified by these several reflections. Interreflections from large areas such as walls can cause *amplification* of the chroma or saturation of the wall color, so that it appears more vivid than the original isolated sample from which it was selected. The color quality of the light reaching a surface is the average of the light that reaches it from all points in the environment.

Surface Finish. Colors of objects often appear to change with surface finish. Specular or mirror reflection from glossy surfaces may, in extreme cases, increase the chroma and darkness of the sample at one angle and wipe out all color at other angles, as well as cause distracting glare. A matte finish reflects light diffusely, and thus gives an object the appearance of its *natural* color. Deeply textured finishes, such as velvet and deep-pile carpeting, cause shadows that make materials appear darker than smooth-surfaced materials such as, satin, silk, or plastic laminates of the same inherent color.

CHAPTER 12

Decorative Lighting

Variation and areas of interest created by interplay of light and shadow are essential components of aesthetically pleasing home lighting design. *Lighting can create decorative accents* to enhance the appearance of art objects and/ or create a focus of interest in a particular area. The following section describes design considerations and suggests techniques for a variety of decorative possibilities.

PAINTINGS, TAPESTRIES, AND MURALS

Design Considerations.

1. Lighting equipment should be placed so that the light rays reach the center of a painting at an angle of 30 degrees with the vertical. This prevents specular reflections in the direction of the viewer's eye from frame, glass, or surface of the picture, and also avoids disturbing shadows of frame, heavy paint texture, and the like.

2. A study of the sight-lines of people seated and standing anywhere in the room should be made, to insure that no unshielded sources are in view.

3. Excessive luminance difference between the lighted object and surrounding areas is undesirable. Higher levels of illumination applied for extended periods may cause deterioration of the paint surface.

4. The primary consideration should be the intent of the artist.

Typical Techniques.

1. Entire picture wall lighted by cornice or wall-wash equipment.

2. Individual frame-mounted luminaires.

3. Individual framing spotlights.

4. Individual spot or floodlamps not confined to picture lighting.

5. Lighting from below by luminaires concealed in decorative urns, planters, mantels, and the like.

SCULPTURE

Design Considerations.

A sculpture is a three-dimensional object in space. A certain amount of specular reflection is often pleasurable. Experimentation with diffused and/or a directional source or sources will help to determine the most acceptable solution for a specific situation.

A luminance ratio between 2 and 6 usually results in a good three-dimensional effect with transparent shadows. If the modeling ratio is reduced below 2, the lighting becomes too *flat* and solid objects appear two-dimensional. If the ratio is above approximately 6, the contrast tends to become unpleasant, with loss of detail in the shadows or in the highlights.

Typical Techniques.

Adjustable spots, floods, individual framing spots, and back lighting.

PLANTERS

Design Considerations.

1. Lighting that provides for enhancement of appearance may not necessarily be suitable for plant growth. Plants may have to be rotated to a growth area from time to time to keep their beauty.

2. Plants tend to grow toward the light. This should be kept in mind when locating equipment.

3. Light sources, particularly incandescent, should not be located too close to plant material because the heat may be detrimental.

4. Backlighting of translucent leaves will often reveal leaf structure, color, and texture.

5. Front lighting of opaque leaves will often reveal leaf structure, color, and texture.

6. Silhouetting of foliage can add an additional dimension.

7. The necessary fire safety precautions should be taken if artificial plants are used.

8. Live plants are not static; they change in height and bulk.

Typical Techniques.

1. Incandescent downlights recessed, surface-mounted, or suspended above planting area.

2. Luminous panel or soffit using fluorescent and/or incandescent sources.

3. Silhouette lighting with luminous wall panel, lighted walls, concealed uplights.

4. Low-level incandescent stake units.

5. Planting racks containing fluorescent and/or incandescent lamps especially selected for plant growth.

6. Luminaires recessed in earth to provide uplight.

NICHES

Design Considerations.

1. The lighting method is determined by what is to be displayed in the niche.

2. The luminances of shielding media and interior niche surfaces should be carefully controlled to avoid excessive difference from surrounding areas.

Typical Technique.

1. Incandescent spotlights from one or more directions.

2. Incandescent or fluorescent sources concealed vertically or horizontally at the edges.

3. Luminous sides, bottom, top, back, or combinations.

4. Lamps concealed behind an object to create a silhouette effect.

BOOKCASES

Design Considerations.

1. The distribution of light should cover the faces of books and other objects on all shelves.

2. Luminance ratios between the bookcase area and the surrounding wall surfaces should be kept within comfortable limits.

Typical Techniques.

1. Adjustable spot or flood lamps or wall washers aimed into the shelves from the ceiling in front of the bookcase.

2. Lighted bracket, cornice or soffit extending in front of the shelves.

3. Tubular sources concealed vertically at the sides of the bookcase or horizontally at the front edge of individual shelves.

4. Lighting concealed at the back of the bookcase for silhouette effects.

5. Some bookcases may require modification such as cutting the shelves back or using glass or transparent plastic shelves, to permit an ample spread of light over the entire bookcase.

FIREPLACES

Design Considerations.

1. Luminance relationships between the surrounding area and the fireplace facing of brick, stone, wood, and the like should be carefully studied.

2. It may be desirable to emphasize the mantel as the focal point of interest in the room. Orientation in the room may be an important consideration.

Typical Techniques.

1. Cornice and wall brackets.

2. Recessed or surface-mounted downlights close to the fireplace surface for grazing light.

3. Adjustable spots or floods (ceiling-mounted) aimed at decorative objects or mantel, or grouped together to light the entire fireplace wall.

4. Recessed or surface-mounted wall-washers for even distribution of light over the entire fireplace surface.

5. Lighting equipment concealed in the mantel to light pictures or objects above and/or surfaces below.

TEXTURED WALLS AND DRAPERIES

Design Considerations.

1. Light directed at a grazing angle emphasizes textured surfaces.

2. Luminaires close to walls usually create patterns of light distribution on the walls. Some of these may be desirable, others unacceptable, depending on circumstances and individual preferences.

Typical Techniques.

1. Recessed or surface-mounted downlights located close to the vertical surface to direct grazing light over the surface.

2. Cornice, valance, wall brackets, or wall-washers.

3. Individual adjustable incandescent luminaires.

4. Concealed up-lighting equipment.

LUMINOUS PANELS AND WALLS

Design Considerations.

1. Luminous elements larger than about 16 square feet (1.5 square meters) should not exceed 50 footlamberts (170 candelas per square meter) average luminance when they will be viewed by people seated in a room.

2. Luminous elements of smaller size (or larger elements seen only in passing) may approach 200 footlamberts (690 candelas per square meter).

3. Variable brightness controls are valuable in adjusting the luminance of a panel to the desired level for specific activities.

4. Uniformity of a luminous surface may or may not be desirable, depending on the spacing of the light sources.

5. Overall pattern, shielding devices, structural members, or other design techniques can add interest and relieve the dominance and monotony of large, evenly luminous panels.

Typical Techniques.

1. Fluorescent lamps uniformly spaced vertically or horizontally in a white cavity.

2. Fluorescent lamps placed at the top and bottom or on each side of a white cavity.

3. Random placement of incandescent lamps (perhaps with color) in a white cavity.

4. Uniform pattern of incandescent lamps in a white cavity.

5. Incandescent reflector lamps concealed at the top or bottom of a white cavity reflecting from the back surface of the cavity.

6. For outdoor luminous walls, floodlamps directing light to a translucent diffusing screen from behind.

OTHER ARCHITECTURAL ELEMENTS

Many homes have especially distinctive architectural features that require special lighting. Because of the variety of problems it is difficult to list design considerations that apply to all. However, the following general principles should be observed in all cases:

1. All light sources should be concealed from normal view.

2. *Unnatural lighting effects* are to be deprecated.

3. Lighting patterns that conflict with or distract from the architectural appearance of the element being lighted should be avoided.

LUMINAIRES AS DECORATIVE ACCENTS

Design Considerations.

1. Luminance of luminous parts can be critical.

2. Luminaires should not introduce distracting or annoying light patterns.

3. Where exposed lamps are used for a desired decorative effect, dimmer controls to provide high or low levels of luminance should be considered.

Typical Techniques.

1. Decorative crystal chandeliers, wall sconces, lanterns, and brackets.

2. Decorative portable lamps.

3. Pendant luminaires.

CHAPTER 13

Insulation

As electric heat continues to be the fastest growing method of home space conditioning in the country, there have been refinements in not only product and equipment, but also in installation practices and in the use of insulation.

Thermal insulation has a singular purpose—to minimize the transfer of heat from a space where it is wanted to a space where it is not wanted. *For example,* keeping heat within a structure during winter, or, for another, preventing heat from entering a structure in the summer.

Where the insulation is effective, significant benefits accrue. The more important benefits to designers, builders, and home owners are as follows.

1. *Savings in heating/cooling equipment costs.* Proper insulation can reduce the input rating of the space conditioning units and, therefore, lower their cost.

2. *Economy of space.* Systems with lower ratings usually occupy less space, easing the space requirements and simplifying installation.

3. *Lower operating costs.* Energy savings return the added cost of insulation in a few years and pay dividends for the life of the structure.

4. *More comfort in winter.* Insulated walls are warmer, lessening down drafts and the chilling effect of cold surfaces.

5. *Reduced noise.* An insulated structure can provide freedom from the distraction and annoyance of external sounds.

It is impossible to eliminate heat losses and heat gains completely. They can, however, be held within specified maximum limits to assure acceptable levels of comfort and energy consumption by the space conditioning system. These limits, and the manner in which the recommended values for thermal resistance may be achieved for various construction methods and materials, are described in this chapter.

MAXIMUM WINTER HEAT LOSS FOR ELECTRICALLY HEATED HOMES

Table 27 lists the widely recognized heat loss limits for electrically heated dwelling units, which are recommended by the Electric Energy Association. The table is based on an assumed infiltration rate of three-quarters air change per hour. Wall, floor, and ceiling construction, including insulation, must have sufficient thermal resistance to limit heat loss to the values shown or less. These values are expressed in watts and Btuh (1 watt equals 3.413 Btuh) per square foot of floor area of the space to be heated, measured to the outside of exterior walls.

TABLE 27

degree days	maximum heat loss values	
	watts/sq ft	Btuh/sq ft
over 8000	10.0	34
7001 to 8000	9.4	32
6001 to 7000	8.8	30
4501 to 6000	8.2	28
3001 to 4500	7.6	26
2000 to 3000	7.0	24

MAXIMUM SUMMER HEAT GAINS LIMITS FOR AIR CONDITIONED HOMES

The curves shown in Fig. 1 indicate the maximum heat gain limits for electrically heated dwelling units, recommended by the Electric Energy Association. They are adapted from the FHA Minimum Property Standards for single and multi-family housing. The values shown do not include heat gains to pipes or ducts.

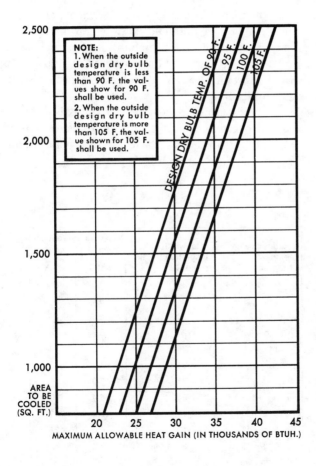

NOTE:
1. When the outside design dry bulb temperature is less than 90 F. the values show for 90 F. shall be used.
2. When the outside design dry bulb temperature is more than 105 F. the value shown for 105 F. shall be used.

MAXIMUM ALLOWABLE HEAT GAIN (IN THOUSANDS OF BTUH.)

Fig. 1.

TYPES OF INSULATION

There are a number of types of insulation in use, ranging from mineral wool to air spaces to glass and wood fiber to pulverized paper to the newer rigid insulations which may be preformed into planks and sheets, sprayed on, or foamed in place. Regardless of the type, there is standard terminology to express insulation effectiveness.

R-Value. The most commonly used expression for thermal resistance.

U-Factor. Overall coefficient of heat transfer, expressed in Btu per hour per square foot of surface per degree Fahrenheit temperature difference between air on the inside and air on the outside of a structural section such as an opaque wall. The U-factor is the reciprocal of the sum of the thermal resistance values (R-values) of each element of the structural section.

Table 28 shows insulation recommendations in terms of R-values, and overall coefficients of heat transfer in terms of U-factors, for major areas of heat loss/heat gain in home

TABLE 28

outdoor design temperature (degrees F)	total width of insulation (inches) rated R-5	heat loss per foot of exposed slab edge	
		watts	Btuh
—30 and colder	24	10.0	34
—25 to —29	24	9.4	32
—20 to —24	24	8.8	30
—15 to —19	24	8.2	28
—10 to —14	24	7.9	27
—5 to —9	24	7.3	25
0 to —4	24	7.0	24
+5 to +1	24	6.4	22
+10 to +6	18	6.2	21
+15 to +11	12	6.2	21
+20 to +16	edge only	6.2	21
+21 and warmer	none	—	—

walls, floors, and ceilings. They are presented by type of construction.

There are two other major heat loss areas which are not included in Table 28—doors and windows.

DOORS

The U-factors range from 0.64 for a 1-inch solid wood door to 0.43 for a 2-inch door. The addition of a metal and glass storm door results in U-factors of 0.39 to 0.28, respectively. A wood storm door with approximately 50 per cent glass reduces these U-factors to 0.30 to 0.24. The combination of a 2-inch solid wood door and wood storm door has a heat loss 3½ times greater than that of an equivalent wood frame wall section. However, modern exterior doors, either acrylic film or metal faced, with a core of rigid insulation, have a U-factor as low as 0.074. Infiltration heat loss, which is determined by the effectiveness of the weatherstripping, must also be considered.

WINDOWS

The U-factor for single glazing is 1.13. Double insulating glass with a ½-inch air space has a U-factor of 0.58 and, therefore, transmits approximately half as much heat. Double glazing is recommended where winter design temperatures are 7°F or lower or degree days exceed 4500. In addition to the economic and thermal comfort benefits, multiple glazing has the advantage of lower noise transmission.

CONTROL OF CONCEALED CONDENSATION

In a well-insulated structure, adequate control of concealed condensation takes on added importance. The presence of water in insulation can impair or destroy its insulating value, may cause its deterioration, and may eventually lead to structural damage by rot, corrosion, or the expansion of freezing water.

An excessive accumulation of moisture in walls, roofs, floor, attics, or crawl spaces can be prevented by one or more of the following: (1) providing a vapor barrier near the inner surface to limit vapor entrance into a structural element; (2) ventilating the living space to reduce relative humidity; and (3) ventilating structural cavities to remove vapor that has entered.

1. *Vapor barrier.* Without a vapor barrier, fibrous or granular insulation would permit transmission of water vapor from inside spaces. Vapor barriers are usually factory-applied on batt and blanket insulation. When none is included in the insulation, the vapor barrier can be provided in the form of materials such as polyethylene sheeting, two mils or more thick, foil-backed gypsum board, or other material with comparable vapor-resistant qualities.

Rigid insulations having a closed cellular structure are relatively impervious to water vapor and provide a suitable barrier when installed near the inside surface of the structural element. However, a vapor barrier is required in roofing systems when the rigid insulation is applied over wood or other material having high moisture permeability.

When the outside material of an opaque wall is of low permeance, such as plastic-faced or metal siding, it is necessary to provide a highly resistant vapor barrier on the winter warm side of the wall and to ventilate with outdoor air to the outside any air space between the insulation and the outside wall material.

The entrance of ground moisture into crawl spaces can be reduced through the use of a suitable ground cover.

2. *Ventilation of living spaces.* This is a necessary accompaniment to a vapor barrier, since, if the barrier blocks entrance into the walls, the water vapor must be removed from the living spaces by other means. The volume of air change necessary is not large and normal infiltration alone is usually all that is required in winter weather. Supplementary ventilation may be necessary in kitchen and laundry areas where moisture levels are high.

3. *Ventilation of structure.* Areas outside the insulation/vapor barrier envelope, but within the structure, except for unheated basements, must be ventilated. Unheated

attics are ventilated at all times. Crawl spaces are ventilated in the summer and may or may not be ventilated in the winter, depending upon local ground water conditions.

OPAQUE WALLS

Frame. Rolls and batts of mineral, glass, or wood fiber insulation are in common use for wood or metal stud walls. Some have an impregnated membrane (or reflective foil facing), which provides an integral vapor barrier. These types incorporate two 1-inch flanges for attachment to wood studs. Unfaced, friction-held rolls and batts stay in place without mechanical fasteners but require the use of a polyethylene film or foil-backed gypsum board to serve as a vapor barrier on the winter warm side of the studs. (*See* Fig. 2.)

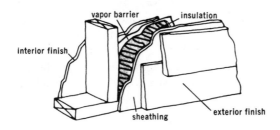

Fig. 2.

If metal studs having a thermal conductance greater than wood studs are used, the thermal resistance of the exterior sheathing should be increased to compensate for the higher rate of heat transfer through the studs.

Sprayed-on or foamed-in-place rigid insulation is effective for insulating walls and lends itself to production line techniques for fabricating wall sections and modules. It is easy to apply even to irregular or hard-to-reach areas and it imparts some additional stiffness to the structure as well. Before application it is low in bulk, minimizing storage problems.

Exterior opaque frame walls and walls between separately heated units should have a maximum U-factor of 0.07, requiring insulation rated at R-11, or greater. Thermal resistance of R-11 can be provided by, *for example,* 3½-inch fiber batts.

Masonry. Masonry is a relatively poor thermal barrier. A masonry cavity wall, *for example,* consisting of 4-inch face brick, an air space, 4-inch concrete block, an air space, and gypsum wallboard has a resistance value equivalent to that of approximately one inch of rigid insulation. A solid concrete wall would have even less resistance value. To bring the thermal resistance of masonry wall sections to an acceptable value, insulation must be added. There are several ways of doing this.

1. Rigid insulation may be applied to the interior side of the concrete blocks by: (a) fastening furring strips to the blocks at regular intervals and applying rigid insulation between them, secured by mechanical fasteners or adhesives; (b) spraying insulation in these same areas to a thickness slightly less than that of the furring strips; or (c) applying preformed rigid insulation to the entire area and securing it by adhesives. The interior wall finish may then be applied by nailing or by adhesives. (*See* Fig 3.)

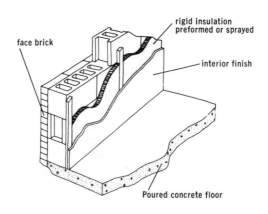

Fig. 3.

2. Preformed rigid insulation may be placed between the face brick or other exterior finish and the concrete blocks during construction, or the space may be left empty and then filled using foamed-in-place rigid insulation before the wall is capped (Fig. 4).

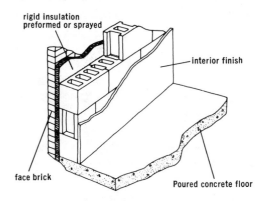

rigid insulation preformed or sprayed

interior finish

face brick

Poured concrete floor

Fig. 4.

3. Where concrete walls are used, preformed rigid insulation may be applied to either side by adhesive bonding, followed by the desired wall finish (Fig. 5).

Exterior opaque masonry walls for housing should have a maximum U-factor of 0.11, requiring insulation rated R-7 or better.

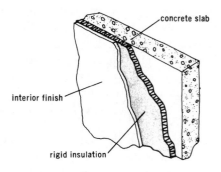

concrete slab

interior finish

rigid insulation

Fig. 5.

Metal. Metal sections used for both load-bearing and curtain walls in house construction consist essentially of the following main elements: (1) a prefinished metal exterior panel reinforced by weldments and integral ribs; (2) a layer of rigid insulation (preformed or foamed in place); (3) an inner wall panel; and (4) a metal supporting frame, where required. (*See* Fig. 6.)

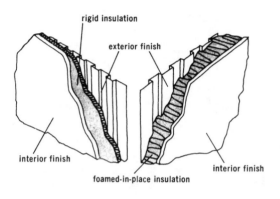

Fig. 6.

Exterior opaque metal walls for housing should have a maximum U-factor of 0.07, requiring insulation rated R-11 or better.

Sandwich. The sandwich-type of wall element is a layered construction formed by bonding two thin facings to a core of lightweight material. The facing can be any strong material such as glass fiber reinforced polyester resin, or metal, and the core any lightweight material, like paper honeycomb, with rigid insulation.

Exterior opaque walls of this type for housing should have a maximum U-factor of 0.07. Since the thermal resistance of sandwich construction depends upon its composition and thickness, the amount of additional insulation required to obtain this U-factor must be calculated in each case.

SPECIAL WALLS

Basements. The exterior walls of heated basements, as well as the walls separating heated areas from unheated areas in partially heated basements, should be insulated. The insulating material may be of any type and should have a thermal resistance of at least R-7. Exterior walls of unheated basements are not insulated since there is no need for reducing heat loss through such walls. (*See* Fig. 7.)

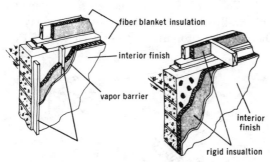

Fig. 7.

Crawl Spaces. The exterior walls of unvented crawl spaces should be insulated (Fig. 8). A thermal resistance of at least R-7 is recommended. Blankets or batts placed between 2-inch X 2-inch studs or rigid insulation, either adhesively bonded to the interior surface of the wall or

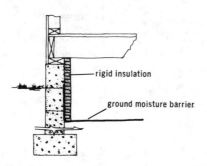

Fig. 8.

foamed in place, may be used. The exterior walls of vented crawl spaces require no insulation.

Walls Between Separately Heated Dwelling Units. The walls between adjacent separately heated dwelling units should be insulated to avoid excessive heat loss from one unit when the temperature in the abutting unit is lowered for any reason and to lessen sound transmission between units.

Insulation with a thermal resistance of R-11 is recommended. When adjacent modular units are joined together a double wall is formed, and the insulation may be placed in either wall or proportioned between them (Fig. 9).

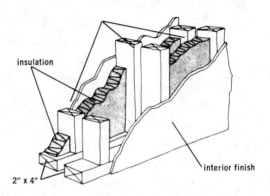

insulation

interior finish

2″ x 4″

Fig. 9.

FLOORS

Frame. Wood or metal joist floors over unheated basements, vented crawl spaces, and between separately heated units should be insulated. The methods used for installing floor insulation depend upon the specific fabricating process utilized in constructing the floor section and includes the following.

1. Blanket or batt form insulation, with the vapor barrier facing the winter warm side, is placed between the joists and supported by sheets of gypsum or other type board at the bottom.

Where the joists are to remain exposed from below, batt

or blanket form insulation with reverse nailing flanges may be applied. This permits the standard inset stapling technique to be used, while keeping the vapor barrier toward the winter warm side.

2. Full-wide roll insulation is placed over the inverted floor section and the supporting board fastened to the joists through the insulation. In this instance, additional insulation would be laid between the joists after the assembly has been turned right side up and before the subflooring is laid.

3. Panels of rigid insulation are fastened to the bottom of the joists. This makes maximum use of the dead air space for added insulating effect.

4. Rigid insulation is sprayed between the joists from above directly on top of the gypsum or similar type supporting board before the subflooring is applied. Or it can be sprayed between the exposed joists from below to the required thickness.

Wood or metal joist floors for housing should have a maximum U-factor or 0.07, requiring insulation rated at R-11 or greater. (*See* Figs. 10, 11, and 12.)

Concrete. Thermal resistance of concrete floors can be brought to recommended levels by one of several methods. Preformed rigid insulation can be bonded to the top or

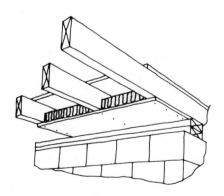

Fig. 10.

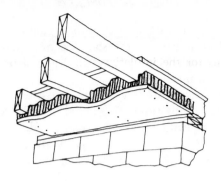

Fig. 11.

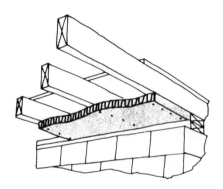

Fig. 12.

bottom surface. Concrete can be formed around a core of rigid insulation to form a sandwich. Or, rigid insulation can be sprayed on the underside of the floor.

Concrete floors for housing over unheated spaces and between separately heated dwelling units should have a maximum U-factor of 0.11 requiring insulation rated R-7 or better.

Slab-on-Grade. Some building methods employ slab-on-grade floors which are poured in place. The FHA Minimum Property Standards specify that perimeter insulation be used to reduce slab heat loss. Maximum heat losses in terms of watts and Btuh per linear foot of exposed slab edge, as recommended by the Electric Energy Association, are shown in Fig. 13. The insulation of the width indicated should be installed on the inside of the foundation wall extending down from the top of the slab. An alternate method calls for the insulation to extend from the top of the slab down the thickness of the slab and then continue horizontally back under the slab, using the total width of the insulation. Insulation rated R-5 is required for this application. (*See* Table 29.)

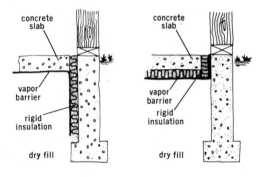

Fig. 13.

OTHER CONSTRUCTIONS

Sandwich. Floors of the sandwich type are similar to sandwich walls previously described. When installed over unheated spaces or between separately heated dwelling units, a maximum U-factor of 0.07 is required.

Crawl Spaces. The use of a ground moisture barrier over exposed earth surfaces in crawl spaces under wood or metal floors is recommended. This barrier may be a concrete slab, heavy roll roofing or polyethylene film, 4 mils or more thick, laid on the graded surface. The edge of the roofing or film should be sealed to the foundation with an asphaltic compound.

CEILING AND ROOFS

Frame. Insulation is recommended for all ceilings below unheated spaces. It is also recommended for both thermal

TABLE 29

Sᴌᴀʙ Eᴅɢᴇ Hᴇᴀᴛ Lᴏss

type of construction	opaque sections adjacent to unheated spaces			opaque sections adjacent to separately heated dwelling units		
	walls	floors	ceilings	walls	floors	ceilings
frame	R-11 U:0.07	R-11 U:0.07	R-19 U:0.05	R-11 U:0.07	R-11 U:0.07	R-11 U:0.07
masonry	R-7 U:0.11	R-7 U:0.11	R-11 U:0.07	R-7 U:0.11	R-7 U:0.11	R-11 U:0.07
metal section	R-11 U:0.07	not applicable	not applicable	R-11 U:0.07	not applicable	not applicabl
sandwich	R-* U:0.07	R-* U:0.07	R-* U:0.05	R-* U:0.07	R-* U:07	R-* U:07
heated basement or unvented crawl space	R-7 U:0.11	insulation not required	insulation not required	*Since the thermal resistance of sandwich construction depends upon its composition and thickness, the amount of additional insulation required to obtain the maximum U-factor must be calculated in each case.		
unheated basement or vented crawl space	insulation not required	see "floors"	insulation not required			

Note: Insulation R-values shown above refer to the resistance of the insulation only.

and acoustic reasons for the ceiling/floor sections between separately heated dwelling units.

When blankets or batts are installed before the finished ceiling is applied, they are stapled to the joists or squeezed between the joists and held in place by their own resiliency. If the finished ceiling is applied first, the insulation is laid in from above. In both cases the moisture barrier goes on the winter warm side of the insulation (*See* Fig. 14).

Loose insulation may be applied from above by blowing, utilizing standard blowing equipment, or manually, by emptying bags evenly between ceiling joists.

Foamed-in-place or sprayed-on insulation is applied from above after the finished ceiling is in place or before the ceiling section is joined to the wall units.

Whatever the type of insulation, it should be extended over the top wall plate, with care taken to avoid blocking any eave vents.

With pitched ceilings, where the ceiling and roof are combined into the same basic structural element, there are two conditions to consider:

1. Where rafters or beams are exposed in the finished structure, a composite roof deck having the required insulating value and vapor barrier can be applied above them. When plank roof deck is used, rigid insulation of the required R-value can be applied to the top surface of the deck, after applying a suitable vapor barrier.

2. Where rafters or beams are not exposed, the insulation is fastened between the rafters in the same manner as for frame walls.

Wood or metal frame ceilings below unheated spaces should have a maximum U-factor of 0.05, requiring insulation rated at R-19 or better. Ceilings/floors between separately heated dwelling units should have a maximum U-factor of 0.07, requiring insulation rated R-11 or better.

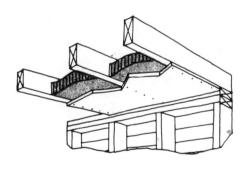

Fig. 14.

Concrete. Concrete ceilings beneath unheated spaces or between separately heated dwelling units can be insulated by adhesively-bonded preformed or sprayed-on rigid insulation applied to either surface, or by forming the concrete around a core of preformed rigid insulation to make a sandwich. Concrete ceilings should have a maximum U-factor of 0.07, requiring insulation rated R-11 or better.

A special case occurs when the concrete ceiling slab also serves as the roof of the structure. In this case, rigid insulation of the required thickness can be applied to the top surface before finishing the roofing system.

Sandwich. Where the sandwich-type of construction (similar to sandwich wall or floor sections) is used in ceilings between separately heated dwelling units, the maximum U-factor of the complete ceiling should be 0.07.

If the sandwich is used in ceilings below unheated spaces or as part of a roofing system, a maximum U-factor of 0.05 should be attained.

CHAPTER 14

Electric Irons and Ironers

Electric hand irons are available in various styles and are generally classified into four types: nonautomatic, nonautomatic with heat indicator, automatic, and steam iron.

Nonautomatic Iron. The nonautomatic iron is heated by a calrod unit embedded in the soleplate. Other nonautomatic irons are heated with mica-insulated heating units inserted between the pressure plate and the soleplate. Both types may be used on either a.c. or d.c. at 115 volts and are rated at 575 watts.

Nonautomatic irons are controlled by plugging or unplugging the appliance cord. If the iron does not heat, use the continuity test lamp in the iron or cord set. If no trouble is indicated, disassemble the iron. Remove the nut from the handle bolt and withdraw the bolt from the handle. Remove the thumb rest and handle and withdraw the top assembly screw. Lift the top assembly off and disassemble other parts if necessary in order to make repairs or replacements. (See Trouble and Remedy Chart.)

To reassemble the iron, reverse the disassembly procedures and test for a wattage reading within 10 per cent of the rating on the nameplate.

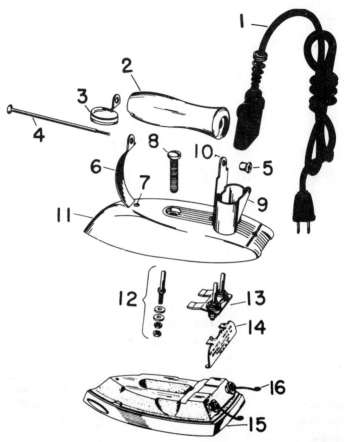

Fig. 1. Nonautomatic iron. (1) cordset; (2) heel stand cover plate; (3) rest thumb; (4) handle bolt; (5) handle bolt nut; (6) front handle support; (7) handle support rivet; (8) top assembly screw; (9) plug receptacle; (10) handle support; (11) top assembly; (12) terminal assembly; (13) terminal support assembly; (14) heel stand cover plate; (15) unit and lead assembly; (16) lead.

Nonautomatic Iron with Heat Indicator. The construction of the nonautomatic iron with heat indicator is similar to that of the preceding iron with one exception: there is an added bimetallic blade indicating the approximate temperature at which the iron is operating. Both irons are operated and tested in the same manner.

To disassemble a nonautomatic iron with heat indicator (Fig. 1), remove the parts in the following order: Remove the handle bolt and nut and lift off the handle. Remove the large screw from the top cover. (To remove the cover, pull slightly toward the rear of the iron and reach under the cover; then pull back on the heel cover plate that is pivoted in the terminal slot.) Take the nuts off the bottom ends and remove the terminals. The heat indicator is secured to the unit and soleplate assembly with one screw. To reassemble, reverse the disassembly procedures; test the iron.

Automatic Iron. The typical automatic iron (Fig. 2) weighs 3 lbs.; it is heated by a calrod unit which is cast

Fig. 2. Automatic iron.

in the aluminum soleplate. This iron has a fabric indicator plate with a thermostat control knob and a pilot light to indicate when it has been heated to the desired temperature. It is approved for use on a.c. only at 115 volts and is rated at 1000 watts.

If it is necessary to disassemble an automatic iron (Fig. 3), proceed as follows: Pull the cord bushing from the cavity in the handle and slide it back onto the cord set. Remove the lamp. Remove the two nameplate screws and lift off the knob and the nameplate assembly. (To replace any part of this assembly, remove the retaining ring.) Withdraw the two top assembly screws and carefully lift the top and the handle assembly up and back while feeding the cord set through the hole in the handle. Further disassembly is a simple procedure but should be undertaken

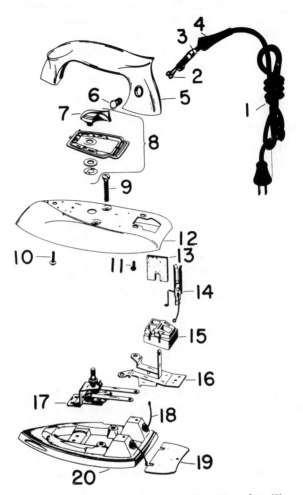

Fig. 3. Automatic iron. **(1)** cordset assembly; **(2)** eyelet; **(3)** cord clip; **(4)** cord bushing; **(5)** handle; **(6)** lamp; **(7)** knob assembly; **(8)** knob and nameplate assembly; **(9)** top screw; **(10)** handle front screw; **(11)** handle rear screw; **(12)** top; **(13)** mica insulator; **(14)** lamp bracket assembly; **(15)** terminal insulator; **(16)** terminal bracket insulator; **(17)** thermostat; **(18)** knob and nameplate assembly; **(19)** heel stand cover plate; **(20)** unit and lead assembly.

only if required. (See Trouble and Remedy Chart.)
Adjustment of Pointer on Knob. For reassembly, reverse

the procedures, but note that the removal of the knob and the nameplate assembly from an iron of this type destroy the relationship between the pointer on the knob and the position of the thermostat shaft. To reassemble the parts in their correct position, connect the *cold* iron to a continuity light circuit and rotate the thermostat shaft to the point at which the contacts close. Then, with the knob pointing between *rayon* and the *Off* position but closer to *rayon*, attach the knob and the nameplate assembly to the iron. Rotate the knob to make sure that it will turn to the full *Off* position at one end and to *linen* in the other direction before encountering the stop at either end. If there is interference, lift the knob and the nameplate and move the knob one or two splines in the indicated direction. Then reassemble the parts.

Adjustment of Thermostat. After reassembling an automatic iron, and if testing equipment is available, test and adjust the thermostat. Using testing equipment for irons, connect the iron to a 115-volt a.c. and make sure that the iron is off when the knob is turned to the *Off* position. Turn the control knob to *silk* and read the *On* and *Off* point temperatures for the second cycle. The adjustment should be such that the *On* point temperature is not less than 200° F. and the *Off* point temperature does not exceed 350° F. Turn the knob to maximum heat and again read the *On* and *Off* point temperatures. The *On* point temperature should not be less than 475° F. and the *Off* point temperature should not exceed 600° F.

To adjust the thermostat, lift the knob and the nameplate assembly clear of the thermostat adjusting stud. Rotate the knob clockwise if the readings were high, counterclockwise if the readings were low. By moving the knob one spline, you can change the temperature setting about 15° F.

Automatic Steam Iron. The automatic electric steam iron (Fig. 4) is equipped with a water tank, a boiler, a valve that controls the flow of water to the boiler, and passages that guide steam from the boiler to the steam ports in the soleplate. The steam control valve permits the iron to be used either as a steam iron or as a dry iron. The iron is filled through a hole in the top of the handle. A knob controls the flow of steam. This iron is approved for use on a.c. only (at 115 volts) and is rated at 1000 watts.

To disassemble and reassemble a steam iron of the type illustrated (Fig. 5), specially designed tools are necessary. The procedure is as follows: Lift off the top cover plate. Withdraw the screw in the cover plate of the handle. Re-

Fig. 4. Steam iron.

move the handle cover plate and free the cord set from the slot in the handle. Remove the bottom cover plate screw and the bottom cover plate. Withdraw two rear handle screws. Remove the steam knob by lifting it while prying up from the back and then raising it from the stop with a pair of thin nose pliers. Turn the valve assembly out of the valve body. Use a bushing tool to remove the filler tube bushing, the retaining ring, and the valve assembly. Remove the two filler tube washers, the filler tube gasket, and the handle from the top and tank assembly. Use the insert tool to lock the prongs of the sleeve in the slots of the valve body. Unscrew the latter part and remove the top and tank assembly. (Some older models of steam irons have bushings and valves with slots too narrow for the bushing and sleeve tools. Remove the bushing with a screwdriver; or break the handle, if necessary, and remove the valves.) Boiler cover screws and steam chamber screws may bind in the soleplate. To release them, place a

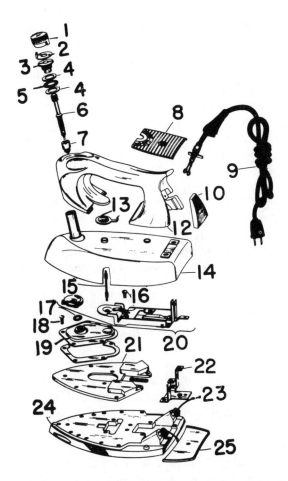

Fig. 5. Steam iron. (1) steam knob; (2) stop; (3) filler tube bushing; (4) filler tube washer; (5) filler tube gasket; (6) assembly valve; (7) body valve; (8) top cover plate; (9) cordset assembly; (10) handle cover plate; (12) assembly handle; (13) control lever; (14) top and tank assembly; (15) control shaft; (16) thermostat attachment screw; (17) valve body gasket; (18) steam chamber screw; (19) boiler cover assembly; (20) thermostat assembly; (21) steam chamber; (22) bracket; (23) wire lead; (24) unit and lead assembly; (25) bottom cover plate.

heavy solid screwdriver in the slot and rap it sharply with a medium-weight hammer.

To reassemble, reverse these procedures. When installing a thermostat, paint the soleplate area under the thermostat and back to the steam chamber cover with silicone resin. Be careful not to get any of the resin in the steam chamber. If necessary, thin the resin with *thinner*. When a steam chamber and a boiler cover are assembled to a soleplate, point the joints with silicone adhesive to prevent leaks. Thin, when necessary, with kerosene. Lubricate all thread screws that enter the aluminum soleplate. Before replacing a soleplate, note its appearance. If the boiler area has a light frosty appearance, the soleplate has been treated at the factory of the manufacturer for improved wetting characteristics. If it does not show this treatment, a tankful of saturate solution of gypsum should be boiled off after the iron has been reassembled. When reassembling the top and handle, be sure that the thermostat cam and the control lever are in corresponding positions. As each rotates 180°, it is possible to assemble them so that the thermostat is locked either at the *high* or at the *low* position. When installing the steam knob, make sure that the valve assembly is turned to the *Off* position; then push the knob onto the shaft, with the stop adjacent to the lug inside the handle. Most of the necessary materials are available at electrical supply dealers or at appliance service stations.

After reassembly, proceed as follows: Place the iron on the iron test stand and connect it to a 115-volt a.c. The adjustment screw can be reached through a hole in the top and tank assembly. Turning it clockwise decreases temperatures, and turning it counterclockwise, increases temperatures. Adjust the thermostat so that all the following conditions are fulfilled: The reading at 115 volts should be between 925 watts and 1050 watts. The iron should be off when the thermostat knob is turned to the *Off* position with the iron at room temperature. The *On* point temperature with the thermostat knob set at the center of the steam position (red marking) should be 310° F. plus or

TROUBLE AND REMEDY CHART
For Electric Irons

Trouble	Cause	Remedy
Inoperative iron.	Defective cord. Blown-out fuse.	Repair or replace. Check cause of blowout. Make necessary repairs. Replace fuse.
	Defective calrod heating unit.	Disassemble iron, following directions given. If unit has a broken lead give it a series test, using a 100-watt bulb in a test set. The test prongs should be placed on each terminal. If the lamp lights at full brilliance, there is a short. Check carefully to be sure that the short is not caused by the broken lead. If the short cannot be corrected, the heating unit cannot be repaired. If the lamp does not light, the calrod coil is open and the heating unit cannot be repaired. If the lamp lights at less than full brilliance, the heating unit is all right and new leads should be installed.
	Bent or dirty calrod terminals.	Straighten bent calrod terminals. Clean out any dirt, oil, or other substances between terminal and sheath to prevent shorts.
	Broken leads.	*Replace.
	Defective mica insulation heating units.	*Test; replace, if necessary.
(automatic and steam irons)	Defective thermostat.	*Test; replace, if necessary.
Unsatisfactory heat. (non-automatic iron with heat indicator)	Defective or faulty adjustment of heat indicator.	Loosen screw of heat indicator. Reset heat indicator at cold edge of slot and retighten. *If heat indicator fails to function after adjustment, replace indicator.
Low heat. (automatic and steam irons)	Incorrect adjustment of thermostat.	*Check with special temperature test equipment and make necessary adjustment.
Failure of pilot light.	Defective lamp.	Replace lamp as follows: Cut about ½″ of the small end of a cord bushing to unscrew the burned out bulb while turning the bushing. If the new lamp does not glow, check the lamp leads and the lamp socket for continuity.
Leaks or sputters. (steam irons)	Actual leak in tank. Leaky steam controls valve.	*Replace. *Check the position of the knob on the steam control shaft in relation to the closed position of the valve.
	Tank filled to the overflow vent.	Watch filling.
	Iron filled with the steam control valve open.	Watch filling.
	Valve opened without the two minute preheat.	Correct procedure.

* Consult service department of manufacturer.

TROUBLE AND REMEDY CHART—*Continued*

Trouble	Cause	Remedy
	Sometimes steam rising up around the iron condenses on the shell and runs off in drops.	If continuous, check iron.
	Thermostat control is set too low, causing water to run through iron without being turned to steam. Thermostat adjusted too high, causing water to be blown through the steam passages so fast that it comes out as droplets instead of steam.	*Reset thermostat.
Iron fails to steam or gives too little steam. (steam irons)	Excessive deposits due to the use of tap water with a high percentage of dissolved solids in the steam passages.	*Test iron for steaming. If the test fails to empty the tank in 25 minutes, inspect the steam passages. Check the valve opening, the valve cleaning rod, and the position of the knob on the steam control shaft.
Spots. (steam irons)	Charred organic material in the steam chamber and passages. This may be caused by lint that works up into the iron or by contaminated vessels or bottles used for filling the iron.	Clean by filling tank and then flooding soleplate while iron is cold. Turn the iron to steam range with valve still open. Boiling water will flush brown material from the iron. Repeat if necessary.
Burns holes in clothes. (steam irons)	So-called "distilled" water obtained at automobile service stations that is apt to be contaminated with sulphuric acid.	Use water specified by the manufacturer.

* Consult service department of manufacturer.

minus 25° F. The *On* point temperature with the thermostat knob set at maximum position should be 510° F. plus or minus 50° F. The amplitude should be greater than 20° F. and less than 80° F. The amplitude plus overshoot should not exceed 120° F.

After reassembly, test the steam iron with the steam valve closed, fill the tank, and preheat the iron for two minutes. Check the steam valve and tank for possible leaks. Turn the control valve to the steam position and note whether the steam flows steadily and without sputtering.

AUTOMATIC IRONERS

A standard type of automatic ironer is shown in Figs. 6 and 7. It operates on 110 volts a.c. only. While some types may differ in outward appearance, the operation and maintenance of all modern automatic ironers are fundamentally the same.

Components and Their Use. Figure 7 shows the components of a typical automatic iron.

A. The *ironer roll* corresponds to an ironing board. It is covered with a soft pad and a muslin cover. The roll can be rotated forward and backward by hand.

B. Each of the *two temperature controls* thermostatically regulates the temperature for half of the ironer shoe. The temperature for the various types of fabrics can be selected by reference to markings on the *control knobs,* and the heat selected is automatically maintained.

Fig. 6. Cabinet rotary electric ironer when iron is not in use.

C. The *ruffler plates* are located at each end of the shoe and are used for ironing small ruffles and gathers.

D. The *ironer shoe* corresponds to the ironing surface of a hand iron, with a rust-proof and scratch-resistance surface. The tip end of the ironer shoe is similar to the point of a hand iron. The top and bottom edges of the shoe are insulated for protection of hands. In most types the ironer

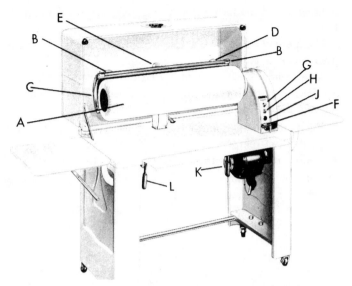

Fig. 7.   Parts of a rotary electric ironer. **A**, roll; **B**, temperature controls; **C**, ruffler plates; **D**, shoe; **E**, shoe release; **F**, finger tip control; **G**, two-speed control; **H**, on and off heat switch; **J**, pilot light; **K**, knee control lever; **L**, knee press lever.

shoe when tilted back in a horizontal position can be used for steaming.

E. The *shoe release* is used to move the shoe away from the roll (if the electric current goes off or is turned off while ironing) by pushing the lever away from the ironer roll. After electric service has been resumed, reset the ironer shoe by pulling the lever forward as far as it will go. For cleaning the shoe or for steaming garments, tip the lever up to a horizontal position.

F. The *finger tip control* is located below the pilot light. It starts and stops the ironing action in the same way as the knee control lever.

G. Many automatic ironers are equipped with a *two-speed control*. The *fast speed* is used for most ironing. The *slow* is a drying speed used for extra-damp clothing which requires a longer application of heat and pressure.

H. The *On* and *Off heat switch* is usually located at the right of an ironer.

J. The *pilot light* indicates when the heat switch is turned to *On.*

K. The *knee control lever* starts and stops the roll and applies and releases pressure automatically. It can be adjusted to a comfortable position for ironing by pressing the two parts of the lever together and sliding it to the position desired.

L. The *knee press lever* also starts the ironer. When using this control, hold the left knee against the lever. The roll will stop revolving until the knee is withdrawn. This additional control is employed to dry seams and extra-damp fabrics or to press heavy garments.

The *master switch* (in the type illustrated) automatically shuts off both the heat and the motor when the cabinet cover is closed.

Installation. To install an automatic ironer, proceed as follows: Uncrate the ironer and connect it to the rated power supply specified on the nameplate by the manufacturer. The ironer cord should be plugged into a permanent receptacle; avoid use of extension cords and do not connect other appliances to the same circuit.

Operation. The three basic factors necessary for the ironer to operate properly are heat, pressure, and time. The heat and pressure are supplied by the ironer. The homeowner should be familiar with the amount of time required.

The correct procedures for operating an automatic ironer are as follows:

Before using the ironer, move the ironer shoe back from the roll, turn it to a horizontal position, and clean its surface with a damp cloth. Preheat the ironer by turning on the heat switch. Then turn the temperature controls on the ironer shoe to .the temperature specified by the manufacturer for the type of fabric to be ironed. During this preheating period, the shoe must be away from the roll. Select the required operating speed by using the knee control or the finger tip control.

Flat work, which is easy to iron, probably constitutes the largest part of the ironing done in the home. Begin with flat pieces, such as towels and napkins, to become familiar with the operation of the ironer. Straighten the clothes on the roll. Never poke the clothes between the roll and the shoe. Turn the roll away from the body by hand so that the edge of the fabric will come under the shoe.

To iron, touch the knee control lever lightly with the right knee to move the shoe against the roll. To straighten the clothes while ironing, stop the ironer with a light touch on the knee control lever. As the garments go through the ironer, keep them straight by smoothing them away from the center with hands. Do not pull the clothes down at the corners because that will curl the corners. When ironing bias-cut garments, always keep the threads parallel to the roll so that the finished shape of the garment will not be changed. Spread and smooth the materials along the full length of the roll. This will keep the padding uniform and make good use of all the heat. If the clothes begin to wrinkle, release the shoe or use the knee press lever to straighten the garment before ironing is resumed. Damp portions of garments, including thick seams and hems, may be dried by using the knee press lever to stop the roll, thus exposing them to more heat and making it unnecessary to put them through the ironer a second time. All garments should be ironed until they are dry; otherwise, they will pucker as they dry.

When ironing is finished, push the shoe release back to move the hot shoe away from the roll. Let the shoe cool before closing the ironer. Wind the electric cord around the hooks that are located under the cabinet at the rear of the ironer.

Care of the Ironer. The following instructions are applicable to most automatic ironers:

Method of Cleaning the Shoe. To remove baked-on starch or other substances from the ironer shoe, use a non-abrasive household cleaner. Then wash the surface with a damp, soapy cloth, wipe off all traces of soap, and dry the shoe thoroughly. Occasionally, when the metal surface is warm, rub paraffin or beeswax over the surface of the shoe so that the wax will spread evenly.

Method of Cleaning and Replacing Ironer Roll Cover. When the muslin cover on the roll becomes soiled, it should be removed and laundered. To remove the cover, untie the drawstrings at each end of the roll and slip it off. To replace it, insert one edge of the roll cover under about 6" of the flannel undercover. Iron the cover on the roll smooth and tight, pull up the drawstrings at each end, tie them, and tuck them under the ends of the roll.

Method of Repacking Roll Padding. Frequent use of the ironer will pack the padding and will cause the ironer to lose some of its efficiency. When the padding becomes hard and packed, remove the muslin cover and unwind the flannel undercover, the cotton padding, and, finally, the burlap which is taped to the roll. Untape and fluff the burlap; let it remain a few hours in the open air and sunlight. Then tape the burlap back on the ironer roll with a common household tape (the adhesive, electrician's, or transparent gummed type). The construction line of the roll is a good guide. Without releasing the shoe from the roll, iron on the burlap; then iron on the cotton padding, the flannel undercover, and, finally, the roll cover. This procedure prevents wrinkles. Use the knee press lever to stop the roll when necessary. Pull the drawstrings until the muslin cover fits smoothly over the ends of the roll and tie them.

Lubrication. All modern ironers are permanently lubricated at the factory. No additional oiling is necessary unless specified by the manufacturer.

Service. Although instructions for making operational and electrical tests, repairs, and replacements have been given, the homeowner who does not have the necessary equipment and experience should arrange for such work to be done by the manufacturer's service department. Nevertheless, every homeowner should know the fundamentals about the ironer and similar appliances. Such knowledge will be useful if it is necessary to call a service man, for his analysis of and charge for the job will be more clearly understood.

TROUBLE AND REMEDY CHART

FOR ELECTRIC IRONERS

TROUBLE	CAUSE	REMEDY
Insufficient heat.	Poor contact between plug and wall outlet.	Test and correct.
	Faulty service cord, plug, or outlet.	Test; repair or, if necessary, replace.
	Use of too long or too light extension cord, with resulting loss of current.	Plug ironer directly into socket, using same size of cord as that supplied by manufacturer.
	Voltage on line too low.	*Consult local supplier.
	Loose connections or broken wires.	Check for breaks and loose or dirty connections. Repair breaks and tighten loose connections. Clean dirty connections, especially at terminals.
	Inoperative heating elements.	*Test thermostat. If thermostat is operating properly, check heating elements for broken or loose connections and make repairs or replace unrepairable elements.
	Inoperative thermostat.	*Replace.
	Excessive moisture in material being ironed.	Dry material.
Motor failure.	Mechanism failure, disconnected motor leads, faulty capacitor, bound motor shaft, or bent fan blade.	*Check mechanism capacitor and current to motor before assuming motor failure is due to faulty motor. Repair or replace motor, if necessary.
Shoe does not heat.	Inoperative thermostat or heating element.	*Remove plate from underside of table and check current up to this point. Remove shoe from ironer and cover from shoe. Short across thermostat to ascertain whether fault is in thermostat or in heating element. If necessary, replace thermostat or heating element or both. Before removing the clamping plate to replace a heating element, mark the exact position of the thermostat mounting plate by scribing around it onto the shoe.
Insufficient shoe pressure on roll.	Incorrect setting of pressure spring and shoe pressure spring bolt.	*Set pressure spring to approximately 1 5/16″ in length by tightening or loosening nut on pressure spring bolt; then tighten locknut. Adjust shoe pressure spring bolt until the bolt head rises from link pin approximately 1/32″ when the shoe is brought against the roll. Recheck the length of pressure spring and tighten the screw at the lower link pin.

* Consult service department of manufacturer.

CHAPTER *15*

Electric Ranges

The electric range shown in Fig. 1 is designed for 115-120 volt and 230-240 volt, single-phase, three-wire a.c. service. Automatic models are equipped with standard 60-cycle timers. Replacement motors are available for converting timers to 50-cycle or 25-cycle service. The fluorescent lamp ballast may be changed for 50-cycle service, but a 25-cycle lamp ballast is not available because of objectionable flicker.

INSTALLATION

The electric range must be connected to the power supply and grounded in accordance with the local electric code. The range service switch or combination range and lighting switch should have a 60-amp. circuit. An approved polarized range receptacle should be installed in the circuit at the range location to permit disconnecting the range for moving or repairing.

Method of Grounding Range. Every range must be grounded at the time of installation. Consult the local utility for grounding requirements. The UL seal of approval on the serial plate means that the range has met all their safety requirements. UL Form O installation means that, according to laboratory measurements of surface temperatures on exterior panels, the range can be installed against combustible material such as wood cabinets or wood lath walls. UL installation Form No. 1 means the range is approved for installation 1″ from such combustible material. The

Fig. 1. Electric range.

electric range shown in Fig. 1 has been approved for Form O installation.

Service Connection. Electrical service connections (Fig. 2) are usually made through a 3' flexible "pigtail" which is equipped at one end with a plug for insertion into a standard range outlet receptacle. This type of connection is generally preferred because it provides a means for easily disconnecting the range. Metallic conduit must be used where the local code requires this type of connection. The wire size of the service cable should be No. 6 AWG or larger for the two line wires and No. 8 AWG or larger for the neutral wire. The service cable or "pigtail" is not supplied with the range. It can be obtained at any local hardware store. The service opening provided is a 7/8" diameter hole scored for a 1⅛" knockout. A strain-relief clamp should be secured to the service cable to protect the terminal connections.

Line connections are made to the three terminals of the terminal block located just above the service opening. Connect the red, white, and black leads of the "pigtail" to the corresponding terminals on the terminal block. After making the service connection and replacing the cover, the

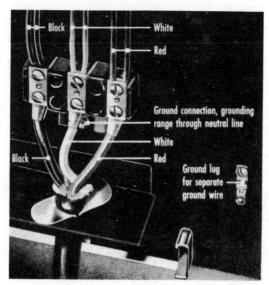

Black
White
Red

Ground connection, grounding range through neutral line
White
Black
Red

Ground lug for separate ground wire

Fig. 2.

range may be pushed into position. Remove one of the lower service drawers or the lower front panel and reach through the opening to insert the plug into the range outlet receptacle.

Method of Leveling. All electric ranges must be level to obtain proper baking results. After the range is in the desired location and has been connected, proceed as follows: For a level indicator, place a spirit level or a pan of water on a rack in the oven. Remove the two lower service drawers or the lower front panel to reach the slotted ends of the four leveling screws located at each corner of the base. Adjust the leveling screws (Fig. 3) with a screwdriver until the level indicator is even. If there are no leveling screws, shims may be placed under the base to raise the range to a level position.

Fig. 3.

INSTRUCTIONS FOR WIRING

There is a wiring diagram on the rear of each range. Figures 4 and 5 show two typical wiring diagrams.

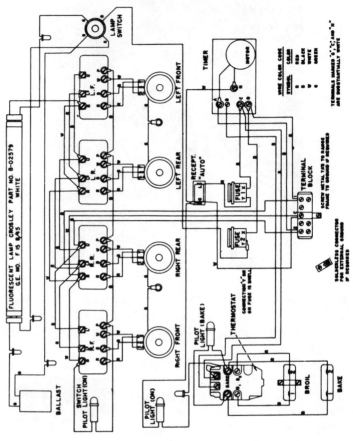

Fig. 4.

Three-Wire, 120/208 Volt, Residential Service. In some locations 120/208 volt, three-wire, residential service is supplied from a three-phase, 120/208 volt, four-wire industrial lighting and power network. Residential service from these networks is usually confined to apartment houses and to a limited number of individual residences situated within the district.

The performance of a standard electric range (rating 118/236 volts) on 120/208-volt service is slightly slower

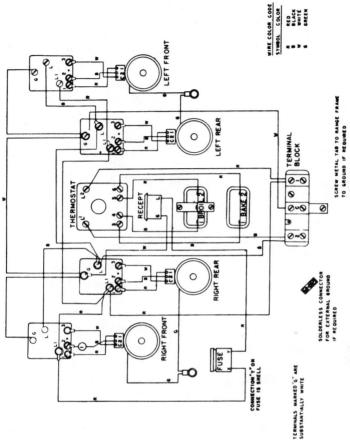

Fig. 5.

when set for the open preheat time, broiling, and the three highest heat positions of the surface units. This is the result of the lower maximum operating voltage. However, baking performance is approximately the same, and the decrease in broiling speed is rarely noticed because broiling is normally a short-time operation. Moreover, the four lower heat positions of the surface units, which are used most frequently, operate at the normal fast-cooking speed. Because of the wide selection of ranges available in the standard rating, many installations have been made on 120/208-volt service and are giving satisfactory performance.

Auto Transformer Installation for Two-Wire, 230-Volt Range. An auto transformer can be obtained from either the General Electric Company or the Acme Electric Company for installing an electric range on a two-wire, 230-volt house service which does not have the third neutral wire required for a range connection. By inserting this transformer in the circuit between the range and the 230-volt fuse box at the meter, the two-wire input circuit is transformed into a three-wire circuit and has the neces-

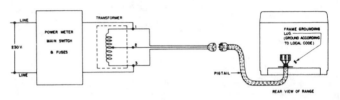

Fig. 6.

sary three range lead-in wires, one of which is a neutral connection 115 volt from each side of the line. Figure 6 shows how the auto transformer should be connected to the range. The three wires emerging from the transformer can be identified as the No. 1, No. 2, and No. 3 leads shown on the wiring diagram by the numbered metal tabs attached to each wire (Fig. 7). When wiring according to the diagram shown in Fig. 7, do not ground the range to the neutral wire by screwing the metal tab beneath the terminal block to the range frame. This would make the range "hot," as there can be considerable voltage between the neutral wire from the transformer center tap and the ground of the power system. Instead, ground the range by running a lead from the grounding lug on the back of the range (Fig. 6) to a good ground connection as specified by the local code.

The transformer is built with convenient bolt holes for mounting (Fig. 7). The best place for the installation depends on individual requirements. *Local code require-*

ments must be followed. The layout of the house and the location of the fuse and meter panel must be considered. The weight of the transformer is also important, as the installation must be strong enough to support the type of transformer used. If the transformer is mounted on the power meter board at a distance from the range, either conduit or standard range cable must be used (depending on local electrical regulations) to make a connection to the wall outlet for the range "pigtail" cable. If the local code permits mounting the transformer directly behind the range, the lead-in cable should be size No. 6 wire or larger.

Fig. 7.

Fuse Location. All parts of the range designed strictly for 115/120 volt operation, with the exception of the pilot lights and the motor for the automatic timer, are protected by a 15-amp. fuse located on the range. If an oven light, top light, or appliance outlet does not function, first check the 15-amp. fuse or fuses. Use standard household fuses of 15-amp. rating for replacements (Fig. 8). All high backguard ranges have the 15-amp. fuse located in the top of the backguard beneath the top lamp shield (Fig. 9). Some electric ranges have the 15-amp. fuses located behind

Fig. 8.

the large storage drawer front panel. Others have the fuse located behind and to the right of the deepwell unit (Fig. 10), and some have the fuse located behind the nameplate on the backguard. On cluster top models, the fuse can be reached by lifting out the left rear unit. To gain access to the fuse on divided top ranges, remove the

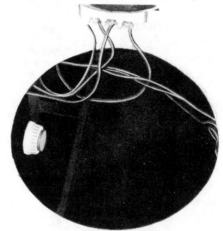

Fig. 9.

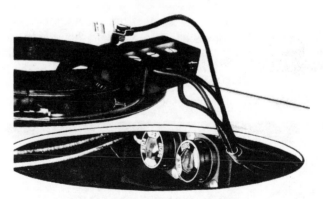

Fig. 10. Fuse location.

large service drawer and the rectangular drip tray. The fuse may be seen through the drip tray slot and easily reached through the storage drawer opening.

Fold-Down Backguard. The backguard may be folded forward to permit any repairs or replacements from the front of the range. To lower the fold-down backguard (Fig. 11) proceed as follows: Disconnect the range from the power supply by unplugging the range "pigtail" or

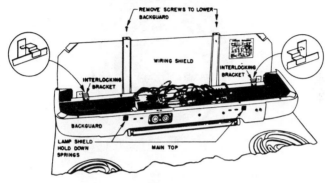

Fig. 11.

by opening the main switch. Spread a protective pad or cloth over the main top. Take out the two screws next to the right and left lamp shield hold-down springs. Lift the backguard about one inch and then pull it forward until it rests face down on the main top.

COMPONENTS AND THEIR USES

The user of an electric range should become familiar with the advantages, operation, proper care, and servicing of the surface units.

Surface Units. There are two common types of surface units: the monotube unit and the triangular rod (two-in-one) unit.

Monotube Surface Unit. The features of the monotube surface unit (Fig. 12) are as follows: There is a large utensil contact area with uniform heating throughout the length of the monotube rod at all heat positions. The heating rod sheath is made of Inconel, a corrosion-proof metal alloy that transfers heat quickly and will not warp, crack, or scale. Nickel-chromium resistance wires are permanently coiled and positioned within the rod for quick, even heating. The monotube rod operates on a swivel to

Fig. 12.

an upright position with a removable center support to permit cleaning. The element rod is not removable and, in case of failure, the entire surface unit must be replaced.

Fig. 13.

Triangular Rod Surface Unit. The triangular rod surface unit (Fig. 13) is constructed as follows: There is a triangular cross section with a mounting that allows for lateral expansion, thus providing a maximum flat contact cooking surface and keeping the unit level. Both the heating rod sheath and the nickel-chromium resistance wire are the same as in the monotube unit. The reflector pan is designed to direct maximum radiant heat to the utensil bottom and, on most models, is removable for easy cleaning. Two separate coils give a versatile heat selection to accommodate every size of cooking utensil. Element rods are easily removed. In case of failure, only the element rod that failed need be replaced.

Operation of the Surface Unit. Heat for cooking is generated by passage of the electric current through the resistance heating element. This heating element is centered and is electrically insulated from the heating rod sheath by a rocklike material, which transfers heat instantly to the rod sheath. Since the flat top of the heating rod is in direct contact with the bottom of the cooking utensil, the heat is conducted directly into the food. In fact, the large contact area of the surface unit enables heat to be transferred to the utensil so quickly that the surface unit will not heat the kitchen air. To provide the various degrees of heat for different cooking operations, each surface unit is made up of two separate resistance heating elements. These are operated either in parallel, singularly, or in series, and on full voltage or one-half voltage to give the user seven degrees of heat (Fig. 14).

Method of Using the Surface Units. A common mistake among new electric range users is to underestimate the cooking speed of the surface units. The six important points to remember about surface-unit cooking are: (1) Only a small quantity of water need be used for cooking fresh or frozen fruits and vegetables. (2) The high heat and second heat positions are used primarily to start cooking operations quickly, depending upon the size of the pan. (3) After foods begin steaming, violent boiling can be avoided by using the lower heat positions until the cooking time is completed. (4) Special cooking utensils are not necessary for satisfactory results; however, pans should be large enough to cover the entire heating area. A utensil with a flat bottom makes good contact with the heating unit and therefore uses the heat most effectively. A tight-fitting lid also shortens the cooking time because it holds the steam. Variations in the flatness of the pan bottoms, the weight of the pan, the material of which the pan is made, the voltage of the electric service and other factors cause variations in cooking speed. (5) If the heat suggested in a recipe seems a little too slow or too fast, the next higher or lower heat should be used. (6) When using glass saucepans, skillets, and percolators on an electric range, place a wire grid between the glass pan and the surface unit.

Suggested uses for the seven heat positions are as follows:

High heat.	For starting most cooking operations; for deep frying.
Second heat.	For starting frying and pan broiling in large utensils; for starting most cooking operations in small utensils.
Third heat.	For continuing frying and pan broiling.
Fourth heat.	For cooking foods with egg or milk bases without a double boiler; for continuing slow frying and frying pancakes.
Fifth heat.	For continuing slow cooking in large utensils after boiling point has been reached; for cooking foods which have been started on second heat position in small utensils; for cooking small quantities of foods with egg or milk bases without double boiler.
Sixth heat.	For cooking average quantities of food after boiling point has been reached.
Simmer.	For long, slow-cooking operations; for keeping foods hot until serving; for maintaining cooking temperatures.

WATTAGES

POSITION	VOLT.	TRI-ROD				COOKER OPEN COIL UNIT	COOKER PLATE UNIT	MONOTUBE	
High	230	1250	1425	1600	2100	1000	800	1100	1900
2	230	675	675	850	1200	600	450	600	1050
3	230	575	750	750	900	400	350	500	850
4	115	312	356	400	525	250	200	194	331
5	115	169	169	212	300	150	112	150	263
6	115	144	190	190	225	100	87	125	212
Sim	115	78	88	100	131	63	50	69	119

Fig. 14.

Deepwell Cooker. The deepwell cooker is used for heating liquids, deepfat frying, steaming foods, and sterilizing. It is usually located, for convenience, in the left rear position. Some ranges are equipped with the elevating or double-duty type of deepwell, which can be used either as a deepwell unit or (when raised to the top) as a surface unit. There is a clearance between the rim of the cooker pot and the main top panel which is necessary for obtaining maximum heat transfer. On ranges using the stationary deepwell, a 1000-watt open-coil type of cooker unit is used to provide maximum heat control in the lower heat positions. A higher-wattage cooker unit (1250 watts) is used in the elevating or double-duty type of deepwell to provide both the necessary high heat positions required for surface-unit cooking and the low heat position required for deepwell cooking. It also permits a shorter starting time in deepwell cooking of large quantities of liquids and in deepfat frying.

MAINTENANCE OF SURFACE AND COOKER UNITS

The enclosed rod-type unit as a rule has a longer life span than the open-coil type cooker unit. The rod sheath and insulation protect the nickel-chromium heating element from food spillage and also retard the low oxidation rate of the heating element wire. The possibility of food spilling onto the open-coil cooker unit is minimized by the design of the deepwell jacket and the cooker pot. If the cooker pot is stored in the deepwell when not in use, it covers the open-coil cooking unit and protects it from spattering grease and from spillage.

The 7-heat switch controls the surface-unit heating rate. The seven heat positions are obtained by contacts inside the switch body that operate the two heating elements either in parallel, singularly, or in series, and on full voltage or one-half voltage.

RANGE OVEN

Electric range ovens are usually insulated by 2¼ lbs. or 3 lbs. (density) of fiberglass insulation. Some ranges have a thicker oven insulation to maintain reasonable surface temperatures on the main top panel. The range shown in Fig. 1 obtains good results because the design of the main top panel has provided for a deep air space between the oven-top insulation retainer and the main top. This design restricts heat transfer to the working surface of the top panel.

Oven Vent. Because air expands when it is heated, all ovens must be vented. This is usually done through the right rear surface unit. In oven cooking, the heated oven air absorbs moisture in addition to expanding. To prevent the moisture-laden oven vapor from condensing on the sides of the oven or on the front of the range, the oven door must be sealed at the top and have a gap of approximately 1/32″ at the bottom so that sufficient air can circulate through the oven and discharge the moist vapor through the oven vent.

On most ranges, an aluminum oven shield seals the opening between the frame and the main top panel to prevent condensation of the hot moist vapors which escape from the side of the oven door. If moisture does condense on the front of the range frame around the oven door, it is an indication of an improper oven door top seal or a restricted or clogged oven vent.

Oven Door Broil Position. In ovens equipped with top heat for broiling, provision is always made for leaving the oven door ajar during the broiling period. This gives the effect of true charcoal broiling instead of high-temperature baking. Some ranges have oven door hinges equipped with the broil stop position. Others have a broil stop mounted on the broil unit. This stop should be turned to the forward position to hold the oven door ajar for broiling. When the oven door is to be left ajar, the drip tray should be pulled out about 1½″ to prevent the oven heat from discoloring the plastic knobs which are directly above the oven door (Fig. 15).

Fig. 15.

Oven Units. The oven units used in electric ranges are generally of the open-coil type. In contrast to surface-unit cooking for which contact-heating is most desirable, oven cooking requires radiant heat for optimum performance.

The bake unit (always located in the bottom of the oven) is covered by a baffle to provide protection from food spillage. This baffle also absorbs radiant heat from the bake-unit heating element and helps distribute and equalize the heat in the oven. The oven units can be pulled straight out of the oven (Fig. 16). They need be removed only when the user desires to give the oven a thorough cleaning. The oven units themselves should never be placed in water, for they require no cleaning. The bake-unit baffle, which may be subjected to food spillage, can be removed from the bake unit for separate cleaning.

Oven Unit Element Wire. Balanced heat distribution in the oven is obtained through uniform spacing of the spiral-wound element wire. If the coils become distorted or lengthened by careless handling, they can be adjusted with a pair of needle nose pliers. The open-coil resistance

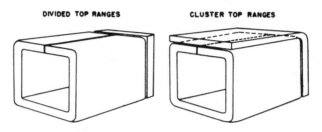

Fig. 16.   Oven insulation.

element wire reaches operating temperature in a matter of seconds and transfers heat quickly to the oven. In open-coil units, some adjustment can be made to correct uneven browning of foods being baked. Close the coil spacing by moving the wires to the areas where additional heat is required, and open the coil spacing where heat is to be reduced. Do this, however, only if no other adjustment or correction will give satisfactory results.

If new element coils are added, they must be stretched uniformly to slightly less than the desired assembly length. A steady pull while the element wire is held firmly at both ends will stretch the element coil evenly.

Oven Unit Terminal Connection. The oven unit electrical heating element wire is terminated at the terminal pins. Because of the high operating temperature of the element wire, the terminal connections must be properly made. If the terminal connections were to develop resistance to the flow of current, energy would be dissipated in the form of heat. This additional heat at the terminal connections would decrease the life of the element wire at that point. To obtain a proper terminal connection and to minimize the possibility of failures due to resistance, note the following points:

The terminal washers used against the element wire must be clean and new in appearance to make good electrical contact. If they show any signs of rust or corrosion, they should be replaced. Monel is generally used for terminal washers. However, nickel-plated washers and also silver-plated washers are sometimes used. Brass (alloy or copper and zinc) cannot be used in terminal washers because the zinc in the alloy would oxidize, thus forming an insulating film which leads to excessive resistance.

When element wire is heated, an oxide coating forms on the exposed surface of the wire. To make a terminal connection with a used element, scrape the oxide coating from the element leads with a knife. Do not nick the wire, for a reduction in diameter will increase the resistance and lead to an early failure. Emery cloth should not be used to remove the oxide coating as emery may be imbedded in the surface of the wire and cause trouble. Always form the element around the terminal in a clockwise direction. The end must be formed into a **U** and placed between the washers (Fig. 17). Be sure the end of the element wire

does not cross or lap over the element lead under the washers because the pressure of the terminal connection may cut partly through the element. The end beyond the washers can be wrapped around the element lead. Avoid having the cut end make contact with the element lead. Be sure the terminal nuts or screws are drawn down tight.

Fig. 17.

Oven Unit Terminal Pins and Receptacle Inserts. The oven bake and oven broil units on most electric ranges are removable. Terminal pins on the heating unit and a receptacle block in the rear of the oven are used to obtain a "plug-in" electrical connection. The terminal contact pressure which can be felt when removing or installing an oven unit is necessary to prevent burning and pitting of the terminal pins and inserts. Defects in the oven unit terminal pins and receptacle inserts that may be noticed by the user include: an oven unit that is difficult to remove;

an element wire failure at the terminal connection; erratic oven temperature regulation or failure of the oven unit to heat. If inspection of the terminal pins indicates arcing or burning, both the terminal pins and receptacle inserts must be replaced. Terminal pins and receptacle inserts are nickel-plated to increase electrical conductivity and resist rust and corrosion.

Oven Thermostat. The primary function of the oven thermostat is to regulate oven temperature. Since the oven units are of fixed wattage connected to a constant voltage supply, temperature regulation is obtained by the action of the thermostat in opening and closing electrical contacts in series with the oven unit electrical supply line. The thermostat used on most ranges is of a hydraulic type. It consists of two basic parts: the motive element and the switch (Fig. 18). The motive element is the mechanism

Fig. 18. Oven thermostat.

which expands and contracts with temperature changes and thereby operates the switch mechanism. The switch mechanism in turn controls the oven temperature by means of snap-acting electrical contacts.

Motive Element. The motive element consists of three parts: a container or bulb, a capillary tube, and an expansible device such as a bellows or diaphram (Fig. 19).

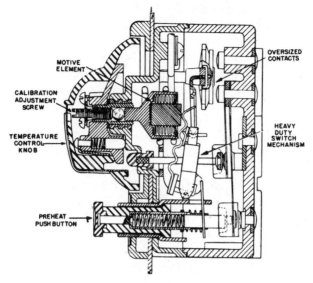

MOTIVE ELEMENT

CALIBRATION ADJUSTMENT SCREW

TEMPERATURE CONTROL KNOB

PREHEAT PUSH BUTTON

OVERSIZED CONTACTS

HEAVY DUTY SWITCH MECHANISM

Fig. 19.

All parts of this motive element are accurately controlled to obtain a temperature range of approximately 150° F. to 550° F.

Switch Mechanism. The switch mechanism consists of a combination of oversized contacts and a heavy-duty snap-action mechanism (Fig. 19) to contribute further to consistent and reliable operation of the thermostat. It is designed and built to provide the two necessary functions of carrying the current and breaking the arc. The switch mechanism is designed so that the contacts tend to perform a wiping action on closing. This action assures clean contacts for maximum current-carrying capacity without appreciable contact wear.

Operation of Oven Thermostats. In addition to its primary function (the automatic regulation of oven temperature) the oven thermostat may serve as the broil unit *On-Off* switch and as the switch for manual or automatic preheat. In broiling, the oven thermostat does not normally cycle. The broiling temperature is determined by the distance between the food and the broil unit rather than by the thermostat setting. The oven thermostat also serves as an over-temperature protective device.

There are a number of baking operations which require the oven to have a specified temperature before the food is placed in the oven. Manual preheat and automatic preheat are provisions within the thermostat which allow the oven to be brought up to the desired temperature more quickly by using both the top heat from the broil unit and the bottom heat from the bake unit during the heat-up period. Manual preheat is used when the oven thermostat must be manually switched from preheat to bake before the food is placed in the oven; an automatic preheat thermostat will switch from preheat to bake when the oven temperature reaches the temperature set on the thermostat dial. "Preheat" saves time and is used for all cooking operations except those for cooking from a cold oven start. The three types of thermostats generally installed on ranges are the Hart, the Wilcolator, and the Robertshaw.

The *Hart* thermostat (Fig. 20) is used for manual pre-

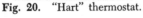

Fig. 20. "Hart" thermostat.

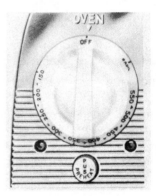

Fig. 21. "Wilcolator" thermostat.

heat. With the selector lever pointer at the center, set the thermostat dial to the desired temperature. Both oven units will heat the oven up to temperature quickly, whereupon the pilot light will cut "off." The selector lever may then be turned to the *bake* position and food placed in the oven. When cooking from a cold oven start, or using the small oven thermostat, set the selector lever pointer to *bake* and turn the thermostat dial to the desired temperature. When

Fig. 22.   "Robertshaw" thermostat.

the pilot light cuts "off," the oven is up to temperature. This pilot light then cycles *On* and *Off* as the oven requires heat, indicating an even baking temperature in the oven. To broil, point the selector lever at the right (broil) and turn the temperature dial as far past 550° as it will go.

The Wilcolator (Fig. 21) is a push-button automatic preheat thermostat. Push the preheat button and set the thermostat dial to the desired temperature. Both oven units will heat to bring the oven up to temperature quickly, the preheat button will "pop out," and both pilot lights will cut "off." The food may then be placed in the oven. The *bake* pilot light will cycle *On* and *Off* as the oven requires heat, indicating that an accurate, even baking temperature is being maintained. When cooking from a cold oven start, or using the small oven thermostat, turn the thermostat dial to the desired temperature. When the bake pilot light cuts off, the oven is up to temperature. To broil, turn the thermostat dial to the broil position.

The Robertshaw (Fig. 22) is an automatic preheat thermostat. Turn the thermostat knob from *Off* all the way around to *broil*, then immediately back to the desired temperature setting. Both oven units will heat until the oven is up to temperature and the pilot lights cut "off." Then food may be placed in the oven. The bake pilot light then cycles *On* and *Off* as the oven requires heat, indicating that an accurate even baking temperature is being maintained in the oven. When cooking from a cold oven start, or using the small oven thermostat, set the thermostat dial to the desired temperature. When the bake pilot light cuts off, the oven is up to temperature. To broil, turn the thermostat knob to *broil;* leave the oven door ajar.

Oven Pilot Lights. On ranges using the Hart thermostat, one pilot light is used to indicate that either a baking or broiling operation is going on in the oven. In a baking operation, this pilot light cycles *On* and *Off* in conjunction with the bake unit, and it is always *On* during a broiling operation because the broil unit does not normally cycle during broiling. Some ranges use separate pilot lights to indicate the baking and broiling operations. The bake pilot light cycles *On* and *Off* in conjunction with the bake unit. The broil pilot light is *On* during broiling and also is *On* with the bake pilot light during an automatic preheat period.

BAKING PROBLEMS

All baking problems fall into two general classifications: mechanical installation and adjustment of the range, and proper use of the range. All parts on the range must be in working order if it is to function properly; the thermostat must be in adjustment, the oven vent must be open, the oven door sealing at the top must have the proper $\frac{1}{32}''$ clearance at the bottom, and the range must be level at the oven racks. (A range must be level to obtain the best baking results. It is a mistake to assume that the kitchen floor is level or to level the range to the top units. Always level the range to the oven racks, as previously described.) A cake baked on an oven rack that tilts will not be of uniform thickness (Fig. 23). The batter flows to the low end of the pan and makes one end of the cake too thick.

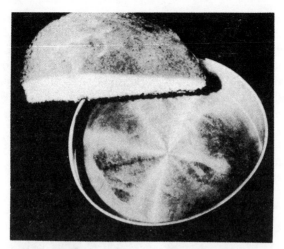

Fig. 23. Cake from uneven range.

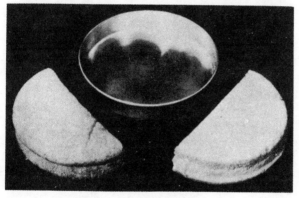

Fig. 24. Cake from standard pan.

Also, the flow of heat under the pan toward the high end browns the thin part of the cake too quickly.

Proper Use of the Range for Baking. The homeowner should be familiar with all parts of the range and their correct uses. Each part has its proper function. Any cake will be affected by a failure to move the thermostat selector lever to *bake* after preheating the oven. Not all thermostats have automatic preheat. On some ranges the selector lever must be switched manually from preheat to bake.

Utensils and Their Properties. One of the most important factors in baking is the use of suitable baking pans and tins.

Standard Pan. The cake shown in Fig. 24 was baked in a standard medium-weight aluminum pan. The sides of the pan are straight and highly polished, and the bottom has a slight dull sand-blast finish. The weight and finish of the material in this type of pan provide for the desired degree of bottom browning. Cake recipes are usually prepared for aluminum pans. If an oven type of glass utensil, or darkened tin, or enameled pan is used for cakes, the

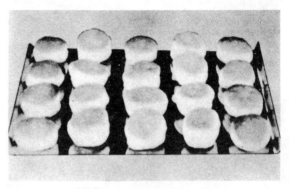

Fig. 25. Standard baking sheet.

oven temperature should be set 25° lower than called for in the recipe. The degree of browning varies with the capacity of the different pan materials to absorb and reflect radiant heat.

Standard Baking Sheets. The biscuits shown in Fig. 25 were baked on a standard baking sheet. Every other row has been turned over to compare the top and bottom browning. The sheet should be 4″ smaller than the oven to allow about 2″ all around for the circulation of heat. An oversized baking sheet will result in poor top browning and overbrowning of the bottom. The oversized sheet restricts the flow of heat to the top of the oven. The best results are obtained when a flat baking sheet of medium-weight aluminum is properly centered in the oven. Biscuits or cookies should not be baked in a utensil with sides. If the utensil has sides, use it upside down. Bake only one sheet of cookies or biscuits at a time.

Arrangement of Food in the Oven. Food should be

Fig. 26. Arranging four pans.

Fig. 27. Arranging two pans in small oven.

arranged in the oven in a central location so that the heat can circulate freely around the pans. The pans should not touch each other, or the sides, or the back of the oven. Figure 26 shows *four cake pans* arranged in the oven in such a way as to permit the free circulation of heat. Rack

Fig. 28. Arranging cookies.

Fig. 29. Complete oven meals.

positions 3 and 7, or 4 and 7, give convenient spacing between oven racks. All rack positions are counted from the bottom. Place *two cake pans* in the oven (Fig. 27) on one rack either on the fourth or fifth rack position or slightly below the center of the oven. If the range has six rack positions, use 2 or 3. For *biscuits* and *cookies,* place the baking sheet on the fourth or fifth rack position (Fig. 28). For *complete oven meals,* foods which will cook at approximately the same temperature and require the same cooking time should be selected. The foods should be placed to allow for the free circulation of heat (Fig. 29).

(*Note:* Consult Trouble and Remedy Charts for electric ranges on pages 386 and 387.)

TROUBLE AND REMEDY CHART

TROUBLE	CAUSE	REMEDY
Complete range inoperative.	Wall plug on cord may be loose in outlet.	Check plug for contact in outlet.
	Fuse in house circuit may be blown.	Check fuse; replace, if necessary.
	Wall outlet may be defective.	Check outlet with another appliance; replace, if defective.
	Open circuit in supply cord.	Check electrical continuity of supply cord; repair or replace.
	Defective oven control.	*Check circuit continuity through oven control, and repair.
	Open circuit in oven control lead wires.	*Check circuit continuity through oven control lead wires, and repair.
Cook-broil or bake unit inoperative.	Incorrect setting of switches.	Use correct switch settings as specified.
	Unit shorted or open circuited.	*Check circuit continuity of unit. Make necessary repairs.
	Circuit containing unit open circuited.	*Check continuity of all components of circuit. Make necessary repairs.
Pilot light does not light.	Lamp loose in socket.	Tighten lamp in socket.
	Lamp burned out.	Replace lamp.
	Lamp circuit open.	*Check continuity of lamp circuit. Repair, if necessary.
Door snaps open when oven is heated. Door sticks and is difficult to open or close.	Door catch not properly adjusted.	*Adjust door catch. Check the door spring adjustment. Adjust, if necessary.
Knobs loose.	Switch shaft compressed.	Spread shaft by inserting screwdriver in split and opening slightly.
Tiny hairline cracks form on porcelain enamel.	Wiping enameled surface with cold damp or wet cloth while enamel is still hot.	Wipe only when range is cold.
Electrical shock received when the body of the range is touched.	Some portion of the electrical circuit is grounded to the non-current carrying part of the range.	*Check range for grounds and correct.
Excessive smoking in oven while broiling.	Door closed while broiling.	Door should be left ajar.
	Placing fat meat too close to broil unit.	Check and correct.
	Not cutting edges of meat to prevent them from curling up.	Check and correct.
	Washing meat.	Wipe with damp cloth.
	Spillage in oven from last usage.	Clean after every usage.
Unsatisfactory baking results: Browning more on the bottom than on the top.	Pans dark on the bottom.	Dark pans absorb heat; correct.
	Not properly preheating oven before placing product in oven.	Check and follow baking procedures specified by the manufacturer.
	Pans with high sides or pans that are too large for the recipe.	Use correct utensil.

* Consult service department of manufacturer.

TROUBLE AND REMEDY CHART—*Continued*

Trouble	Cause	Remedy
	Product placed too close to the bottom of the oven.	Place on correct rack.
	Defective control staying on too long.	*Check; repair or replace.
Oven bakes more or less in back than in front.	Range or oven rack not level.	Level.
	Unclean oven (causing oven lining to absorb more heat).	Check and clean.
	Placement of pans.	Move large pans or sheets away from back of lining.
Cakes are uneven in size and browning.	Control out of adjustment (oven too hot causes cakes to brown in center and uneven rising).	*Check and correct.
	Range or oven racks not level.	Level.
	Warped cake pans or discolored pans.	Do not use.
	Control set too low (causes uneven browning and longer baking time).	Check and adjust.
	Control set too high (overheated oven gives porous and dry cake).	*Check and adjust.
	Not properly preheating oven when called for.	Check instructions.
Baking period too long even though thermostat control is correct.	Food placed in oven directly from refrigerator.	Bring food to room temperature.
	Proper oven temperature not used.	Check and correct.
Oven bakes too fast.	Control calibrated too high.	*Check and adjust.
	Line voltage extremely high.	*Check on range outlet and correct.
Moisture around oven door.	Improper door seal.	*Check the range for proper insulation, especially at the front frame of the oven. If faulty, make corrections. Check and adjust the oven door for proper seal to insure a close fit between the oven frame and door lining, especially across the top and down the sides as far as possible.
	Partially obstructed vent tubes.	Inspect the oven vent tube for obstructions (insulation or foreign objects) and clean.
Timer does not operate properly.	Faulty or loose connection.	Check and correct.
	Time-set shaft binding at the crystal of the timer.	*Check and correct.
	Burnt-out coil or rotor, or bent metal disc.	*Check and replace.
	Slight warpage of back splash plate.	*Correct.
No heat on any of the surface units.	Blown-out fuse.	Check and replace.
	Loose connection.	Check and replace.
	Broken wires.	Check, repair, and replace.
No heat on one of the surface units.	Broken lead wire to the unit.	Check and repair or replace.
	Grease on the lead assembly.	Clean and check leads.
	Loose connection at the switch.	Check and tighten.

* Consult service department of manufacturer.

CHAPTER 16

Automatic Washers

Successful operation of any automatic washers of the type shown in Fig. 1 is dependent largely upon the following factors: (1) An adequate supply of hot water (150° F.) is necessary. (2) The water heater must have sufficient storage capacity or hot water recovery to operate the washer through at least three complete cycles. Usually a fast-recovery gas-, oil-, or coal-fired heater having a 30-gal. storage capacity is adequate. If an electric or other slow-recovery heater is used, it must have a storage capacity of at least 60 gals. (An excessively low recovery or inadequate capacity may prevent continuous operation of the washer through a normal family washing.) (3) Hot and cold water lines must be equipped with individual shut-off faucets and should be located within 6′ of the left rear bottom corner of the washer. (4) Waste water outlet facilities may be a sink, set tub, stand pipe, or floor drain—properly vented and trapped to prevent disagreeable odors from entering the washer. If you use an extension cord to supply electricity to the washer, be sure the wire is of the gauge specified by the manufacturer.

Wiring should be checked to see that the addition of the washer will not overload the circuit. Any alterations to the plumbing or electric connections should be made prior to installation of the washer and in accordance with existing local plumbing and electricity codes.

INSTALLATION

It is not necessary to bolt the washer to the floor, and

it may be installed on any type of floor which is reasonably solid and level. (Two types of washer installations are shown in Fig. 2.) Choose a location that has separate hot and cold water faucets within 6′ of the left side of the washer where the inlet hoses are connected.

When a washer must be installed on a weak or "springy" floor, reinforcement may be necessary to prevent excessive

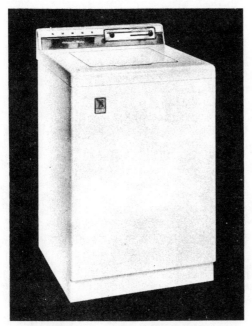

Fig. 1. Automatic washer.

vibration. Reinforcement can be secured by attaching a ¾″ plywood panel to the existing floor with screws (Fig. 3). The panel must be of sufficient length to rest on the floor over at least one joist on either side of the washer, and it must be attached firmly by screws, preferably to the floor joists. In cases of extremely weak floors, install a steady brace under the floor where the washer is to be placed (Fig. 3) by using post jacks (or wood supports on a firm footing). Any bracing used under the floor should

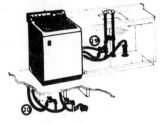

FIRST FLOOR INSTALLATION

BASEMENT INSTALLATION

Fig. 2.   Types of automatic washer
installations.

also support at least one joist under each side of the washer.
A laundry tray, sink, waste pipe, or other suitable outlet
should be provided into which the pump hoze nozzle can be

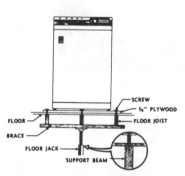

Fig. 3.   Installation of washer on
weak floor.

placed to discharge waste
water. The pump outlet
hose need not be placed
at any particular height
from the floor, since the
pump is not directly con-
nected with the basket
which contains the water
during washing and rins-
ing operations. Do not install the outlet end of the
drain hose more than six
feet above the floor on
which the washer is installed. (Figure 4 shows a rear view
of the washer.)

Plumbing Connections. Do not attach the drain hose

directly to the waste water drain line by means of a clamp or similar device as disagreeable odors and waste water may "back up" and enter the washer. Attach the hose to the pump in such a position that it will not kink or twist when the washer is placed in its final location. The end of the drain hose is molded into a curve and to prevent kinking the hose, the hose must be connected to the pump outlet with the curve toward the drain (Fig. 5). Secure the drain hose to the pump outlet with the clamp provided and place the nozzle over the sink, set tub, or drain.

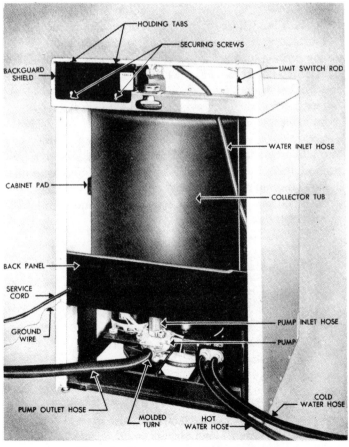

Fig. 4. Rear view of automatic washer.

To make proper hose connections, proceed as follows: Place one thin rubber gasket in each inlet hose connection,

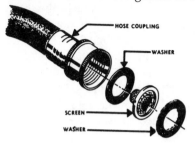

followed by a screen and a second thin gasket (Fig. 5). Place the screens at the faucet ends of the hoses. (They prevent foreign matter from entering the water mixing valve.) Place one thick gasket in the op-

Fig. 5. Hose washers and screen.

posite end of each hose and attach one hose to the cold water (or right) side of the mixing valve and tighten it securely. Attach the remaining hose to the hot water (or left) side of the valve and tighten it securely (Fig. 4). Attach hot and cold hoses to their respective faucets.

An extension of hot and cold water fill and drain hoses, (while not recommended) can be provided as follows: Use additional hoses as specified by the manufacturer and connect them with double male hose couplings. Assemble hoses in desired lengths by attaching a female hose coupling at each end and clamping the couplings securely. Standard screw-type clamps (Fig. 6) can be used successfully if

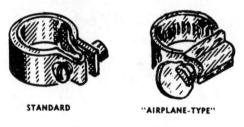

STANDARD   "AIRPLANE-TYPE"

Fig. 6. Hose clamps.

properly tightened. Clamps with thumb screws (Fig. 6) known as "Airplane Type" clamps are also used. When extending drain hoses, cut the drain hose supplied with the washer at approximately 3′ from the end so that the molded turn or the nozzle will not be disturbed. To install the extension, push a 6″ length of 1″ tubing into each end of the extension to a distance of 3″, and push the cut-off ends of the drain hose over the remaining 3″ section which protrudes from the extension. Clamping the hose to the tubing may not be necessary when short extensions are made. Standard hose clamps can be used in cases where long extensions create a pressure condition in the hose. Do not use a pipe nipple or other similar heavy material for a coupling. It will reduce the inside diameter of the hose, and will also create a "step" inside the hose which will

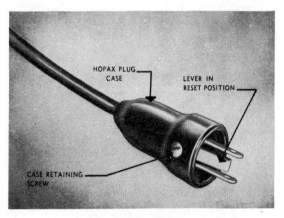

HOPAX PLUG CASE

LEVER IN RESET POSITION

CASE RETAINING SCREW

Fig. 7. Hopax plug.

accumulate lint and cause obstructions. When preparing the coupling, be sure that all burrs are removed from the ends to provide free passage of waste material.

Electrical Connections. An electrical outlet should be provided within reach of the service cord. The use of an extension cord is not recommended. At the start of extraction spin, the motor draws approximately 1300 watts; therefore, the circuit to which the washer is attached must not operate additional electric apparatus which draws power in excess of a total of 300 watts unless the circuit is equipped with a time-delay fuse. If the wiring is of sufficient current carrying capacity, a slightly heavier fuse

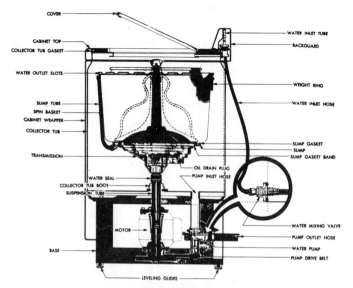

Fig. 8. Sectional view of washer and components.

Fig. 9. ⅓ h.p. motor.

can be used if allowed by the local electric codes. Connect the service cord; make sure that a "Hopax" overload plug is properly set by raising the red lever up between the prongs of the plug (Fig. 7). Attach the ground wire to the cold water pipe with a clamp and tighten it securely. Test the vibration limit switch by pushing the top of the agitator toward the left rear corner of the washer to release the switch. Reset the switch by centering the agitator and pushing down on the reset button on the backguard before starting the motor.

Instructions for Leveling. Washers must be leveled before being used. To level a washer, proceed as follows: Install one of the rubber pads (furnished by the manufacturer) on leveling glides (Fig. 8). Be sure the pad fits the glides properly. All the pads must be installed to prevent the washer from moving out of its position. Adjust the leveling glides to allow equal support at all corners of the washer. Test by "rocking" the washer diagonally from front to rear corners and adjust the glides until all movement is eliminated. Tighten the lock nuts to prevent the glides from turning in the base. Check the space between the washer cabinet and the floor. The cabinet must clear the floor at the lowest point. In cases where the floor has a decided slant, it will be necessary to raise one corner or side to level it. All glides should be turned in as far as possible to maintain proper cabinet clearances and effect reasonable leveling of the washer. The lock nuts must be tightened securely. The washer should be leveled in both directions for proper operation. The basket (Fig. 8), being mounted on rubber, has a tendency to "hang" toward the low side and will not be centered in the collector tub unless the washer is leveled. In cases where the front of the washer is low, this "off-center" position will cause some water to flow outside the basket in both fill and rinse periods. It may also delay operation of the vibration limit switch, under an unbalanced load condition. In cases where the washer is low at the back, the limit switch may stop the washer during the normal spin period.

COMPONENTS AND OPERATION

Figure 8 shows a sectional view of washer and components.

Washer Motor. The motor (Fig. 9) is a special ⅓ h.p.

reversible type fitted with a capacitor and protected by a "Hopax" plug at the receptacle end of the service cord. It operates in two directions: counterclockwise when the transmission is in operation, and clockwise when the basket is spinning to extract water from the clothes. The direction is determined by observing the pulley travel from above. The wattage drawn by the washer during the wash cycle is as follows:

Fill	20— 30 watts
Wash	300— 400 watts
Spin (start)	1250—1350 watts
Spin (running)	225— 275 watts
Agitator overflow	300— 350 watts
Spin rinse	235— 285 watts

Washing Operation. There are very few special washing factors necessary for successful operation of automatic washers. The printed instructions furnished with the washer, however, should be closely observed.

The cycle time and approximate water consumption of the automatic washer shown in Fig. 1 are as follows:

CYCLE TIME	APPROX. WATER CONSUMPTION
4 min.—Fill (adjust 3 to 6) (wash water).	11 gals. hot (or warm).
10 min.—Wash.	
½ min.—Off.	
1 min.—Extract.	
½ min.—Spin rinse (rinse water).	1⅜ gals. warm.
3½ min.—Fill (rinse water).	11 gals. warm.
4 min.—Agitated overflow rinse (rinse water).	11 gals. warm.
1 min.—Agitated rinse without overflow.	
½ min.—Off.	
1 min.—Extract.	
½ min.—Spin rinse (rinse water).	1⅜ gals. warm.
6 min.—Extract.	
32½ min.	Total ......{ 11 gals. hot. 24¾ gals. warm.

LUBRICATION

The transmission of an automatic washer is charged at the factory with 1 gal. of oil, and it is not necessary to oil it at regular intervals. See that the transmission is filled with 1 gal. of new oil whenever it is being overhauled (as a result of normal use) by the service department of the manufacturer.

(*Note:* Consult Trouble and Remedy Charts for automatic washers on pages 398-400.)

TROUBLE AND REMEDY CHART

TROUBLE	CAUSE	REMEDY
No electric power to motor.	Fuse blown.	Replace with time delay fuse or correct circuit for overload.
	Loose or broken wiring harness.	Replace or make proper connections.
	"Hopax" plug open from overload.	Reset by raising red lever until click is heard.
	Vibration switch open.	Redistribute clothes load and reset switch.
	Inoperative cycle control.	*Repair or replace cycle control.
Motor has power but will not turn in either direction.	Tight pump impeller due to foreign object in pump.	*Remove obstruction and free impeller.
	Foreign object lodged in pulley belt groove.	*Remove object and adjust belt tension if necessary.
	Excess end play in motor shaft keeps starting switch open.	*Remove motor for necessary repairs.
Motor has power but will not spin basket.	Brake solenoid burned out.	*Replace solenoid.
	Brake latch misaligned or binding.	*Replace latch or suspension leaf assembly or both. Check brake band for rough spot on holding tabs.
	Loose belts.	*Adjust tension.
	Cycle control points not contacting.	*Adjust points or correct cycle control.
	Collector tub full of water due to plugged or inoperative pump.	*Drain tub and correct pump.
Hot water will not enter tub.	Hot water not available.	Provide hot water source.
	Water temperature selector switch wires reversed.	*Check leads and make necessary repairs.
	One or both water shut-off faucets closed.	Open both supply faucets.
	Hoses reversed at mixer valve or supply faucets.	Change as necessary.
	Wires reversed on mixer valve solenoids.	*Check leads and correct.
	Water temperature selector switch defective or disconnected.	Replace switch. Check connections of wires.
	Screens plugged (faucet or mixer valve).	Clean screens and replace if necessary.
Agitator will not oscillate.	Drive block loose on shaft.	*Replace drive block if necessary and securely tighten.
Rinse water too hot.	Wires reversed at mixer valve.	*Check leads and correct
	Thermostat element in mixer valve defective.	*Change mixer valve.
	Hoses reversed at mixer valve or supply faucets.	Change as necessary.
Rinse water too cold.	Dirt, corrosion, sand, or lime deposit around mixer valve piston.	*Remove piston and clean.
	Hot water supply exhausted.	Correct hot water source.
Circuit breaker opens.	Pump plugged and outer tub filled with water.	*Remove obstruction.
	Brake solenoid burned out or disconnected.	*Replace solenoid or connect wires.
	Defective circuit breaker.	*Change assembly.
Noise, rapid knock, or rattle.	Loose set screw in brake hub, motor, pump, or drive pulley.	*Tighten set screws securely.

* Consult service department of manufacturer.

TROUBLE AND REMEDY CHART—*Continued*

Trouble	Cause	Remedy
	Pump bushings worn or excess vertical play in impeller shaft.	*Replace bushings or adjust end play.
	Flexible coupling shaft loose in support beam bushing.	*Replace bushing or shaft, or both.
	Basket drive tube loose in suspension leaf hub.	*If parts are scored or damaged, replace the suspension leaf assembly and drive tube, and check the lower water seal for leak.
Slow knock in wash or agitator rinse.	Loose transmission gears or linkage.	*Check and make necessary repairs or replacements.
	Vertical motion in agitator shaft.	*Remove end play by adjusting thrust collar.
Basket turns counterclockwise during agitation.	Brake latch spring broken or weak.	*Replace spring.
	Brake latch disconnected, unhooked, or worn.	*Reassemble and align latch. Replace broken pin, clip, or other parts.
	Key broken or missing at lower end of drive tube.	*Replace key or brake hub assembly.
	Cycle control inoperative or short circuited.	*Check terminals; repair or replace as needed.
Motor operates, but basket will not spin.	Broken clutch spring.	*Install new spring.
	Motor rotation reversed.	*Check wiring harness against wiring diagram at motor leads.
	Outer tub flooded with water. Pump inoperative, hose kinked or plugged.	Remove restriction.
	Loose pump belt.	*Adjust tension.
Washer moves on floor.	Rubber pads missing or in bad condition.	Replace all pads and readjust feet.
	Not resting on four feet.	Adjust leveling feet.
Agitator operates while basket spins.	Key at lower end of drive tube broken or missing.	*Replace key or brake hub assembly.
	Broken clutch spring.	*Install new spring.
	Defective cycle control.	*Change or correct cycle control.
Washer vibrates.	Floor too weak for efficient operation.	Reinforce floor.
	Transmission drive shaft bent.	*Straighten shaft or replace assembly.
	Washer not resting squarely on all four feet.	Adjust feet.
	Motor mounting rubber pads hardened from age.	Replace all rubber pads.
	Worn or high spots on pump or transmission belt.	*Replace belt.
	Pulley "out of round" or loose set screw.	*Replace pulley or tighten set screws.
	Motor shaft bent.	*Remove motor and have shaft straightened by motor repair shop.
Basket "hangs" off center in outer tub.	Washer not level.	Level washer in two directions.
	Suspension spring broken.	*Replace spring.
	Suspension leaf bent.	*Replace suspension lead and pad assembly.
	Suspension leaf tube not square with leaves.	

* Consult service department of manufacturer.

TROUBLE AND REMEDY CHART—*Continued*

Trouble	Cause	Remedy
Washer will not start when reset button is pushed down.	Upper "barrel" clip out of position or spring broken.	*Adjust the barrel clips on the reset rod. The distance from the bottom of the rod to the working surface of the barrel clip should be ¾" and 1⅜", respectively.
	Reset knob set too low on rod.	Raise knob by turning it counterclockwise; tighten lock nut.
Water leak from top of collector tub.	Gasket between tub and cabinet top out of position or broken.	*Position or replace gasket.
	Collector tub low at one corner.	*Raise collector tub by placing washer under corner or corners.
	Collector tub loose on base.	*Tighten all screws securely.

* Consult service department of manufacturer.

CHAPTER 17

Automatic Dishwashers

Dishwashers of the type shown in Figs. 1 and 2 are built into the sink unit or work surface, with a permanent drain and power and water connections. Portable units that require no installation are also manufactured. Both types are available in the 24″ or 48″ size. The permanently installed dishwashers have either a front or a top opening for loading and have either a cabinet design or an undercounter design. The dishes are cleaned by forcing hot water over them. A detergent is added to the water to remove food and grease. The dishes are rinsed with clear water and are dried by the heating element.

COMPONENTS

The five basic components of all types of automatic dishwashers are: an impeller; upper and lower loading racks; an automatic timer; a heating unit; and a timer motor.

Impeller. The impeller is usually located in the bottom center of the dishwasher tub (close to the bottom rack) with its two blades scientifically pitched for effective and thorough water action. It is connected directly to the motor shaft and is actuated by the impeller motor. A circular water diverter and the blades of the impeller direct the hot water upward for thorough washing of dishes in both racks.

Loading Racks. The two removable loading racks are constructed of steel wire, electro zinc plated, and coated with a durable resilient plastic. They are unaffected by hot water, grease, or any of the recommended detergents.

Timer. The timer automatically controls the flow of

Fig. 1. Automatic dishwasher.

Fig. 2. Another type of auto-matic dishwasher.

water and the operation of the impeller, the drain pump, and the heating element. It is actuated by the impeller motor. After completion of the washing, rinsing, and drying cycles, it automatically shuts off the dishwasher.

Heating Unit. The heating unit is rated at 1000 watts and operates during the washing, rinsing, and drying cycles.

Motors. The *impeller motor* is a $\frac{1}{3}$ h.p., 60-cycle a.c. motor and operates on 110-120 volts. The *timer motor*

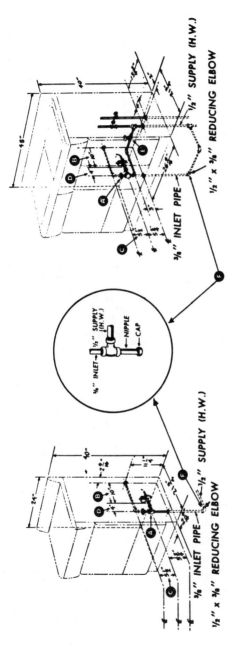

Fig. 3. Hot water inlet connections to 24″ dishwasher. **Fig. 4.** Hot water inlet connections to 48″ dishwasher. **(A)** Install the ⅜″ angle shut-off valve upside down if water inlet comes through wall. **(B)** 10″ from cabinet back to inlet side of water valve. **(C)** 5½″ height required to provide space for service replacement of inlet water valve solenoid coil. **(D)** 4″ maximum plumbing build-up provides access for service replacement of motor. **(E)** Shut-off valve can be installed at this location for convenience. **(F)** A ½″ x ⅜″ reducing tee with a nipple (approximately 4″ long) and cap may be installed at this point instead of the elbow. This will form a trap to catch foreign matter that can be removed periodically from the supply line by removing the cap.

is a $\frac{1}{30}$ h.p., 60-cycle, a.c. motor and operates on 110-120 volts.

INSTALLATION

Automatic dishwashers must have water supplied at a temperature of approximately 150° F. with a minimum pressure of 20 lbs. per square inch and a maximum pressure of 120 lbs. per square inch.

Plumbing Connections. All plumbing connections must comply with local plumbing code regulations. Copper tubing may be used if this conforms with local codes. Water connections should be made to the inlet valve of the dishwasher with $\frac{3}{8}''$ standard pipe fittings. The 4″ maximum

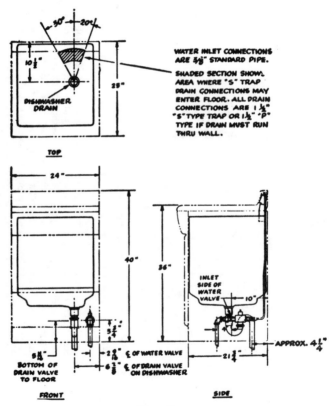

Fig. 5.

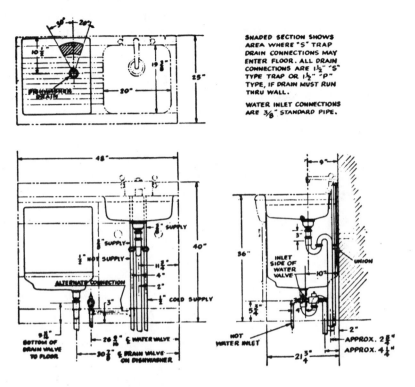

Fig. 6.

build-up of plumbing (Figs. 3 and 4) from the center line of the water inlet hole in the floor to the inlet side of the water valve should be held by the use of two ⅜″ close nipples, one ⅜″ union, and one ¾″ angle shut-off valve or one ¾″ elbow if the valve is installed in the basement. The location for the shut-off valve in a 48″ dishwasher-sink should be on the sink side of the cabinet if a water inlet line is built from the faucet hot water pipe (Fig. 4). With a 24″ dishwasher, the shut-off valve may be located inside the cabinet (Fig. 3) or in the basement.

Drain connections usually consist of 1½″ slip joints, using a "P" or "S" type of trap. In a dishwasher-sink combination, install the dishwasher drain and sink drain with independent connections to the soil pipe (Figs. 5 and 6). Before making connections, taper the ream pipes and remove loose chips or scale from the line. Do not allow white lead or other lubricant to get into the water valve when making final connections.

Electrical Connections. The only electrical connections to be made are those shown in Fig. 7. The dishwasher is completely wired, ready for connection to its power supply. It is designed to operate from a 115-volt 60-cycle a.c. power source. The current rating of the dishwasher alone is 16.5 amps. In connecting power to a dishwasher or dishwasher

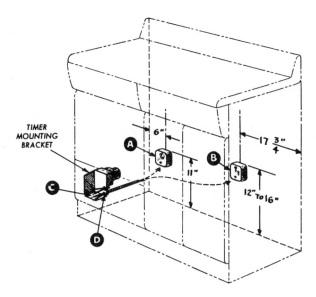

Fig. 7. Typical electrical connections. (A) Location of junction box for a 24″ dishwasher installation. (B) Location of junction box for a 48″ dishwasher-sink installation. (C) Connect 115 volts-60 cycle power across these leads. Connect white lead to grounded side of power supply. Solder connections and insulate each by wrapping with tape. (D) Power leads enter through hole in timer mounting bracket. Provide a strain relief clamp. Connections must be made in accordance with local electrical codes.

disposer combination, use a separately fused branch circuit with fuse ratings as follows:

Dishwasher unit alone .... 20 Amps.

Dishwasher plus disposer .... 25 Amps.

It is advisable that with a dishwasher-sink combination the junction box be located on the sink side of the dishwasher-sink unit, thus making provision for a disposer. When wiring a branch circuit to a dishwasher-sink, provide wire with additional capacity to handle the disposer load. A schematic wiring diagram of the dishwasher unit is attached to the back of the access panel.

Instructions for Installation. After making preliminary plumbing and electrical connections, install the dishwasher

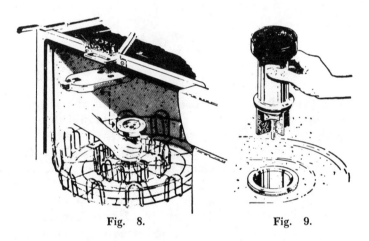

Fig. 8.   Fig. 9.

as follows: Place the upper rack into the carriage. Grasp the rack around the center, lift it upward and forward until the rim of the cup engages the flanges on the sides of the carriage, move it forward, and snap it into place (Fig. 8). Insert the drain plug assembly into the drain hole of the tub. Be sure the black cap and the sealing ring are in place (Fig. 9). Set the lower rack into position on the open door and roll it into the tub (Fig. 10).

Using special parts furnished with the appliance, proceed as follows: Fasten the sink (or top) panel to the frame with a sink clamp and a hex head screw (Fig. 11). Attach any adjoining base cabinet to a 24″ dishwasher with sheet metal screws (Fig. 12). For a 48″ dishwasher, attach any base cabinet to adjoin the left-hand side of the dishwasher.

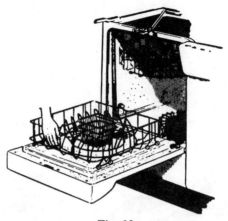

Fig. 10.

After the dishwasher is connected, leveled, and aligned with adjoining cabinets, use a No. 31 drill to drill through the dishwasher side panel (the left-hand panel on the 48″ dishwasher, both side panels on the 24″ dishwasher). Insert screws from the adjoining base cabinet side. The machine screw and the hex nut (Fig. 13) are used to fasten the right-hand side of the 48″ dishwasher to an adjoining base cabinet. Drill $\frac{7}{32}$ clearance holes for machine screws. Pull off the timer control knob and remove the screws at the bottom of the access panel in order to remove the panel. Place the dishwasher in its location and make

Fig. 11.

Fig. 12.

Fig. 13.

any remaining plumbing and electrical connections after leveling and aligning (shims may be necessary) the dishwasher with any adjoining cabinets. Run the unit through several cycles to test all connections. Replace the timer control knob and the access panel.

PROPER LOADING

The automatic dishwasher (Figs. 1 and 2) has been de-

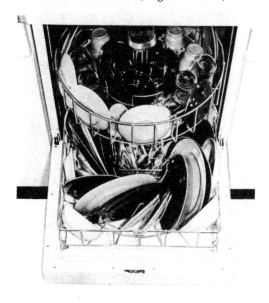

Fig. 14. Top rack loaded.

signed to hold all the dishes and silver necessary for serving six people.

Method of Loading Top Rack. When loading the top rack, pull the revolving rack forward (Fig. 14). Place small juice glasses in the innermost circle. If preferred, they may be hooked over the wire loops. Use whichever method works best. Each glass must be placed mouth down so that the water can reach inside and properly drain. Small sauce dishes fit into the next narrow circle. Place glasses and cups in the widest ring. Alternate them by two's if de-

sired, or you may place half the glasses on one side and
half on the other, with cups between the halves. Be sure
the cups and glasses are mouth down. Saucers and bread
and butter plates fit right into the outer circle (Fig. 14).
Throughout the washing cycle, the revolving rack turns
about 20 times a minute, with a swirling action that cleans
the dishes in the top rack. Every dish in the revolving rack
is exposed to the thorough washing action of the dish-

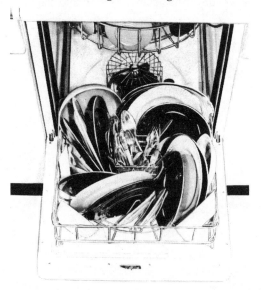

Fig. 15.   Bottom rack loaded.

washer. Even if a large bowl or a skillet in the lower rack
forms a partial shield, the revolving rack allows for complete cleaning.

Method of Loading the Bottom Rack. Pull the bottom
rack out on its rollers so that it rests on the open door.
Place dinner plates, salad plates, soup plates, and other
flat plates into the ring of plate spaces that surrounds the
silver basket. Alternate the large and small items to allow
for the best water action. Large dinner plates will aid
water distribution if they are placed at the corners of the
lower rack. Put deep dishes, mixing bowls, skillets, or pans

in the corner.spaces, tilted towards the silver basket so that water can reach inside and so that they will drain properly. Medium-sized platters can be placed along the sides of the lower rack (Fig. 15).

CYCLE OF OPERATION

The complete cycle of operation is shown in Fig. 16, and the automatic cycle is shown in Table 1.

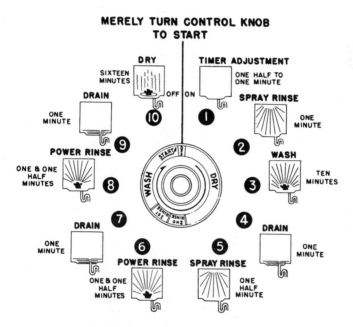

MERELY TURN CONTROL KNOB TO START

COMPLETE CYCLE 34 MINUTES, REQUIRES ONLY 6 3/4 GALLONS OF WATER

Fig. 16.   Cycle of operation.

Excess food soil should be removed by careful scraping and occasional rinsing. Provide plenty of hot water at a temperature of at least 150° F. If the water is too hard, use a water softener in the water supply system. Dishes must be loaded properly. When dishes are being washed once a day, rinse them to dampen and loosen soil. Never

TABLE 29A

AUTOMATIC CYCLE

Timer Adjustment (Fig. 16, 1)	To get timer to zero or starting postion. (a) Timer on. (b) Heater on. (c) Drain valve closed.	½ minute
First Spray Rinse (Fig. 16, 2)	(a) Timer on. (b) Heater on. (c) Impeller inoperative. (d) Drain valve open. (e) Inlet valve open, permitting 1½ gals. of water to be sprayed over the dishes, thus removing the larger particles of soil from the dishes and clearing the cold water from the supply line.	1 minute
Power Wash (Fig. 16, 3)	(a) Timer on. (b) Heater on. (c) Impeller operating; thus washes the detergent from the container and throws the water against the dishes at a high pressure. This force of water causes the upper rack to rotate throughout the power wash.	10 minutes
First Drain (Fig. 16, 4)	(a) Timer on. (b) Heater on. (c) Drain valve open, allowing water to drain from tub by gravity.	1 minute
Second Spray Rinse (Fig. 16, 5)	(a) Timer on. (b) Heater on. (c) Drain valve open. (d) Inlet valve open, permitting ¾ gallon of water to spray over the dishes for ½ minute, floating away any soil remaining in the tub.	½ minute
First Power Rinse (Fig. 16, 6)	(a) Timer on. (b) Heater on. (c) Drain valve closed. (d) Inlet valve open for 1 minute, allowing 1½ gals. of water to enter the tub. (e) Impeller on, forcing the hot water over the dishes.	1½ minutes
Second Drain (Fig. 16, 7)	(a) Timer on. (b) Heater on. (c) Drain valve open 1 minute.	1 minute

TABLE 29A—*Continued*

Second Power Rinse (Fig. 16, 8)	(a) Timer on. (b) Heater on. (c) Inlet valve open 1 minute only. (d) Drain valve closed. (e) Impeller operating.	1½ minutes
Third Drain (Fig. 16, 9)	(a) Timer on. (b) Heater on. (c) Drain valve open for 1 minute.	1 minute
Drying (Fig. 16, 10)	(a) Timer on. (b) Heater on. (c) Drain valve closed. (d) Impeller operates for 16 min., forcing warm, clean, dry air over the dishes. This air is drawn from beneath the tub and past the motor drive shaft.	16 minutes
	Total Cycle	34 minutes

Note: (By A & B in each operation) the timer and heater operate throughout the cycle.

The heater maintains approximately 160° water temperature throughout the entire cycle.

overload the dishwasher as it reduces the effectiveness of the water action. Silver is loaded with the handles down. Do not allow silver to "nest" and do not overload the silver basket. The recommended amount of detergent must be used. The indicator control dial should be set at *Off* at the beginning of a complete washing and drying cycle. Always turn the indicator control dial in a clockwise direction; if turned in a counterclockwise direction, the timer mechanism may be damaged.

The tub and racks of the dishwasher are self-cleaning. If necessary to give the interior an extra cleaning, add the proper amount of detergent and start the machine. When the rinse is completed, turn the control knob to *Off*.

(*Note:* Consult Trouble and Remedy Charts for automatic dishwashers on pages 414 and 415.)

TROUBLE AND REMEDY CHART

TROUBLE	CAUSE	REMEDY
Dishwasher inoperative.	Door handle not turned completely to *On* position.	Turn handle to right as far as it will go to engage strike assembly plunger. A slight click will be heard if the plunger is engaging the micro-switch properly.
	Strike assembly engages micro-switch, but machine still does not operate.	*Defective micro-switch. Replace.
	Loose connections.	Check and tighten.
	Timer will not run.	*Replace timer motor only. If it is still inoperative, replace complete timer. Also check for loose wires at the timer or micro-switch.
Dishwasher continues to run when handle is moved to *Off* position.	Strike assembly plunger stuck.	*Remove plunger and clean plunger as well as plunger hole. Inspect for burrs and bent plunger.
	Micro-switch shorted.	*Replace micro-switch.
Timer control knob not illuminated.	Bulb burned out.	Remove timer knob. Replace bulb.
Loud hum heard during drain cycle.	Solenoid arm twisted and binding on solenoid plunger.	*Check and realign arm to assure solenoid plunger is free operating and in perpendicular position to solenoid and bracket assembly. If solenoid arm is bent, replace.
	Solenoid and bracket assembly bent.	*Straighten bracket if possible or replace.
	Defective solenoid coil.	*Replace solenoid and bracket assembly.
Motor only will not run.	Loose wiring.	Check lead connection at motor and at timer. Use test light across motor terminals to see whether voltage is getting to motor.
	Foreign object makes it impossible for impeller to rotate.	*Remove impeller guard and foreign object.
	Motor burned out.	*Replace motor.
Insufficient water in tub.	Defective wiring.	*Check inlet valve solenoid connections and determine whether voltage is getting to solenoid by using test light.
	Defective solenoid coil.	*Check and replace if necessary.
	Water inlet valve filter screen clogged.	Remove filter screen and clean.
	Pipe line plugged by mineral deposits.	*Remove valve and clean pipe line.
	Low water pressure.	Check local water supply.
	Water inlet tube crimped.	Remove crimp if possible; otherwise, replace complete tube.
	Faulty timer.	*Check and, if necessary, replace timer.
Water will not shut off.	Bleed hole in inlet valve diaphragm clogged.	*Remove diaphragm and clean bleed hole.
Dishes not washed clean.	Not enough water.	See remedy for insufficient water in tub (above).
	Impeller turning too slow, tight motor bearings, defective motor windings, or low voltage.	*Replace motor.

TROUBLE AND REMEDY CHART—*Continued*

Trouble	Cause	Remedy
	Improper quantity of detergent.	Check directions specified by manufacturer.
	Water not hot enough.	Check water heater tank to assure water supply at approximately 150° F.
Water does not remain in tub.	Leaking drain valve.	*Check for damaged "O" ring and foreign particles around drain that hold it open.
	Improper timer operation (opens drain at improper time).	*Replace timer.
	Leaks around drain nut, motor body, and tube assembly, or heater terminals.	*Tighten nut at the leak or replace gaskets.
Water does not drain from tub properly.	Drain solenoid inoperative.	Replace solenoid.
	Improper timer operation.	Check and, if necessary, replace timer.
	Loose wiring to drain coil.	Check and tighten connections.
	Impeller guard legs over flange of drain plug to prevent it from opening.	*Move heater coil to left to its proper position and tighten. This permits guard legs to fit properly.
Dishes do not dry.	Heater element burned out.	*Replace.
	Water not hot enough.	Check water heater tank to assure water supply at 150° F.
Water leakage around door gasket.	Check for gasket damage.	*Remove and replace gaskets.
	Insufficient gasket compression.	*Adjust gasket forward.

* Consult service department of manufacturer.

CHAPTER 18

Disposers

The disposer shown in Fig. 1 is shipped as a completely assembled unit and requires only conventional electrical and plumbing connections. It can be installed in most sinks having a 3½″ to 4″ opening. A space 10″ x 16½″ is adequate in which to make the installation, plumbing, and electrical connections and to provide sufficient ventilation for the motor. The disposer is equipped for a slip-joint connection to any conventional 1½″ trap which conforms to local plumbing codes, and it can be installed in conjunction with an electric dishwasher. On long drain lines, separate traps should be provided for the dishwasher and the disposer.

ELECTRICAL COMPONENTS AND INSTALLATION

The electrical components of the disposer are a motor control or unit switch, a water flow control in some models, and a ⅓ h.p. split-phase drive motor. All electrical installations must conform to local electrical codes governing motors and equipment of this type. The power source must conform to the data on the motor rating plate. To make the necessary electrical connections proceed as follows:

Remove the combination name plate and switch housing cover. Connect the power line as shown in the wiring diagram (Figs. 2 and 3) applicable to the installation. Use an approved conduit fitting. Tape and solder all connections. Attach a suitable ground to the disposer. Fuse the circuit with a 20-amp. fuse. Place the disposer cover in the "drain" position and rotate clockwise to the "grind" position. The motor should start just before the handle reaches the "grind" position. If the cover fails to operate the switch, it will be necessary to make a minor adjustment.

Loosen the two hex screws holding the switch to the mounting bracket. Move the switch in or out until the proper setting is reached. Tighten the two hex bolts and try the actuating cover. Replace the combination name plate and switch housing cover.

PLUMBING INSTALLATION PROCEDURES

Before installing drain line connections to the disposer, inquire whether municipal authorities have approved garbage disposer installations for the community. All local sanitary practices, plumbing codes, and restrictions on vents and traps should be followed. Make a thorough survey of the drainage system of the house before starting plumbing installation. The system will probably conform to the plumbing codes, but your survey may disclose that modifiications are necessary for trouble-free operation of the disposer. For example, grease traps may be found in older drainage systems. Unless the local code requires such traps, they should be removed because they are sure to cause clogging. If a grease trap is required, clean the trap often. Thoroughly clean any old sections of drain lines, fittings, or house traps when installing the disposer.

Fig. 1.   Disposer.

Plan plumbing connections as shown in Figs. 4, 5, 6, and 7. The base of the disposer should be free to swivel horizontally 180° to accommodate existing installations or to facilitate new plumbing connections. You can make the

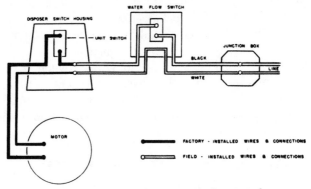

Fig. 2. Wiring diagram with flow switch.

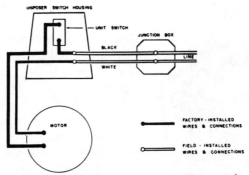

Fig. 3. Wiring diagram without flow switch.

base movable by loosening the six ¼″ nuts at the bottom of the top housing. The lower housing and motor can be shifted to suit the desired connection. If a full 180° turn is required, change the cable connection in the switch housing. The disposer requires a clean, correctly pitched drain line for proper operation. All horizontal runs should have a ¼″ to ½″ pitch per foot. Old lines must be mechanically cleaned before installation.

The following instructions should be followed to insure trouble-free operation of the disposer: Disposer branch drain lines should be a minimum of 1½″ in diameter, pre-

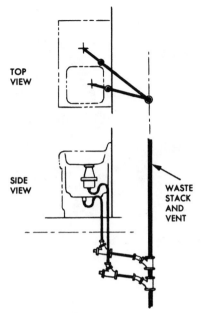

TOP
VIEW

SIDE
VIEW

WASTE
STACK
AND
VENT

Fig. 4. Suggested method of "S" trap
installation.

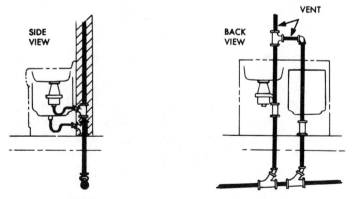

SIDE
VIEW

VENT

BACK
VIEW

Fig. 5. Suggested method of "P" trap installation.

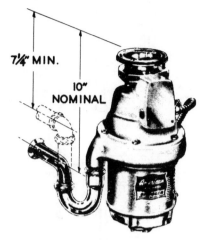

Fig. 6. Wall waste connection using "Hi-Boy" or "P" trap.

ferably 2″, and they must conform to local codes. The lateral drain line must be less than the 7¼″ minimum (Fig. 6). Use approved directional fittings. Do not use square T or X fittings. Bell and spigot fittings with re-

Fig. 7. Floor waste connection using "S" trap.

cessed threads are acceptable. Avoid grease interceptors whenever possible as they require frequent cleaning and are a source of trouble. Wherever possible, use a standard 1½" P trap. When a disposer is installed with a dish-

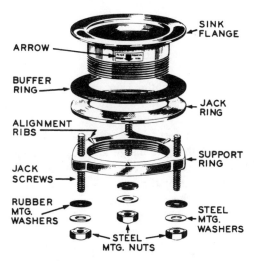

Fig. 8. Sink flange assembly.

washer, it is best to trap and vent each unit separately. Figures 4 and 5 show two suggested methods of installing the units.

GENERAL INSTALLATION PROCEDURES

After making electrical and plumbing connections, install the disposer as follows:

1. Loosen and remove the 3 ⁵⁄₁₆" mounting nuts, washers, and rubber mounting washers which hold the sink flange assembly to the top of the disposer.

2. Lift the sink flange assembly from the top of the disposer.

3. Loosen the three jack screws and unscrew the sink flange from the supporting ring. (A hex wrench for the jack screws is provided by the manufacturer.) The sink flange

assembly (Fig. 8) consists of a sink flange, a buffer ring, a jack ring, a support ring, three jack screws, three rubber mounting washers, three steel mounting washers, and three steel mounting nuts.

4. Prepare a cushion of plumber putty in the sink opening recess.

5. Insert the threaded end of the sink flange through the sink opening and seat the flange in the putty cushion. Align the arrow on the body of the flange with the front of the sink.

6. Place the rubber buffer ring and jack ring around the neck of the sink flange and against the bottom of the sink basin.

7. Screw the support ring onto the sink flange. The lower surfaces of the sink flange and support ring must be even. One of the three curved flats on the support ring must be to the front of the basin.

8. Align one of the three raised ribs on the top surface of the support ring with the arrow on the neck of the sink flange.

9. Tighten the three jack screws progressively and carefully to seat the sink flange in the putty cushion. The screws should be snug. Do not overtighten. Use the hex wrench to tighten the jack screws. Remove excess putty from the area around the sink flange.

10. Check to see that the assembly is level before proceeding with the installation.

11. Place the disposer in position below the sink flange assembly with the switch housing to the front. Raise the unit up to the sink flange assembly by means of the disposer jack (or blocks), forcing the jack screws through the holes in the rubber flange gasket.

12. Install the rubber mounting washers, steel washers, and nuts on the jack screws. Tighten the nuts carefully and uniformly to secure full bearing on the rubber flange gasket. Do not overtighten these nuts, for the resiliency of the rubber mounting will be destroyed.

13. Insert the cover in the sink flange opening and check its position.

The disposer must be suspended in a vertical plumb position from the sink basin to avoid vibration. There should be no strain on the waste outlet when the trap connection is being installed. If all the preceding instructions have been followed, make the final plumbing and electrical connections.

TROUBLE AND REMEDY CHART

Trouble	Cause	Remedy
Disposer fails to start.	Blown fuse in circuit.	Replace fuse.
	Disposer overloaded.	Remove overload. (Motor is equipped with overload protection switch. It will shut the motor off temporarily when overloaded, then reset itself.)
	Disposer jammed.	Check and make necessary adjustments.
	Defective switch.	Check and, if necessary, replace switch.
Disposer leaks water.	Leaking around mounting flanges and rings.	*Examine and tighten if necessary.
	Leakage around motor.	*Disconnect electrical connections, remove plumbing connections, and make necessary repairs and replacements.

* Consult service department of manufacturer.

CHAPTER 19

Dryers

—————————————•—————————————
•

The automatic dryer shown in Fig. 1 is designed and built for flush wall mounting in the kitchen or utility room.

COMPONENTS AND GENERAL OPERATION

The operating temperature of the dryer is controlled by a thermostat so that it does not exceed 160° F. This temperature is not attained until the clothes are almost dry. During the major part of the drying cycle, much of the heat is absorbed as water is evaporated from the clothes. The average operating temperature is between 125° F. and 140° F.

A one-pass system is used whereby the air moved by the blower is exhausted to the outside each minute, and an equal supply of fresh air is drawn in over the heating element and through the tumbling clothes (Fig. 2). There is a flow of approximately 120 cu. ft. of air per minute. The fresh air comes into the unit through an opening below the service door and moves around the junction formed by the top panel and the back panel bridge. As the air enters the back of the unit, it passes between the moving cylinder and the panels, thus keeping the panels cool. The air is then preheated, as it moves over the cylinder and the heater box, to increase the efficiency of the unit.

Heater Assembly. The heating element is rated at 4400 watts, 220 volts, and delivers approximately 15,000 B.T.U.'s per hour. The heater assembly is mounted in a double-walled box assembly consisting of a box frame and a reflecting shield. Screws, clamps, and nuts are used to hold the parts together. The box and the reflecting shield are so designed that a uniform supply of fresh air is drawn in over all sections of the element wire. This gives "even" heating and increases the life of the element. Only two leads are used with this heater: one is attached to the high-limit switch, mounted directly above the heater, and the other is attached to the solenoid switch. A door

Fig. 1. Automatic dryer.

switch placed in series with the solenoid switch interrupts the heater circuit whenever the door is opened, thus eliminating overheating.

Thermostat. Drying temperatures are maintained by a thermo-disc type thermostat mounted in the exhaust section of the blower housing. A gasket prevents the thermostat from contacting the blower housing. The thermostat is preset at the factory and must be replaced if it fails to function properly. Since the evaporation of water from the

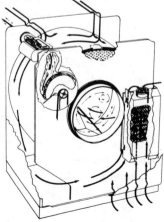

Fig. 2. Air circulation.

clothes keeps the operating temperature (usually between 130° F. and 140° F.) below the thermostat cutoff point, the thermostat comes into operation only during the last quarter of the drying time. The speed with which a dryer attains the thermostat cut-off temperature is controlled by the type and size of load, voltage input to the heating element, and amount of air circulated. All thermostats are designed to cut off at approximately 160° F. Temperatures can be checked by placing a thermometer in the exhaust stream at the rear of the unit.

Timer. The timer used has a three-hour maximum cycle which is necessary if the unit is connected to 100 volts. It exercises control over all the electrically operated parts of the dryer except the door light. Since the solenoid switch controls the major portion of the wattage, the timer is called upon to switch on only about 300 watts used in operating the motor and solenoid. A maximum of 75 min. automatic running time may be obtained by turning the dial clockwise until it reaches the stop tab. For a drying time of less than 10 min., turn the dial at least 10 min. into the cycle; then rotate the dial in the opposite direction to the desired time. This clocks the points and allows the current to pass on to the remaining controls. The timer is a spring-wound mechanism that is automatically activated when the dial is turned to the desired drying time. It is possible to turn the dial in either a clockwise or counterclockwise direction rapidly without damaging the spring motor. The maximum operating time of 75 min. may be increased to 180 min. by removing the timer dial and bending the metal tab.

Cylinder. The cylinder has a 28¼″ outside diameter and rotates clockwise at approximately 43 r.p.m. It has six short baffles that are staggered in order to alternate the wet clothes from front to rear, exposing all parts to the hot moving air. It is supported by a cylinder drive belt and two idler pulleys which are attached to the front plate. The front of the cylinder is a perforated sheet of steel which allows air circulation at all points. A ⅓ h.p., 115-volt, 1725 r.p.m., a.c. motor drives both the cylinder and the blower.

So that air will be directed into the cylinder without loss, a felt seal is attached to the outer circumference of the cylinder. It is crimped to a seal-retaining ring located directly under the 7⁄16″ flange extended from the perforated sheet steel front of the cylinder. The ring and seal are secured to the cylinder with six tabs.

In case the cylinder shaft becomes worn, it may be replaced by cutting off the heads of the rivets which attach it to a reinforcing plate. Remove the old shaft and plate and attach the new one with screws. The screw heads should be inside the cylinder and should be smoothed off carefully so as to leave nothing to catch the clothes.

In order to maintain a good seal and assure proper unit operation, it is necessary to space the cylinder properly in

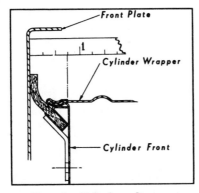

Fig. 3. Cylinder adjustment.

relation to the front plate. The cylinder can be moved forward or backward by means of the rear bearing. Loosen the bearing locknut and insert a screwdriver in the bearing slot; turn the bearing *clockwise* to reduce the cylinder spacing and turn the bearing *counterclockwise* to increase the distance between the cylinder front and the front plate. The proper spacing is ¾″ from the front plate to the end of the cylinder front crimp (Fig. 3). If it is necessary to move the rear bearing for adjustment, let the cylinder make at least two revolutions before taking a measurement. This will allow the cylinder and drive belt to position themselves properly.

To remove the cylinder assembly, it is necessary to remove the top panel. Open the service door to remove the cylinder drive belt. Also remove the front panel collar and felt-seal assembly. Remove the seven Phillips head screws and slip the collar and felt seal out of the front panel. Remove the back panel screws. Remove the cylinder shaft screw, flat washer, and lock washer. Loosen the bearing locknut and remove the bearing assembly while manually supporting the rear of the cylinder. Hold the back of the cylinder up with the left hand and tip the rear panel back until it clears the cylinder shaft. Then remove the cylinder from the unit by working the throat out through the felt seal which surrounds the opening in the front plate.

To reinstall the cylinder, turn the felt seal attached to the cylinder throat opening toward the front of the unit and slide the cylinder throat through the opening. Be sure to coat the shaft with lubricant. Reverse the disassembly procedure to complete the installation.

Top Panel. The top panel has a section at the rear that extends approximately 6″ above the top of the unit. The timer shaft extends through the top panel of the left side of this section, and the nameplate and escutcheon are attached to the raised section. The lint trap is located directly under the top panel on the left side.

To remove the top panel, first remove the timer dial and the chrome plug bottom on the right side of the escutcheon. Remove the screw and washer located behind the plug button; also remove the thin hex nut and special washer located behind the timer dial. Then pull the panel forward until the return flange on the front of the panel is free of the front panel flange. Lift up the front of the panel approximately 3″; pull forward and lift off. To replace the top panel, check the six short rubber gaskets to be sure they are all attached to the return flange of the panel (two in front and two on each side). Place the back of the panel on side panels approximately 6″ forward of the back panel bridge. Hold the front of the panel up about 4″ and slide it back until the timer shaft enters the hole in the panel and the door hinge is behind the lint trap. Lower the front of the panel until it is level and slide it into place. Secure it to the back panel bridge with the screw and nut and replace the dial and plug button.

Back Panel. The back panel is a functional part of the

unit, for it contains the cylinder rear support bearing. A heavy steel plate is welded to the inside of the back panel at the center and is drilled and tapped so that the cylinder-bearing assembly can be threaded through this hole. The bearing assembly, in turn, is attached to the cylinder by means of a screw, lock washer, and flat washer threaded into the cylinder shaft. By turning the bearing assembly in or out of the back panel, the cylinder can be properly spaced in relation to the front plate. A locknut is placed over the bearing assembly so that the proper position can be held.

Front Panel. The front plate acts as a cover for the front of the cylinder and is a mounting plate for the heater box, blower assembly, idler pulley shafts, solenoid switch assembly, speed reducer tongue, fuse receptacle, and belt switch. A felt seal is fitted into a ring groove around the cylinder throat opening to keep the air from leaking in or out. A felt seal attached to the cylinder front rides against the front plate to complete the air-seal on the cylinder. A flange has been formed on the section of the front plate near the left side of the heater-box opening in order to reduce heat loss from the heater box to the blower.

Blower and Pulley Assembly. A suction-type blower pulls air out of the cylinder and directs it along the inside of the blower housing to an exhaust opening which is slightly above the intake opening in the front plate. It is designed to move 120 cu. ft. of air per min. under normal load conditions. The wheel and shaft are dynamically balanced in order to hold vibration to a minimum. The blower wheel is assembled into an "oilite-type" bearing pressed into the blower housing. A belt leading directly from the motor furnishes power to drive the blower wheel at approximately 2500 r.p.m.

EXHAUST CONNECTIONS

Moisture-laden air, very fine lint, and heat are exhausted through a lint trap located under the top panel. These exhaust products should be vented to the outside, if possible. Several exhaust accessories are available from the manufacturer. If the dryer is not to be exhausted to the outside, a deflector can be installed in the lint trap.

Outside Exhaust. The dryer shown in Fig. 1 is equipped with a blower of sufficient power to force the exhaust products through a considerable length of duct work to the outside. This is highly desirable where the dryer is installed in a small utility room, a kitchen, or even a basement. The heat is sufficient to raise the temperature of an average kitchen by several degrees. Since the dryer exhausts well over 1 gal. of water per hr., it tends to cause excess moisture and condensation on walls and windows. All heat, moisture, and lint may be vented outside by attaching a 3″ diameter flue pipe to the dryer and extending it through an outside wall or window. This flue must be very smooth on the inside because any rough places tend to collect lint which will clog the duct and prevent the dryer from exhausting properly. The elbows must be smooth on the inside and have at least a 2″ radius on the inside bend. All joints must be made so that the exhaust end of one pipe is inside the next pipe. The addition of a vent pipe tends to reduce the amount of air the blower can exhaust but it does not affect the dryer operation if held within practical limits. Do not use more than four right-angle elbows and not more than 20′ of straight pipe. Two feet of straight pipe may be added for each elbow less than four used. If the vent passes through a wall, it should be placed in a metal sleeve of slightly larger diameter. This practice is required by some local codes and is recommended in all cases to protect the wall from possible discoloration due to the high temperature of air passing through the vent pipe. If the vent extends through a window, the pane of glass should be removed and a sheet metal or plywood plate put in its place. The vent should not exhaust directly below a window, for this will tend to "steam" the window under certain weather conditions. A deflector of some sort must be placed over the end of the vent to prevent rain and high winds from entering the vent when the dryer is not in use. The lint screen should not be removed from the lint box when a dryer is being exhausted. If the back of the dryer is against an outside wall, the exhaust pipe should be used on the outside to prevent cold air and dirt from blowing in.

Lint Trap. All material exhausted from this dryer is filtered through a lint trap. A removable screen placed inside the trap catches the lint and should be cleaned at the end of each drying day. The screen should be cleaned after each load if linty materials, such as chenille bedspreads or tufted throw rugs, are being dried. To remove the screen, first remove the top panel. Slip the lint-trap assembly back until it clears the mounting ring on the front plate and lift it up and to the right. If the mounting bracket on the lint trap interferes with the return flange on the left side panel, use a slight amount of force to spring the side panel away and allow the lint-trap bracket to clear the panel. Lift the front of the lint trap up and to the right until it is in the triangular area formed between the cabinet brace and the front plate, and remove the lint trap. It may be hard to align the two prongs on the latch with the two holes in the lint trap housing. This difficulty usually arises because the hinge, which is spot welded to the housing, is out of line. To correct it, grasp the front of the lint-trap cover (the end which contains the latch) and twist it slightly either to the right or to the left, depending upon the direction in which the cover is out of line.

MOTOR AND MOUNTING

The drive motor is located on the left side of the unit at the base. All service on the motor can be accomplished through the service door. The motor itself is a ⅓ h.p., 115-volt, 60-cycle, 1725 r.p.m., split-phase type. The two motors illustrated are the General Electric and the Delco, (Fig. 4) both of which have a starting switch. On the Delco motor, the terminal board, which is located on the pulley end of the unit, is protected by a cover plate. There are three terminal posts (Fig. 4). On the General Electric motor, the terminal board is located on the end of the motor opposite the pulley and is protected by a board. Be sure to connect the wires on either motor properly; otherwise, the solenoid switch will fail to energize or the starting winding will not shut off when the motor

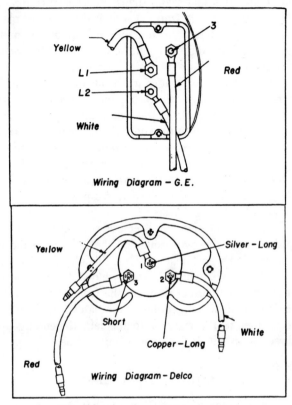

Fig. 4. Motor terminal connections.

reaches 1350 r.p.m. A Delco motor cannot be substituted for a GE motor unless the three motor lead wires are made longer. This must be done because the terminal board on the GE motor is on the rear side at the end opposite the pulley.

A hinged mount is used to secure the motor to the base. A mounting bracket is held to the motor by means of two latches. Note that the oil holes should always be in the upright position. The counterpart of the bracket is welded to the base of the dryer. The two brackets are joined together with a long hinge pin. Thus, the motor is free to pivot and give the blower belt proper tension. There is also a spring which hooks onto the front mounting latch and runs to a small hole in the front plate directly above the motor. This spring relieves some of the weight of the motor on the blower belt. To remove the motor, pull the hinge pin, disconnect the three motor lead wires, and remove the spring.

Motor Pulley. The motor pulley is made of cast iron and has two diameters to give the desired speeds to the cylinder and the blower. The small diameter drives the cylinder 43 r.p.m., and the larger diameter runs the blower 2500 r.p.m. The pulley is fastened to the motor shaft by means of two set screws secured at right angles. These screws must be tightened to insure quiet operation.

Belts. Three belts, specially made to withstand high temperatures, are used to operate the speed-reducer pulley, cylinder, and blower. Blower belt tension is controlled by the weight of the motor which is on a hinged mount. Cylinder belt tension is controlled by attaching a pair of springs from the speed-reducer pulley shaft to the base. The springs pull the speed-reducer pulley downward to tighten the cylinder belt and toward the right to tighten the speed-reducer belt. Worn spots on any of the belts will cause noisy operation; worn spots on the cylinder belt will cause the cylinder to bounce up and down excessively. The belts can usually be aligned properly by adjusting the motor pulley. This also aligns the motor-to-reducer belt. The blower belt can be replaced by opening the service door and removing the belt from the motor pulley. To replace the motor-to-reducer belt, remove the blower belt and the front belt tension spring on the speed-reducer pulley. Remove this belt slowly so that the speed-reducer tongue does not drop out of the left-hand angle bracket. Also check the belt-switch arm to make sure that it stays behind the tongue. Replace the belt, suspension spring, and blower belt. To remove or replace the cylinder belt, remove the speed-reducer pulley, then slip the belt off the cylinder throat.

To reassemble, place the cylinder belt around the throat of the cylinder. Work the belt between the two idler pulleys. Place the bottom of the belt around the small pulley on the speed-reducer pulley. Put the pulley on the shaft by removing the tongue from the left bracket. Bear down on the pulley until the left end of the tongue re-enters the bracket. At this point note the position of the belt switch. Make certain that the switch arm is behind the tongue. Check the position of the belt switch from time to time during reassembly. Replace the motor-to-reducer belt. This will hold the speed-reducer tongue in place as the rear belt tension spring is secured to the base. Replace the flat washer and the front belt tension spring. Again check the position of the belt switch. Replace the front panel collar and felt-seal assembly. Start the seven screws before tightening. Tighten the screws holding the doorlight window if they have been loosened. Replace the top panel and timer dial.

Idler Pulley. The two idler pulleys act as rollers which support the front end of the cylinder and allow it to turn freely: For quiet operation, the cylinder drive belt is run between the pulleys and the cylinder collar. Thus, the cylinder actually turns on a rubber surface. The current-production pulley is made of a special plastic compound with an "oilite" bearing molded into the hub.

Replacement procedures are as follows: Remove the top panel. Remove the seven Phillips head screws mounting the front panel collar to the front panel. The collar and seal assembly will now turn freely. Before removing this assembly from the unit, loosen the two screws securing the plastic doorlight window. The entire collar and felt assembly can now be taken from the unit; compress the felt as it is worked through the door opening. Remove the belt tension spring and flat washer located in front of the speed-reducer pulley. Remove the motor-to-reducer belt to allow the speed-reducer tongue and shaft to drop down toward the right. Unhook the rear belt tension spring from the base of the unit. The pulley can now be brought forward and worked off the shaft. Either idler pulley can be removed. Remove the hairpin clip and flat washer in front of the pulley. Lift the cylinder throat to relieve the pressure from the pulley and, at the same time, pull the pulley forward until it is free of the shaft. Behind the pulley there is another flat washer and then a spring washer to take up end-play.

Before reassembling the idler pulley, put a coating of lubricant on the shaft. Replace the spring washer, the flat washer, and then the pulley. Install the pulley right-side out so that the outer edge is at a maximum distance from the front plate. Lift the cylinder throat so that the pulley can be placed on the shaft. Replace the large flat washer and hairpin clip. Push the belt toward the rear of the pulley so that it will not slip off the pulley when the unit is started. Place the cylinder belt around the small pulley on the speed-reducer pulley. Put the pulley on the shaft by removing the tongue from the left bracket. Bear down on the speed-reducer pulley. Put the pulley on the shaft bracket. Note the position of the belt switch. Make certain that the switch arm is behind the tongue. Check the position of the belt switch several times during reassembly. Now replace the motor-to-reducer belt, which will hold the speed-reducer tongue in place as the rear belt tension spring is secured to the base. Next, replace the flat washer and the front belt tension spring. Recheck the position of the belt switch. Also check the position of the cylinder belt to make sure it is in line with the pulleys all the way around the cylinder throat. Replace the front collar and felt-seal assembly. Get the seven screws started before tightening any of them. Tighten the screws holding the doorlight window. Replace the top panel and timer dial.

WIRING CONNECTIONS

Every dryer should have a separate circuit from the meter board or fuse box because the unit draws a heavy

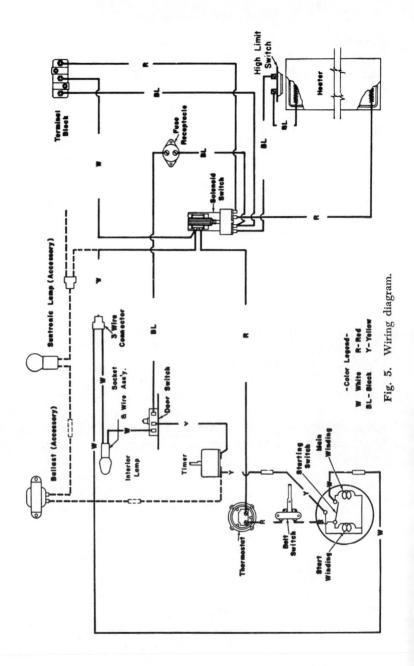

Fig. 5. Wiring diagram.

—Color Legend—

W White   R—Red

BL—Black   Y—Yellow

load. The wiring should always be heavy enough to deliver the necessary current with no appreciable voltage drop. Do not use wires smaller than No. 10. Both sides of the circuit should have 30-amp. fuses. Recommended wire sizes are as follows: zero–60', No. 10 wire; 60'–100', No. 8 wire; over 100', No. 6 wire. Special heaters are available for unusual voltage conditions. The unit should be properly wired and grounded in accordance with local and national electrical codes.

When hooking up the electric dryer for 110-volt operation, it is necessary to move the red wire on the terminal block to the center post which also carries the white wire. The 110-volt line should be attached to both the terminal post carrying the black wire and the terminal post carrying the white and red wire. No further changes need be made when installing this unit for 110-volt operation. A separate 110-volt circuit should be obtained, and it should be fused for no more than 15 amps. The same heating element is used for both 110-volt and 220-volt operation. The unit may be wired directly to the fuse box with No. 12 wire or it may be plugged into a separate 15-amp circuit. A power cord obtained from the manufacturer (or its equivalent) must be used if the unit is to be plugged into an existing power outlet. Extension cords are not recommended; however, if used, they should be at least No. 12 wire. In some areas it is permissible to use an electric range cord as a connection from the wall plug to the dryer when operating on 220 volts. In other cases, conduit, Romex, or BX cable can be used.

The terminal block is located in the upper right-hand corner at the rear of the unit. When connecting wires only to the terminal block, the top panel need not be removed. Conduit, Romex, BX cable, or range cord may be connected to the terminal block. Be sure to match the color coding of the wires on the terminal block; the center connection is "ground." A decal on the back of the unit states the proper wire sizes for 220-volt installation. A diagram pasted to the front plate can be seen by removing the top panel.

ELECTRICAL CIRCUITS

The electrical circuits (Fig. 5) on this dryer are arranged to provide the utmost protection and economy in operating the unit. A 15-amp. fuse is located on the front plate directly above the heater box. The fuse receptacle extends in front of the plate, but the fuse is screwed into the receptacle from the rear of the plate. Removal of the top panel provides access to the fuse. It is a standard-type fuse similar to that used in most house fuse boxes and is placed in the black wire that leads from the solenoid switch terminal to the door switch (Fig. 5). This is the "hot" lead supplying power to the main motor, interior lamp, and solenoid switch. It does not fuse the heating element. If this fuse blows, it shuts off the entire unit by interrupting the flow of power to both the main motor and the solenoid which controls the power to the heating element. By wiring the solenoid through the motor starting switch, it is possible to start the main motor (which requires considerable current during the starting process) without having the heating element turned on. If both were to start at the same instant, there would be danger of blowing a house fuse, especially if some other load were on the fuse at the same time.

The starting switch in the motor is a spring loaded, centrifugal-type which is *On* from the time the motor is turned on until it reaches approximately 1350 r.p.m. At this time it turns *Off*, thus interrupting the flow of power to the motor starting winding.

TROUBLE AND REMEDY CHART

TROUBLE	CAUSE	REMEDY
Unit runs but does not heat.	Insufficient voltage at terminal block.	*Check with neon test lamp on black and white terminals. If the lamp lights, test on red and white terminals. If the lamp does not light at this point, 220 volts are not being delivered to dryer. Check the main fuse. If voltage is satisfactory, faulty operation may be due to other causes.
	Burned or open solenoid cord.	*Test and replace if necessary.
	Thermostat contacts open.	*Check for loose connections. Repair, or replace if necessary.
	Belt switch contacts open.	Same as above.
	Motor starting switch contacts fail to open.	*Check. Repair, or replace if necessary.
	Broken wire in heating element.	*Replace element.
Unit does not run.	Insufficient voltage at terminal block.	Same as remedy for trouble listed above: Unit runs but does not heat.
	Blown fuse.	Check and replace. Ascertain cause.
	Door switch contacts open.	*Check and repair.
	Defective timer.	*Check and replace.
	Burned winding in motor or broken starting switch.	*Check and replace.
Slow drying.	Drop in voltage caused by inadequate size of the connecting wire from fuse box to dryer.	Check and replace with size of wire recommended by manufacturer.
	Insufficient voltage at junction box.	*Test while unit is in operation; ascertain cause of insufficient voltage and correct.
	Air leak.	*Check all felt seals, especially cylinder felt against front plate. Replace if necessary.
	Lint cloggage.	Check the lint trap and clean it.
	Bound blower.	*Check. Disassemble, clean, and re-oil.
	Improperly set or faulty thermostat.	*Check. Reset, or replace if necessary.
	Overloading.	Load only as specified by manufacturer.
Noisy unit.	Vibrating blower.	*Replace.
	Loose service door.	Tighten.
	Loose cabinet or mounting screws.	Tighten.
	Lumpy belt.	Replace.
	Solenoid chatters or stutters.	*Check to ascertain cause, and make necessary adjustments.
Unit will not shut off.	Jammed or defective timer.	*Check. Make necessary adjustments or replacement.

* Consult service department of manufacturer.

CHAPTER *20*

Refrigerators and Freezers

The prime function of a refrigerator (Fig. 1) or a freezer (Fig. 2) is to remove heat from the space to be refrigerated. Any substance absorbs heat while changing from a solid form to a liquid form. Ice melts in an ice box because it absorbs heat from the food and from the interior of the food storage compartment. Likewise, any liquid when changing from a liquid form to a vapor form absorbs heat. For example, the water in a tea kettle over the flame of a gas burner absorbs heat from the gas flame until the temperature of the water rises to 212° F., after which the continued application of heat changes the water to steam or vapor without a change in temperature. If a liquid were to be placed in an open container and the temperature of the surrounding air were higher than the boiling point of the liquid, the liquid would boil. For example, if a pan of water were placed in a room in which the temperature was over 212° F. (the boiling point of water), the water would boil and change to vapor. The heat required to bring about this change (latent heat of evaporation) would be absorbed from the air in the room, resulting in a reduction in room temperature. This principle is used in all refrigeration and freezing processes.

Heat is absorbed and removed from the interior of the refrigerator or the freezer by means of a suitable refrigerant. A refrigerant must have a boiling point much lower than 212° F. to be used in the preservation of food. Freon (F-12), a freon which is widely used as a refrigerant, has a boiling point of 22° below zero (−22° F.) If,

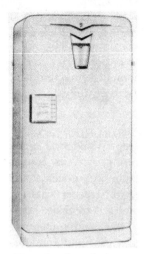

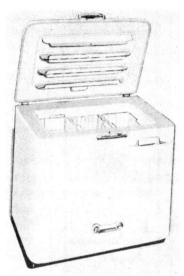

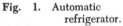

Fig. 1. Automatic refrigerator.

Fig. 2. Freezer.

instead of the pan of water, a pan of liquid freon were used, a room temperature much lower than 212° F. could be obtained. Since the boiling point of freon is −22° F. at atmospheric pressure, heat could be absorbed from the air in the room by the "evaporation process" until the room temperature fell to −22° F. If a constant supply of liquid freon were on hand to replenish the liquid freon that had evaporated or boiled away, the room temperature would eventually drop to 22° below zero (−22° F.)

In refrigeration, the evaporator or freezing compartment absorbs heat from the space to be cooled. If refrigerants such as freon were inexpensive, the evaporator or freezer compartment alone would be sufficient for refrigerating purposes inasmuch as more refrigerant could be added to replace the amount which, in boiling, changes to vapor and escapes into the outside air. The liquid freon in the open jar (Fig. 3) will boil violently and change to a heat-laden vapor as it absorbs heat from the interior of the cabinet. Thus, heat is removed from inside the refrigerator or freezer and carried to the outside air by the escaping vapor. Refrigerants, however, are expensive. It would be neither economical nor practical to refrigerate by constantly re-plenishing the refrigerant. Thus, the condensing unit of all mechanical refrigeration units functions to liquefy (by compressing and cooling) the heat-laden refrigerant vapor so that it can be used repeatedly.

Fig.   3.

REFRIGERATING SYSTEMS

Refrigerating systems consist of the following components: the refrigerator or freezer cabinet, the evaporator (freezing compartment), the condensing unit (motor-compressor and condenser), and the capillary tube.

The *refrigerator* or *freezer cabinet* is the insulated space in which the lower temperature required for preserving or freezing food can be maintained. It is insulated to retard the rapid transfer of heat into its interior.

The *evaporator* contains the boiling liquid refrigerant (Fig. 4). In a freezer it is the tubing which encircles the food liner (Fig. 5). The refrigerant boils inside the evaporator and thus absorbs heat from the walls of the evaporator or freezer liner which in turn, absorbs heat from the warmer air in the cabinet. This reduction in air temperature reduces the temperature of the food stored in the cabinet. Because metals are usually good conductors of heat and are capable of rapidly transferring (by conduction) the heat from the cabinet to the colder boiling refrigerant, all evaporators are made of metal. The metal evaporator is connected to the condensing unit by two pipes, or tubes. One is the liquid supply line (capillary tube) which carries the liquid refrigerant to the evaporator; the other is the suction line which carries the heat-laden refrigerant vapor back to the condensing unit.

The *condensing unit* consists of two main parts: the

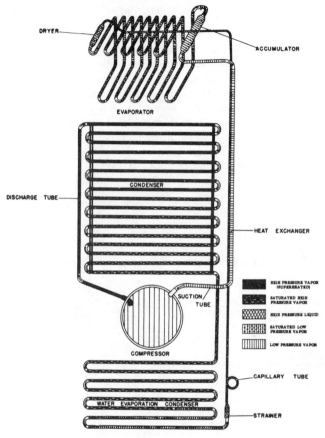

Fig. 4.

motor-compressor and the condenser (Figs. 4 and 5). Its function is to condense (liquefy) the heat-laden refrigerant vapor which is drawn from the evaporator. In condensing processes the vapor is first compressed by the motor-compressor and then cooled sufficiently to change to the liquid state as it passes through the condenser, thus giving up the latent heat absorbed during the evaporation process. This liquid is then returned to the evaporator to be re-evaporated.

The *capillary tube* is a long coiled tubing (with a very

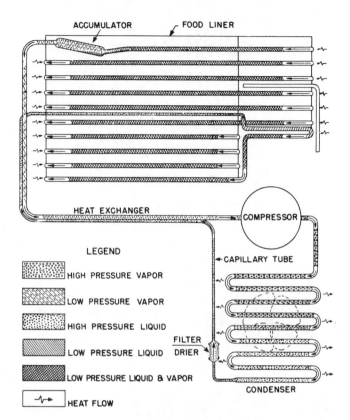

Fig. 5.

small bore) extending from the condenser to the evaporator (Fig. 4). It carries the liquid refrigerant to the evaporator. The long narrow passage of this tube provides sufficient frictional resistance to the flow of the liquid refrigerant to build up a head pressure that is high enough to produce condensation of the gas in the condenser.

A *filter-dryer* is placed in front of the capillary tube to catch foreign matter and to remove any moisture that might freeze into ice at the evaporator end of the tube. If the tube becomes plugged, the evaporator will defrost and continuous operation of the unit will result (Figs. 4 and 5). An *accumulator* which acts as a surge tank and separator is used at the evaporator end of the suction line to prevent liquid freon from passing through the suction line to the compressor (Figs. 4 and 5). The suction line and the capillary tube make contact for a portion of their length and thus create a *heat-exchanger* section where heat passes from the high temperature liquid in the capillary tube to the low temperature vapor in the suction line (Figs. 4 and 5). The heat-exchanger improves operational efficiency by cooling the liquid refrigerant in the capillary tube and at the same time acts to vaporize the liquid and prevent spillover into the suction line.

ELECTRICAL CIRCUITS

Most units use a split-phase motor designed to operate on a 60-cycle, 115-volt, single-phase current. The motors employ a relay for starting purposes and are equipped with an automatic reset thermal overload cutout to assure satisfactory operation. Voltage conditions measured at the unit should not vary more than plus or minus 10 per cent of the nameplate voltage. Wherever possible, a separate circuit should be employed. Do not connect through long extension cords or drop cord outlets that are switch controlled. Extension cords with a maximum length of 12′ may be used, providing that the wire is No. 16-2 rubber-covered or heavier. A wiring diagram of the complete electrical circuit of a unit is usually attached to the rear cabinet panel by the manufacturer. (A typical wiring diagram is shown in Fig. 6.)

Relay Operation. As a general rule, one of the two wires from the incoming supply cord leads through the junction box to terminal L on the relay (Fig. 6); the other wire leads through the junction box to terminal 2 on the temperature control. When the control contacts close, current flows through the junction box to terminal 3 on the thermoguard (overload device). When operating conditions are normal, the electrical flow continues through the thermoguard to the *common* terminal on the compressor assembly, which is the common terminal for both starting and running

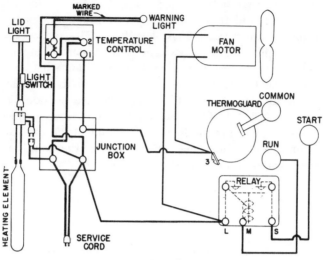

Fig. 6.

windings. The other lead of the service cord which leads to L on the relay supplies a current source to the relay coil, to the relay terminal marked *M* (Fig. 6), and to the *run* terminal on the motor compressor, the other terminal of the *running* winding of the motor. Since the starting torque of the running winding is out of balance with the starting load of the compressor, it is necessary to incorporate an additional heavier winding for starting purposes. The relay (magnetic type) places this winding into the circuit when required and takes it out of the circuit when the motor reaches normal speed. When current is supplied to the running winding, which does not have power to start the compressor, the excessive current drain causes a magnetic field to be set up within the coil of the relay. Thus energized, it operates as a solenoid and lifts a connecting bar which closes the circuit between terminals L and S, temporarily shunting the current to the starting winding through the *start* terminal on the motor compressor. The starting winding circuit remains closed only until the motor is brought up to approximate operating speed, at which time the current flowing through the relay coil to the running winding is no longer sufficient to support the bar which closed the circuit between L and S and it drops, taking the starting winding out of the circuit.

Thermoguard. The current flow through the thermoguard is from the temperature control lead through the junction box thermoguard to the *common* terminal of the motor compressor assembly. Basically a thermoguard is a bimetallic, disc type of switch wired in series with a resistance element. The assembly is mounted so that one side of the disc is sufficiently near the motor-compressor housing to be affected by the latter's temperature. On the other side of the disc the resistance element is located in such close proximity that any abnormal temperature induced in it by excessive current requirements will directly affect the action of the bimetallic disc. The resistance elements employed in a thermoguard are selected on the basis of the maximum allowable current established for the motor with which it is used. Similarly, the bimetallic discs are further factory-adjusted in assembly in direct relationship to the resistance element with which they will be used. Thermoguards, therefore, that have failed cannot be repaired and must be replaced. A thermoguard will automatically reset itself upon cooling; consequently, a unit may cycle on thermoguard operation indefinitely. Where the air surrounding the refrigerator is cold, such an occurrence may be detected by noting a gradual increase in cabinet temperature. Under all conditions, however, close attention to the thermostat through a running cycle should disclose whether or not the thermoguard is controlling operation of the unit.

Temperature Controls on Fully Automatic Units. On fully automatic refrigerators, the timer knob and the temperature control knob are usually located at the top center of the trim frame, with the former at the left and the latter at the right. Under normal usage and weather conditions, the temperature control knob, in the type shown, is usually set at the 5½ position. This position will provide proper refrigeration with greatest efficiency. Position 1 on the dial is the warmest setting and is generally used as a vacation setting. Colder settings are obtained by turning the knob toward position 10 (clockwise). Position 10 is the coldest position and is generally used for fast freezing. For periodic cleaning and manual defrosting, the knob should be turned to the *Off* position; in this position the compressor will not run.

Defrost Controls on Fully Automatic Units. Defrost controls are electrically and thermostatically operated. When installed the control should be set to the correct time of day by turning the knob clockwise. The defrost cycle will then occur at 3 A.M. daily. At that time it will not interfere with the food preparation plans of the user, the defrost cycle and the recovery of freezing temperatures after defrosting will be most rapid, and the freezer plates will be relatively free from frost during the daytime hours when the refrigerator is used the most. If manual defrosting is desired for periodic cleaning (on the automatic defrosting types), turn the temperature control to the *Off* position and rotate the timer knob clockwise until you hear an audible click at the 3 A.M. (night) position. This will energize the heaters and defrost the refrigerator. The compressor will remain off until you rotate the control knob from the *Off* position to the desired temperature setting, after which you should reset the timer to the correct time of day.

Push Button Power Defrost Control. When the defrost button in the center of the temperature control knob is pressed, a lever is operated between the temperature control and the defrost control which turns off the compressor and energizes the defrost heater. The heaters remain on a sufficient length of time for the evaporator to defrost.

REFRIGERATING CYCLE

The refrigerating cycle of the principal components of refrigerators or freezers (Figs. 4 and 5) works as follows: When the cabinet temperature is above the required temperature of the automatic temperature control of the unit, the operating control contacts close and start the compressor. The compressor draws off the evaporated vapor from the evaporator. This reduces the pressure on the liquid refrigerant which begins to boil and evaporate. During this process heat is absorbed from the cabinet interior, whereupon the cabinet temperature begins to drop. The low-pressure vapor is drawn through the suction line into the compressor where it is compressed and passed on to the condensor as a high-pressure vapor. As this high-pressure vapor (loaded with heat picked up from the cabinet interior, plus the heat from compression) passes through the coils of the condenser, it gives up heat to the cooling medium (the room air surrounding the condenser) and reverts to a high-pressure liquid. This liquid then passes through the filter-drier into the capillary tube, from which the refrigerant enters the evaporator inlet tube. As it proceeds, there is a drop in pressure because of the frictional resistance of the small bore in the capillary tube. When it reaches the end of the tube, the refrigerant expands into the comparatively wider opening of the evaporator as a low-pressure liquid and immediately begins to evaporate. As it evaporates the liquid travels through the coils of the evaporator, boils, and absorbs heat from the cabinet, again becoming a low-pressure vapor. Thus, the refrigerant has circulated completely through the system. This process continues and heat is removed from the cabinet until its temperature drops to the required temperature of the temperature control.

When the required refrigeration temperature is reached, the control contacts open and stop the compressor. The refrigeration unit remains *Off* until the flow of heat that penetrates the cabinet insulation raises the temperature of the cabinet interior, whereupon the unit starts and a new refrigeration cycle begins.

INSTALLATION

All refrigerators and freezers are thoroughly tested, checked, and assembled at the factory. Do not place the refrigerator or freezer in a confined area where there is little or no ventilation or in excessively damp places. Be sure the voltage is constant within ten per cent of the prescribed voltage. Extension cords and outlet boxes should be heavy enough to maintain proper voltage during the starting period and should be protected by a 20-amp. to 30-amp. fuse. Be sure that the unit is level, with all four corners resting on the floor. To be certain that the door gasket is making a good seal, check it by placing a piece of thin paper on the flange of each corner, then closing the door. A definite pressure should be noted when the paper is pulled out.

TROUBLE AND REMEDY CHART

Trouble	Cause	Remedy
No refrigeration. Motor will not run.	No voltage at outlet. Defective service cord. Loose electrical connection. Defective thermostat. Defective relay. Defective thermoguard. Open motor winding.	*Make complete electrical continuity tests. Repair or replace parts as required.
No refrigeration. Motor will not run, or will hum, attempting to start, but overload will open circuit.	Low voltage. Loose connections. Defective relay. Defective capacitor. Short circuit in motor windings.	Check with voltmeter at plug-in point. (If low, report to utility company.) *Make continuity test for other possible causes.
No refrigeration. Motor runs continuously. Discharge service valve is open.	No refrigerant in system. Plugged screen in drier. Plugged capillary tube.	*Locate and repair leak. Evacuate and recharge. *Clean out capillary tube or replace.
Excessive running. Motor runs too much or continuously. Temperature too high.	Short refrigerant charge. Partial restriction in drier or in capillary tube. Excessive load of warm food being frozen. Partially blocked condenser. Fan not operating properly. Inefficient compressor. (Check other possibilities first.)	Test for leaks and make repairs accordingly. *Evacuate and recharge. Correct use of unit as per directions of manufacturer. *Clean condenser. *Replace fan motor. *Replace compressor.
Excessive running. Motor short cycles. Temperature too high.	Voltage too high or low. Fan motor inoperative. Air in system. Blocked condenser. Overcharge of refrigerant. Defective discharge valve in compressor. Restriction in liquid line (capillary tube or drier.) Overload protector has burned, or has pitted the contact points, or is improperly adjusted. Defective relay.	Check voltage. Report to utility company if necessary. *Replace overload (thermoguard). *Replace relay.
Excessive running. Motor runs continuously. Temperature too low.	Short charge. Partial restriction (drier or capillary tube.)	*Add refrigerant. *Check and, if necessary, replace filter-drier or capillary tube assembly.
Excessive running. Motor runs too much or continuously. Temperature too low.	Temperature control capillary not far enough in well or not in well at all. Range adjustment of control is too low. Control contacts stuck in closed position due to fused contact points or faulty mechanical action.	*Insert capillary in temperature control as far as it will go. *Readjust or replace. *Replace control.

* Consult service department of manufacturer.

CHAPTER 21

Small Appliances

Small electrical appliances are divided into two general groups; small heat appliances that are either automatically or nonautomatically controlled, and motor-driven appliances. Heating coils on small appliances are of two general types; totally enclosed coils that cannot be repaired and must be replaced as a unit, and open-type coils in which temporary repairs can be made.

HOTPLATES

The hotplate shown in Fig. 1 is equipped with an enclosed coil and a thermostatic control. Other types of hotplates are equipped with open coils and are nonautomatic. Hotplates must be used on a.c. current unless otherwise specified by the manufacturer.

The circuit continuity of the unit thermostat, wiring, or complete unit may be checked by referring to the wiring

Fig. 1. Hotplate.

diagram (Fig. 2) and using a 110-volt series test light, an ohmmeter or any series-type circuit tester.

The thermostat assembly is so constructed that the bimetallic thermostat lever bearing against the actuator assembly of the thermostat opens and closes the thermostat contacts to maintain the temperature of the stove unit. Turning the control knob to any specified position closes the contacts and moves the actuator arm to a position that is a predetermined distance from the bimetallic lever. As the unit comes up to the required temperature, the lever

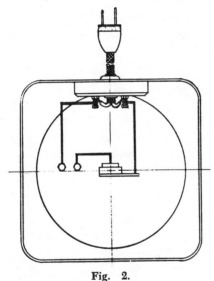

Fig. 2.

moves toward the actuator and opens the circuit when the required temperature is reached. The temperature is then maintained by the cycling of the thermostat. The transfer of heat from the unit to the bimetallic lever takes place by means of the thermostat mounting bracket. The mounting bracket must fit tightly against the unit casting, and the thermostat assembly and bimetallic lever should fit tightly against the bracket.

All functional parts of the thermostat are pretempered at 800° F. to withstand fabrication stresses, thus eliminating the possibility of any change in the parts during their life.

Care should be taken to see that they are not bent. The control screw and all movable parts of the thermostat should move freely. A sticky control screw may be loosened up by the application of a small amount of oil. The flexible jumper from the movable contact on the thermostat actuator arm to the connection stud on the unit must be formed so that it clears all non-current-carrying parts of the thermostat. It must be formed smoothly and not be unduly kinked. When any part of the thermostat assembly or enclosed coil is found to be defective, the complete assembly must be replaced by the manufacturer.

Repair of Open-Coil Heating Elements. The following are two methods commonly used for repairing open heating coils of nonautomatic small appliances.

(1) The broken coil may be reattached to the terminal. This repair can be made only when the break is close to the terminal. Clean the wire at least ½″ from the end, and fasten it around the terminal in the direction in which the screw or nut tightens. (2) The broken ends may be fastened together with a metal sleeve. Clean the broken ends, place them in the sleeve, and crush the sleeve on the wire with pliers.

TROUBLE AND REMEDY CHART

TROUBLE	CAUSE	REMEDY
Units do not heat.	Wall plug on cord may be loose in outlet.	Check to see that plug makes contact in outlet.
	Fuse in house circuit may be blown.	Check fuse.
	Wall outlet defective.	Check outlet with another appliance or lamp.
	Open circuit in unit or wiring.	*Check unit and wiring for circuit continuity.
	Defective thermostat.	Check thermostat for proper operation.
No *Off* position.	Incorrect thermostat adjustment.	*Readjust thermostat.
Units do not heat fast enough.	Cooking vessel used is too big, or too large a quantity of food is being cooked.	Correct.

* Consult service department of manufacturer.

PERCOLATORS

The two types of percolators are the nonautomatic and the automatic. The nonautomatic type consists of a con-

Fig. 3. Electric coffeemaker.

ventional glass or metal percolator equipped with an open-coil or closed-coil heating attachment similar in construction and operation to the hotplate coil. The automatic coffeemaker shown in Fig. 3 is the type generally used; it is rated at 400 watts for use on 110-120 volts a.c. current only. It must be used on a.c. current because the thermostat switch used is of the slow make-and-break type.

The interior of the body of the type shown, including the heater plate nose, is completely tin lined. This keeps the brew from coming in contact with the copper body, which would give the coffee a bitter taste. The pump, tube, basket, and spreader plate are made of aluminum.

OPERATION

The pump or siphon is the heart of the percolator. It consists of the pump chamber (Fig. 4, A), valve washer (Fig. 4, B), and valve seat (Fig. 4, C). When the inset is in position, the base of the pump rests on the heater plate nose (Fig. 4, D). The surfaces that rest against each other are very carefully machined to insure that the pump is seated securely on the heater plate nose. The valve seat and valve washer are lapped to insure a tight seal between these two parts. The proper seating of the pump on the heater nose and the valve washer against the valve seat is necessary for proper operation of the percolator. When the pump is placed on the heater plate nose, water already being in the percolator, the water enters the pump chamber and rises in the inset tube to the level of the water in the percolator. The small amount of water in the heater plate nose is heated quickly and a small amount of steam is formed.

The valve washer, being seated tight against the valve seat, prevents the steam and water from escaping into the main body of the percolator. The steam pressure pushes some of the water in the heater plate nose and pump chamber up the tube. This reduces the pressure in the pump chamber, allowing the valve washer to be raised by the pressure of the water in the main body of the percolator so that the water enters the pump chamber and heater nose. This cycle is repeated; the heated water is forced up through the tube and out onto the spreader plate (Fig. 4, E). The spreader plate is designed to give an even distribution of the water over the coffee grounds. The perforations in the spreader plate are raised approximately $\frac{1}{32}''$ above the surface of the plate, allowing the water to collect on the spreader plate until a depth of approximately $\frac{1}{32}''$ is reached. The water then flows through all the perforations, with even distribution.

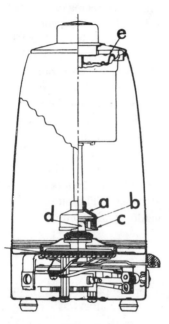

Fig. 4.

The percolation time is controlled by the control in the base of the percolator. The control lever is set at the desired position between *mild* and *strong,* and the percolator is then connected to the outlet. When the coffee has reached the strength corresponding to the setting, the current is automatically switched from the main unit to the warming unit and the pilot light comes "on." The warming unit will then remain on until the percolator is disconnected. If a second pot of coffee is to be made immediately after the first, the interior of the percolator should be thoroughly rinsed with cold water. It should be disconnected

TROUBLE AND REMEDY CHART

Trouble	Cause	Remedy
Percolator will not heat.	Wall plug on cord may be loose in outlet.	Check to see that plug makes contact in outlet.
	Fuse in house circuit may be blown.	Check fuse.
	Wall outlet may be defective.	Check outlet with another appliance or lamp.
	Open circuit in supply cord.	Check electrical continuity of supply cord.
	Loose connection in appliance plug on cord.	Check appliance plug for loose connection.
	Open circuit inside of percolator.	*Check percolator for circuit continuity.
Percolator heats but will not percolate.	Inset not properly seated on heater nose.	*Check to see that base of the pump rests securely against the heater nose.
	Valve washer not seated properly on valve seat in inset pump chamber.	*Check valve washer and valve seat.
	Valve seat not securely screwed into inset pump chamber.	*Check to see that valve seat is securely screwed into inset pump chamber.
	Small holes or flaws in valve seat that allow water to flow out of pump chamber.	*Inspect valve seat for small pinholes and flaws. If holes are found, valve seat must be replaced.
	Coffee grounds or other material in the heater nose.	Clean out heater nose.
Coffee begins to percolate again after having been switched over onto the warming unit.	Bimetallic heating unit burned out.	*Replace control assembly part.
Pilot light will not light.	Bulb burned out or part of bimetallic heater unit is shorted.	*Check and replace.
Unit operates erratically.	Loose bimetal.	*Check and adjust.
Coffee not strong enough.	Adjustment lever not set high enough.	*Increase setting.
	Insufficient coffee used for number of cups of water used.	Normal amount of coffee is one rounded tablespoon for each cup of water.
	Internal adjustment set too low.	*Adjust control.
	"Pumping time" incorrect.	
Coffee tastes bitter.	Stale coffee used.	Use fresh coffee.
	Percolator not properly cleaned after each use.	Clean percolator regularly as described by manufacturer.
	Tin lining on the inside of the percolator worn through, allowing the coffee to come into contact with the copper body.	Percolator must be returned to factory for relining.
Stale odor in interior of percolator.	Percolator not properly cleaned after each use.	Clean percolator regularly as described by manufacturer

* Consult service department of manufacturer.

while being rinsed. The percolator should never be connected unless there is water in the pot, nor should the warming unit be left "on" after all of the coffee has been served.

MAINTENANCE

Coffee should never be allowed to stand in the percolator (and coffee grounds should not remain in the basket) from one use to the next or for any long period of time. This practice will cause the interior to discolor and tend to impart a rancid taste to the coffee. After each use, the inset should be removed, emptied, and washed thoroughly in hot, clean water. Unscrew the valve seat, remove the valve washer, and make sure that the valve seat, valve washer, and pump chamber are clean. The interior of the percolator should be thoroughly washed out in hot soapy water and rinsed in hot clean water. Be careful not to allow any water to get into the base of the percolator. The appliance should never be immersed in water after being used, it should be left open to the air. The well in the heater nose should be cleaned out with a small brush periodically. Do not scratch or mar the heater nose, for this will affect the operation of the percolator. Remove stale odors by occasionally operating the percolator with two tablespoons of baking soda dissolved in the water.

TOASTERS

The three general types of toasters are nonautomatic, semiautomatic, and automatic.

Nonautomatic Toasters. Nonautomatic toasters are equipped with either mica-insulated heating units or open-coil units. In mica-insulated units the heating unit consists of ribbon-shaped nichrome wire wound on sheet mica which is bound top and bottom with metal. Nonautomatic toasters consist primarily of a base, top, and two end shields which may or may not be in one piece with the heating unit mounted vertically between the top and the center line of the base. In open-coil units the heating unit consists of a length of nichrome wire wound in a helical coil, strung vertically from top to bottom, and supported by porcelain posts. Some models have a toggle switch located in the base to control *Off* and *On*.

Removal and Replacement of Heating Units. When it is necessary to remove and replace a defective heating unit, proceed as follows: Remove screws or spring clips from the underside of the base. Separate the base from the sides and the top shell assembly. Disconnect ends of the unit wire from the terminal nuts. Pull out the heating unit and guard wires, and remove the wires. Insert heating and guard wires. For replacement and reassembly of the unit, reverse the foregoing procedures.

Semiautomatic Toasters. Most types of semiautomatic toasters have three heating elements; one is located in each of the doors and the other (open on both sides) is in the center. These elements, so arranged, toast two slices of bread on both sides. A bimetallic thermostat responds to the surface temperature of the bread where it lies against the thermostatic strip. The color of the toast depends on how far the end of the thermostatic strip moves before it separates the switch points and shuts off the current. A coil of wire in the top of the toaster is connected in series with the main circuit. When this breaks loose, the toaster becomes inoperative.

AUTOMATIC TOASTERS

Automatic toasters (Fig. 5) are equipped with a variable-speed clock mechanism that automatically adjusts the toasting time.

The following are causes of common difficulties in the operation of automatic toasters:

Toaster fails to heat. A fuse is blown, there is a break in the cord or lead, the main switch does not make contact, or all heating elements are burned out.

Toaster heats, but toast fails to "pop up." The auxiliary switch does not make contact, the timer release lever binds, or there is a broken operating lever spring.

Operating lever will not stay down. The timer release lever binds, the timer release lever spring is broken, or the main trip lever is not properly adjusted to the timer release lever.

Toast is too light or too dark. The timer mechanism is not properly adjusted.

Successive slices of toast are not uniform. The timer

Fig. 5.   Automatic toaster.

mechanism is not properly adjusted, or the trouble is due to the texture and type of bread.

Adjustments, Replacements, and Repairs. The following are the most common adjustments, replacements, and repairs for automatic toasters:

To Remove Case and Crumb Pan Assembly. Remove the Bakelite operating handle and the timing button. Unlatch the crumb tray and remove the two crumb-plate attachment screws. Remove only the four self-tapping screws, located in each corner. Do not remove the other five screws as they fasten the Bakelite base to the toaster base plate. Work off the case by rocking it slightly from left to right. Be careful not to injure the lead wires on the left side of the toaster. Tilt the operating lever upwards and backwards into the case assembly while removing the case. It may be put into this tilted position when you reassemble the toaster, and by reaching under the front of the case with a finger you can drop it out through the slot in front of the case.

To Replace the Heating Element. Remove the case. Then remove all the bread guard wires. If a center element is burned out, the outside element must be removed to reach the center element. Use a socket wrench or flat-nosed pliers to remove the two hexagonal nuts located at the bottom of the outside elements. After removing the screws, you can slip the outside element out by pressing the top of the element inward, until you can remove it through the top of the bread slot. Be careful not to tear the element wire on the bread rack while removing it. If, after removing the outside element, it is necessary to remove the adjacent center element, remove the two screws and two bus bars which hold the center element in place. This element should also be removed through the oven slot. When installing the new element, be sure that the bottom edge of the element fits into the four slots provided in the base of toaster and that the side of the element on which the wire is wrapped faces the bread slot. When installing the bread guard wires, place the hooked ends on the inside of the

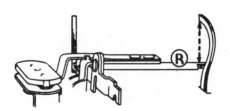

Fig. 6. Switch closed.

elements to hold them firmly in place against the frame. The center elements and the outside elements do not have the same amperage rating. On the metal clip at the top of the center elements the amperage rating of 2.49 is stamped, and the outside elements are rated 2.73.

To Repair Switch and Trip Lever Assemblies. Figure 6 shows the operating lever depressed, the switch closed, and other parts in actual operating positions. The operating lever assembly operates both the upper switch contact assembly and the reset arm which are integral parts of the lever assembly. The far end of the reset arm, as it rises, applies pressure against the curved portion of the shunt lever, which resets the timing mechanism (Fig. 7).

Upper switch assembly. Remove the case. Then depress the operating lever until it locks at the bottom. File off or, with a small power emory wheel, grind off the heads of the two rivets which secure the contact spring arm to the main lever assembly. Either an anvil or some small supporting block should be used under the bracket (through which

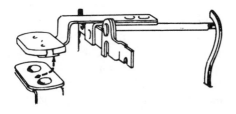

Fig. 7. Switch open.

the rivets penetrate) to keep the supporting bracket at a right angle to the main lever assembly. In reassembling, eliminate the rivets and replace them with two ¼ x 5-40 R.H. screws, two ¼ x 5-40 hex. nuts, and 2 lock washers.

The switch assembly must be correctly adjusted to insure proper operation. Correct adjustment is indicated as follows: When the operating lever is pushed downward, the upper switch assembly should touch the rear contact point first. At this time there should be a space of about ¹⁄₁₆″ between the front contact point and the upper switch assembly. After the upper switch assembly has touched both of the lower contact points, the outward sweeping motion should take place as shown in Fig. 7. When the operating lever locks into toasting position, the upper switch returns by a sweeping action to a horizontal position as shown in Fig. 6. To obtain this switch action, pressure may be applied with the finger on the upper switch assembly upward or downward, as may be necessary.

Lower switch assembly. First remove the case. Then remove the complete timer base and the auxiliary element assembly by taking out the two screws which secure this assembly to the base. To remove the lower switch assembly, knock out the rivet with a sharp punch. Re-rivet the new lower switch assembly to toaster base. (Do not attach the new assembly with nut and screw.)

Trip lever assembly. Remove the case, the entire timer base, and the auxiliary element assembly. Working from the opening in the base plate directly under the bimetallic operating arm assembly, spread the horseshoe washer open, remove it, and slide the main trip lever pin out. Replace any broken part.

To Repair or Replace the Timer Release Lever and the Release Lever Spring. If the timer release lever binds, the bread may not stay down or the toast may not pop up. One reason may be that the release lever drags on the toaster base plate. The never should be freed by bending it toward the center of the slot. Another reason may be that the entire mechanism does not operate freely. It should be removed and, if it has not been damaged, should be cleaned with naphtha and oiled. If this does not remedy the trouble, the entire release lever and timer bracket must be replaced. If the operating lever will not stay down but the timer release lever operates properly, the difficulty may be caused by a broken release lever spring. Replace the release lever spring. Be sure that the trip release lever and the main trip lever are properly engaged when the operating lever has been pressed down. Engaging clearance can be made from the underside of the toaster by bending the "elbow" of the release lever up or down to change the position of that portion of the release lever which projects above the toaster base plate.

To Change the Main Operating Spring. If the toast does not pop up, the cause may be a broken main operating spring located inside the hinged crumb tray. Remove the broken spring and install a new one. Prongs in the center of the base plate permit tension adjustment of the spring.

To Install Bread Racks. To replace broken or damaged bread racks, remove the case. With long-nosed pliers twist the metal lugs on the bread rack support bracket located at the rear of the toaster. Then remove the bread rack through the slot in the rear frame. When replacing a bread rack, be sure not to mutilate the retaining lugs; do not leave the bread rack dangling in the toaster oven (insert in the slot provided in the front frame) or damage a heating element.

Operation of Timing Mechanism. The bimetallic strip in the timing mechanism is heated by an auxiliary element connected in series with the main toaster elements. The flexing of the bimetallic strip controls the toasting cycle. It is voltage-compensating and will automatically increase or decrease the toasting time if used within the limits of 105 volts to 120 volts. Since the bimetal is affected by toaster temperatures, the toaster should be at room temperature before any tests are made. A heat-up, cool-off cycle controls the timing operation. During the heat-up part of the cycle, the auxiliary element heats the bimetallic strip, causing it to flex. This action forces the operating arm slowly forward until it hits the timing shaft. The bimetallic strip, still heating, then flexes in the opposite direction until it has moved off the shunt lever trigger. This causes the auxiliary switch to close and shunt out the auxiliary element. The release link then drops into the path of the operating arm. At this point the cool-off part of the cycle begins. The operating lever moves back toward the starting position until it comes in contact with the release link, moving the release lever until the toaster trips. The toast automatically pops up, opening the main switch and shutting off the current. The entire timing mechanism is now reset for the next operation.

To Replace Auxiliary Element. It is not necessary to remove the case to replace the auxiliary element. Remove the hinged crumb plate, then detach the bimetallic spring, and remove the two screws which fasten the auxiliary element terminal leads to the long and short bus bars. Re-

move the nut and the washer from the end of the bimetallic
pivot and remove the entire bimetallic operating arm. Slide
the auxiliary element off the free end of the bimetal. Check
the new auxiliary element to see that it is not out of shape,
for the curvature of a bent element will affect the mechanical operation of the bimetallic assembly. When installing
the new auxiliary element, be sure that the free end of the
bimetal is inserted between the two center pieces of the
element mica and that no part of the element wire is touching the bimetal or any other part of the toaster. Slide the
element on until it reaches the center rivet at the other end
of the bimetal and see that it is centered properly. Reinstall the entire assembly and connect the leads. Be sure

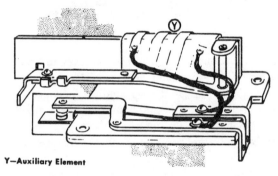

Y—Auxiliary Element

Fig. 8.

they do not put a tension on the movement of the bimetallic
strip or that they do not touch each other. Then adjust
the timer mechanism and test the toaster. Figure 8 illustrates the auxiliary element properly installed as part of the
timing mechanism.

To Adjust General Timer Mechanism. Disconnect the
toaster, turn it upside down with the timer end to the left,
and remove the undercover. The toaster should be at room
temperature.

1. Pull the operating lever handle into toasting position.
2. See that the bimetallic strip is straight and parallel
with timer operating arm. Then unhook the bimetallic
spring from the bimetal to make sure that the tension of

the auxiliary element leads does not affect the movement of the bimetallic strip. (Figure 8 shows the approximate position of the leads.) Then reconnect the spring and reset it.

3. Slowly push the bimetallic strip to the right until it

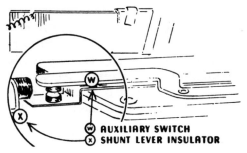

W AUXILIARY SWITCH
X SHUNT LEVER INSULATOR

Fig. 9.

just trips the shunt lever. The release lever link (under the operating lever adjustment arm) should snap up approximately $\frac{1}{64}''$ to $\frac{3}{64}''$ and stop against the operating lever adjustment arm. At this point, inspect the contacts of the auxiliary switch, but do not close them.

4. If contacts are not closed, push the operating lever farther to right (if necessary spring the bimetallic strip slightly) until the release lever link just clears the bottom of the operating lever adjustment arm. At this point, the release link should move approximately $\frac{1}{64}''$ to $\frac{3}{64}''$ and the auxiliary switch should be closed. The shunt lever should not rest against the underside of the bimetallic strip. If the auxiliary switch is still open, bend the long bus bar and contact assembly closer to the other contact point.

To Adjust Auxiliary Switch. If the auxiliary switch (Fig. 9) is not making contact, the auxiliary element will continue to heat and the operating arm will not complete the cool-off part of cycle, thus causing the auxiliary element to heat too long. This overheating results in permanent damage to the bimetallic strip, which would have to be replaced. If the bimetal is not damaged, adjust the contact points.

Timer Adjustment. Two consecutive toasting operations should be performed to determine whether the switch adjustments are correct. If the toast produced is satisfactory, the adjustments should not be disturbed. Before any test can be made, all necessary repairs should be completed and the proper adjustments should then be made as specified below. Do not tamper with the internal adjustments after the toaster has been assembled. Only the bimetallic arm and the trip arm adjustments should be made at this time.

1. The timing button should be set at "darker" when measurements of the trip arm adjustments are made.

2. The distance between the timing shaft (Fig. 10, *C*) and the trip arm (*A*) should be $3\%_{64}''$.

3. The distance between the timing shaft (*C*) and the release link (*B*) should be $4\%_{64}''$ to $2\%_{32}''$.

4. Clearances in both cases are measured at right angles to the timing shaft (*C*).

The following figures give the approximate time required for the heat-up and cool-off cycles, with the timing button set at medium position and the toaster connected to 115 volts. This time applies only when bread is actually being toasted; it will not be correct unless bread is used and the toaster is at room temperature at the start of test.

Heat-up Time	Cool-off Time	Total Time
1 Min., 30 Sec.	30 Sec.	2 Min.

If the toaster operates within these time limits and the toast is too light, it may be necessary to increase both clearances. To increase clearance *A* to *C*, loosen the adjustment screw (Fig. 8) and move the operating arm away from the timing shaft not more than $\frac{1}{64}''$. To make the second adjustment, remove the bimetallic arm and with an end cutter or knipper cut off about $\frac{1}{32}''$ from the shoe (the point that protrudes through the bimetallic arm) of the operating arm to provide for a larger clearance between *B* and *C* (Fig. 10). After cutting, file the area smooth to prevent a drag

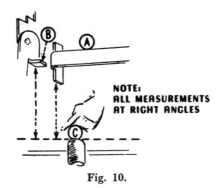

Fig. 10.

or bite against the release link. Do not bend the release link. If the second slice is darker than the first, the distance between A and C should be increased; when the second slice is lighter, only the second adjustment is made.

Voltages higher or lower than standard (approximately 115 volts) will increase or decrease the total time in the time chart. The stated ratio between the heat-up time and the cool-off time should be maintained so far as possible.

ROASTERS

Combination electric roaster-grilles can be used in much the same way as the electric range oven, but on a somewhat smaller scale. They are available in oval, round, and rectangular shapes. The rectangular shape gives the advantage of greater usable space. Efficient roasters should comply with the following specifications: They should have double walls and a bottom (which contain heating elements) and an inch or more of insulation on the outside. There should be a lightweight cover which is easy to remove and is polished to reflect heat back into the roaster without undue loss. Other requirements include good design, a perfect fit at joints, a rigid body and rigid handles, heavy plating and enameling, supplementary cooking utensils, and thermostatic controls (Fig. 11).

Operation. The roaster has one side and one bottom element connected in series with an adjustable thermostat.

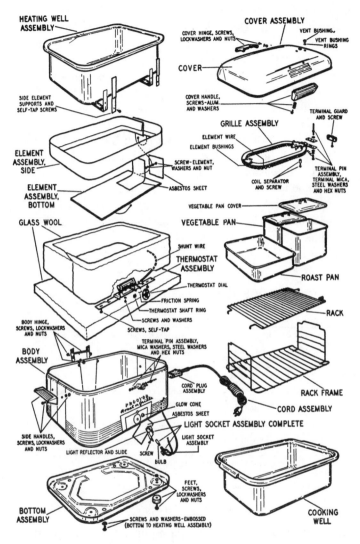

Fig. 11. Parts of a roaster.

When the adjusting dial is turned to the required temperature, the circuit is completed and the roaster is heated. The thermostat automatically shuts off and turns on the circuit to maintain even heat at the selected temperature. Roasters featuring the "Glow-Cone" will produce a "glow" while the roaster is heating. The glow will disappear when the thermostat shuts off the current.

The roaster and the grille must not be in operation at the same time because a house fuse will blow, with possible damage to the roaster. For this reason only one cord is supplied for both parts of this appliance. Do not connect the roaster to a drop cord or to an extension cord because the wires will be of too small size for satisfactory heating. Connect the roaster directly to a wall outlet.

Maintenance. Clean the outside of this appliance with a damp cloth. Whiting and soap may be used to remove grease spots. Do not place the roaster in water for cleaning. The inset pan can be washed in water. Store the roaster in a dry place.

TROUBLE AND REMEDY CHART

TROUBLE	CAUSE	REMEDY
Roaster is not heating.	Thermostat set on *Off* position.	Set thermostat to desired baking temperature.
	Fuse blown; cord or plug needs repair.	Replace or repair.
Roaster does not heat properly.	Thermostat set too low.	Change thermostat to higher setting.
	Connected to drop cord or long extension cord.	Connect to wall outlet.
	Lid may be warped or sprung.	*Repair.
Roaster gets too hot.	Thermostat set too high.	Change thermostat to lower setting.
	Thermostat operating improperly.	*Replace.

* Consult service department of manufacturer.

WAFFLE IRONS

Most waffle irons or bakers are equipped with automatic temperature controls. The automatic temperature control of the waffle baker shown in Fig. 12 differs from controls used on other waffle bakers. Instead of an ordinary bimetallic thermostat responding to air convection and conduction through its supporting members, it employs the grid casting itself as the active member of the thermostat control.

Fig. 12. Combination waffle baker and grill.

Aluminum has a high expansion per degree of temperature change, as contrasted with the length of the procelain segments. This difference in the rate of expansion of the two materials with changes in temperature makes the silver contact switch open or close the circuit to the heating elements. The switch has a manual control knob that permits setting the controls to any desired temperature. Waffle bakers will operate on a.c. current only; they must not be used on d.c. current.

Care of Waffle Baker. Since the grids of the waffle baker are made of cast aluminum or cast iron, they are porous and should me smeared with grease before they are used. Brush the grids of a new iron with unsalted fat, then close the grids, and heat them 8 min.; open them and press slices of bread between them to absorb the surplus fat. After grids have been so treated, they should not be washed. To clean them, leave them open until they are cool; then brush them with a stiff dry brush. If food burns on the grids, moisten a cloth with ammonia and place it between them for several hours. Then brush them with a steel brush and re-apply grease. The outside may be cleaned with whiting and ammonia mixed into a paste. The waffle baker should never be immersed in water. Always connect it to an appliance outlet, never to a drop light socket. About 8 min. is long enough to preheat the iron. Either overheating or underheating will cause waffles to stick.

Replacement of Pilot Light. To replace a burned-out pilot lamp, proceed as follows: Remove the undercover. Pilot lights usually are No. 46—6.3-volt Mazda lamps con-

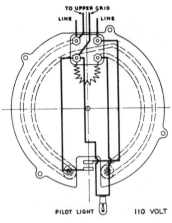

Fig. 13.

nected across a short length of resistance wire in series with the heating element (Fig. 13). Replace the bulb by loosening the screw that holds the pilot light bracket in place. The hole is slotted; therefore, when the screw is loosened, the entire bracket can be slid out, the lamp replaced, the bracket put back into position, and the screw tightened down. On some of the earlier types of bakers these holes are not slotted. In such cases, the best way to reach the light with the fingers is to bend the bracket down, replace the light, and then bend the bracket back into position. The soft metal of the bracket can be bent without injury.

Adjustment of Thermostat Assembly. If the thermostat should get out of adjustment, proceed as follows: While the baker is at room temperature, turn the control screw with a pair of pliers counterclockwise until the contacts are opened. Plug in the baker, preferably to a series test circuit. (In this way the grids will not heat up but the pilot light will go on, indicating that the contact is closed.) Turn the control screw clockwise until the light goes on; then continue to turn the control screw clockwise 1⅛ turns. Without disturbing the control setting, push the Bakelite control knob tight onto the screw with the pointer in a vertical position (at 12 o'clock). The thermostat should now cut off at approximately 410° F. The setting can now be adjusted manually up to approximately 50° higher and lower by moving the control button clockwise or counterclockwise as far as it will go. The control button will move a little less than a semicircle. If it is necessary to raise or lower the temperature setting beyond the stop position, the Bakelite control knob must be pulled off, the shaft must be turned in the desired direction, and the control knob must be replaced for the new setting. The pointer on the control knob should be in a vertical position (at 12 o'clock).

MOTOR-DRIVEN APPLIANCES

The two most common motor-driven small appliances in the home are vacuum cleaners and mixers.

Vacuum Cleaners. The two main types of cleaners are the tank type (Fig. 14) and the upright type (Fig. 15). If the floor coverings are deep pile or hard-to-clean weaves, the upright cleaner with a motor-driven beating brush is

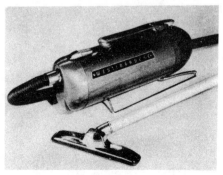

Fig. 14. Tank vacuum cleaner.

TROUBLE AND REMEDY CHART

TROUBLE	CAUSE	REMEDY
Cleaner refuses to pick up dirt.	Bag may need emptying.	Empty bag, turn wrong side out, and brush or clean.
	Improper adjustment of cleaner nozzle.	Adjust nozzle, (some cleaners do not have an adjustment) so that a quarter slides easily between cleaner nozzle and rug. On cleaners with a rotating brush or roll, the clearance should be the thickness of a half-dollar. Hold the machine with the nozzle up, and put a ruler or piece of stiff cardboard over the nozzle. The bristles should not quite touch the ruler, leaving a space of 1/32″. Make the adjustment by moving the pin, screw, or lever at each end of the brush.
	Improper adjustment of brush.	
	Broken belt or belt not revolving.	Examine the belt. Make sure it is around both the brush and the motor shaft and is not tangled with hair and string. If the cleaned brush still will not turn with the motor, the belt is probably stretched and should be replaced with a new belt. To replace the belt, slip it off the motor shaft, remove the brush and the belt from the cleaner, slip a new belt over the brush, and replace the brush in the cleaner.
Motor does not operate.		See Chap. 7.

* Consult service department of manufacturer.

preferable and will do a faster cleaning job with less effort. The tank cleaner is recommended for homemakers who require a cleaner that will do a variety of dusting jobs as well as clean rugs and carpets. Compared with the upright cleaner, it will clean scatter rugs more easily and will clean closer to baseboards in rooms that have wall-to-wall carpeting. Its strong suction whisks dust from the surface and cracks of bare floors. The rug nozzle of a tank cleaner reaches under low modern furniture, making it unnecessary to move heavy pieces, and in this way reduces cleaning time to a minimum.

Mixers. The two main types of mixers are the stationary and the portable types. The stationary mixer remains in a set location on a mixer stand. The beater moves around in

Fig. 15. Upright vacuum cleaner.

the bowl, constantly scraping the food mixture from the sides. The portable mixer permits the motor head and the beaters to be used independently of the mixer stand.

Specifications. A mixer (Fig. 16) should conform to the following specifications:

Deep, full-mix, rust-proof beaters, easily removed from the motor head without force or pressure.

Dough guards to prevent the dough from climbing the beater shafts.

A speed-dial-finder to help the user to select the required mixing speeds.

Easy-to-wash beaters, easily adjusted, with no corners to lodge food particles.

A handle which can be manipulated conveniently.

A bowl platform that revolves at reasonable speed.

Full power delivered automatically for every speed. Low speeds break up solid foods; high speeds whip air into the mixture.

UL approval as to safety.

A number of attachments are available, such as a juice extractor, drink mixer, meat grinder, slicer and shredder, and potato peeler.

Fig. 16. Mixer.

Maintenance. Many mixers have sealed bearings which are packed with grease at the factory. Other mixers require lubrication. If the latter type of mixer is used frequently, it should be oiled biweekly with 3 or 4 drops of light machine oil. If not used often, it should be oiled once a month. Remove the front cover at intervals of 6 months and examine the grease in the gear case. Clean out the old grease and add new grease if necessary.

The motor should be wiped off with a damp cloth after use. Care should be taken not to get water in the oil holes or electrical parts. Cover the motor with a cloth or paper when not in use.

TROUBLE AND REMEDY CHART

TROUBLE	CAUSE	REMEDY
Motor does not operate.		*Repair or replace. See Chap. 7.
Beaters strike the mixing bowl.	Beaters not placed firmly in their sockets. Improper adjustment of the motor head.	Push beaters up until they are firmly in the sockets. Most mixers are provided with an adjustment mechanism to raise and lower the beaters in the mixing bowl. Locate the adjustment mechanism, loosen the lock nut, and turn the screw until the beaters just touch the bowl. Tighten the lock nut.

* Consult service department of manufacturer.

CHAPTER 22

Fans

⋮

Fans are an integral unit of the ventilating system in a home. They range in size from 8″ to 36″ and may be oscillating or nonoscillating. They have either single or variable speed and are equipped with a.c. or d.c. motors. Any of the various kinds of fans (which are classified into portable and ventilating types) may be installed in walls, windows, or attics.

SELECTION OF THE CORRECT SIZE OF FAN

To select the correct size of fan, determine the number of cubic feet (length × width × depth) in the room or area to be ventilated. The result will be the required air delivery per minute. The fan selected should have a capacity equalling or exceeding this figure. Use the following table as a guide:

TABLE 29B

SELECTION OF SIZE OF FAN

Size of Fan	Cubic Feet Per Minute Replacement (C.F.M.)
16″	2000
20″	3500
24″	5000
30″	6500

PORTABLE FANS

Portable fans (Fig. 1) are available in sizes from 8″ to 16″. The fan base on some models can be replaced with a

Fig. 1. Oscillating or non-oscillating fan.   Fig. 3. Automatic window fan.

Fig. 2. Twin-fan ventilator.

VENTILATING FANS

Three types of ventilating fans used in wall, window, and attic are shown in Figs. 2, 3, and 4.

Wall Fans. Wall fans are designed to fit into a wall and discharge the air toward the outside. Some wall fans have an outside shutter arrangement controlled by two lengths of bead chain that opens and closes the shutters and starts the motor. Other models have a separate switch mounted on the inside grille plate for starting and stopping the motor. (See cutaway view of wall fan, Fig. 5.)

Installation of Wall Fans. The installation of a wall fan requires a wall opening, electrical wiring, and mounting of the unit.

Fig. 4. Wall ventilator.

Fig. 5. Cutaway view of ventilator.

Wall Opening. Make a 10¼" x 10¼" opening in the outside wall so that the center of the opening will be at least 6½' above the floor (Fig. 6). If the kitchen range is against the outside wall, the fan should be installed 12" to 18" above the top of the range.

Plan the opening in the wall before starting to cut it. Check the basement to determine whether there are pipes

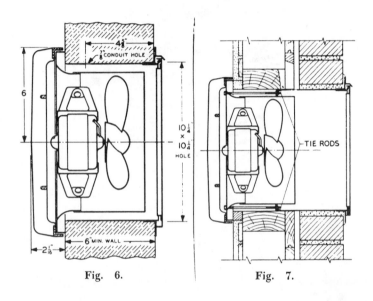

Fig. 6.   Fig. 7.

or wires between the outside wall studs at the desired location for the fan. If there are obstructions, select a location between two other studs. (Note that if the obstructions are to one side of the studding space, however, there may be enough space left for the fan.) After determining the location, locate the studs by tapping the wall lightly with a hammer or by measuring. Drill or cut a small hole to locate the stud and lay out the opening. In the case of frame construction, it may be possible to use a stud as one side of opening. This gives a more rigid mounting. A frame in the wall is not necessary, since the tie rods clamp the wall box and fan section firmly to the wall (Fig. 7).

If a wall ventilator is being installed in a house under

construction, and the outside wall is stone, brick, or block, grout the wall box into the masonry in the same manner as a door or a window frame. The wall box may be either recessed in the wall or flush with the outside of the wall (Figs. 8 and 9). If the box is recessed, make allowance for free opening of the wall box door. If the outside wall of the house is frame, shingle, or clapboard, frame the 10¼″ x

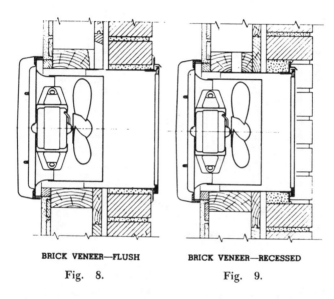

BRICK VENEER—FLUSH   **BRICK VENEER—RECESSED**

Fig. 8.   Fig. 9.

10¼″ opening between the wall studs (Fig. 10). If the location of the fan will allow, use a stud as one side of the frame. After the siding has been nailed in place and finished, mount the wall box in the frame and fasten it securely with nails or screws through holes in the side. Figure 11 shows the ventilator mounted flush in a frame house.

If the walls are more than 11″ thick, or if there is an air space between the wall box section and the fan section, the space should be lined with a No. 30 gauge metal sleeve soldered along its seam. A metal or roofing contractor can supply this material.

Electrical Wiring for Wall Fans. In the case of a completed house, run in the wires or conduit after the opening in the wall has been made. Allow enough length of wires or conduit so that they can easily be connected to the conduit box. When connecting the wires in the conduit box, be sure to connect the white wire to the other white wire, and the black wire to the other black wire. Use the nuts supplied by the manufacturer to fasten the conduit box into the wall box. A switch operated by the opening and

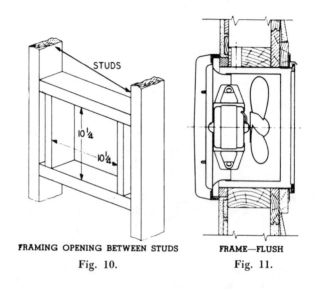

STUDS

10¼

10¼

FRAMING OPENING BETWEEN STUDS

Fig. 10.

FRAME—FLUSH

Fig. 11.

the closing of the outside door controls the fan. Test the switch and the wiring by plugging in a lamp or a 60-watt to 100-watt appliance.

In new frame construction, the wire or conduit should be run through when other house wiring is being done. In new masonry construction, the wires or conduit should be placed when the stone, brick, or block is being laid. Unless grounded metal conduit is employed, No. 14 copper wire should be used to ground the wall box to a water pipe.

Mounting of the Unit. To install the fan, inspect the wall box to see that the door opens and closes easily and completely. Figure 12 shows parts of the wall box section. Remove any mortar or waste material which might interfere with the operation of the door. Turn off the current until the fan is connected in the box. Install the door spring in the proper hole (Fig. 12, F). To install the pull chain (Fig. 12, D) remove the pendant, hook the spring in the proper hole in the door flange, and thread the chain around the pulley (Fig. 12, D, I). Do not let the chain run over the spacer pin of the fan section (Fig. 12, H). Test the door operation.

Figure 13 shows part of the fan section. To install the

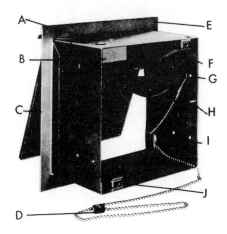

Fig. 12.   Wall box section. **A,** conduit opening; **B,** conduit box; **C,** door;
D, pull chain; **E,** outside wall
flange; **F,** door spring; **G,**
mounting holes; **H,** fan section
spacer pin; **I,** inside pulley; **J,**
spring nuts.

fan section, remove the shipping brace and replace the nuts on the motor bolts. Bring up the fan section and run the chain through the guide (Fig. 13, O). Plug the motor (Fig. 13, N) into the receptacle shown in Fig. 12, B, making sure that the wires do not interfere with the fan wheel. Start the tie rods into the spring nuts (Fig. 12, J) with slotted ends toward the body. Guide rods through the holes in Fig. 13, K and start the nuts on the slotted ends. Hold the fan section against the wall and run the rods in with a screwdriver until the nuts are flush with the fan plate, with

Fig. 13. Fan section. **K,** tie rod holes; **L,** spring nuts; **M,** corner holes; **N,** cord and plug; **O,** chain guide.

the screw heads flush with the nuts, and the fan firmly in place against the wall. Put the pendant on the end of the chain.

The grille shown in Fig. 14 can be mounted with the bars in a horizontal or vertical position. For horizontal mounting, remove the clip and the plug in the lower right-hand corner and drop the chain through the hole. For vertical mounting, leave the clip and the plug in place and drop the chain between the bars. The spring nuts on the fan plate are movable for lining up with the grille screws. Do not run the screws up too tight.

Fig. 14.

The ventilator shown in Fig. 3 may be installed in the ceiling if the duct work is not too long or too narrow. The wall box and the fan section should be in line, just as they were when installed in the side wall. The chain should hang freely between the grille bars.

General Maintenance of Fans. Maintenance suggestions for wall fans are as follows:

Motor. The motor in the wall or ceiling ventilator is suitable for either horizontal or vertical operation. It is constant speed for 50-cycle or 60-cycle, 110-120-volt single-phase operation, and for alternating current only. The power consumption is 60 watts. Direct-current motors are also available from the manufacturer on special order.

Method of Oiling. At intervals of six months, remove the grille and add a few drops of machine oil in each bearing. Do not lubricate excessively, nor while the fan is running. If there are no holes or tubes for the oil, the motor has sealed-in lifetime lubrication and never needs oiling.

Methods of Cleaning. The grille may be washed in soap and water after the excess grease and dirt have been removed with a paper towel. Do not use boiling water. Follow the manufacturer's directions printed on the back of the grille.

Window Fans. Window fans are available in different models and in sizes of 16″ and 20″. The 16″ fan can be installed in a window with or without adjustable panels; or it can be used as a daytime circulating fan. It is equipped with a spiral safety guard inserted in the discharge opening. The 20″ fan is designed for window installation only. The all-purpose 24″ and 30″ fans may be installed in a window, mounted to discharge through a wall opening, or installed in the attic with a plenum chamber. Figure 2 shows a portable twin fan ventilator with a thermostat control that turns *On* and *Off* automatically as the room temperature varies. It is reversible and fits sash and casement windows. Window fans 20″ or larger are belt-driven. They are usually encased in a sheet-metal enclosure that can be adjusted to the size of the window. Necessary screws, cotter pins, chains, and installation fixtures are furnished with these fans.

Installation of 16″ and 20″ Window Fans with Adjustable

Mounting Panels. The window in which the fan is to be installed should be large enough for the purpose; an opening that is too small reduces air delivery and prevents satisfactory operation of the fan. When installing a 16″ fan in a window, remove the guard by pushing down on the supporting wires and pulling it out. Each panel consists of a panel bracket and an extension panel (Fig. 15, A, B). To attach the panels to the fan frame, remove the three screws on each side of the frame, and attach the panel brackets to the frame with the same screws and washers. Secure the extension panels (Fig. 15, B) to the window stop (using the wood screws

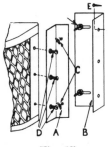

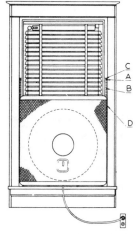

Fig. 15.   Fig. 16.

and washers shown in Fig. 15, E). The face of the panels should be approximately ⅛″ from the inside of the window sash, and the bottom of the extension panels should be ⅜″ above the window sill. Place the square shanked adjusting screws in the extension panel slots. Secure the slots with the four hexagon nuts . Place the fan (with panel brackets attached) on the window sill and secure the brackets to the extension panels with wing nuts (Fig. 15, C.).

Installation of 16″, 20″, 24″, and 30″ Window Fans without Mounting Panels. To secure the fan cabinet to the window, remove the screw on the top of both sides of the cabinet and attach the chains to the cabinet with the screws illustrated in side view in Fig. 16, D. Insert the two screw eyes (Fig. 16, A) in the inside edges of the window stops as follows:

25" above the window sill for a 16" fan
30" above the window sill for a 20" fan
36" above the window sill for a 24" fan
42" above the window sill for a 30" fan

Place the fan on the window sill in the center of the opening so that front edge of the fan clears the inside of the window sash ⅛". If there is a bottom sash, lift and remove the handle. Thread the chain (Fig. 17, B) through the screw eyes and insert cotter pins (Fig. 17, C) through the chain links above the screw eyes to prevent the chains from slipping back through the screw eyes. Adjust cotter pins in the chain links until the fan is secured parallel with the window sash. Spread cotter pin ends to secure the fan firmly. Raise the bottom window sash to the top of the fan-blade opening. Plug the fan into an outlet of the same voltage and frequency specified on the nameplate of the fan.

Installation of 24" and 30" Fans in Upper Section of Window. To mount a 24" or 30" fan in the upper part of a window, use the two hanger hooks and screws supplied with the fan. After locating the center of the trim at the top

Fig. 17.

of the window, place hanger hooks at the following distances to each side of the center: 12⅝" for the 24" fan, and 15¾₁₆" for the 30" fan. Secure the hanger hooks with screws. The bottom of the hooks should be 1" above the bottom of the window top. Hang the fan on hanger hook slots near the top of the fan housing on the discharge side. Lower the upper sash to the window at the bottom of the fan-blade opening. Plug the fan into an outlet of the same voltage and frequency specified on the nameplate of the fan.

Attic Fans. The larger types (36", 42", and 48") are generally used as attic fans. They are encased in wooden cabinets which are open at the front and back. These models are also belt-driven with the motor suspended directly beneath the fan-blade assembly.

Selection of Attic Fan. The major factors to consider in the selection of an attic fan are quietness of operation, simplicity of installation, and the capacity to move a large volume of air economically. Fans of varying sizes are available with the correct speed and design for successful air circulation through the attic. Table 29C shows necessary data for attic fans.

TABLE 29C

DATA FOR ATTIC FANS

Volume of House (Cubic Feet)	Fan Size (Inches, Diameter)	Free Air Discharge (Cubic Feet per Minute)	Normal H.P.	Recommended Fan Speed	Recommended Size of Opening to Attic (Square Feet)			from Attic (Square Feet)	
					Free Air Opening	Wood Grille 60%	Metal Grille 80%	Wood Louver 50%	Metal Louver 80%
7,400	24	3,700	1/6	675	6.2	10.3	7.8	12.4	7.8
11,000	30	5,500	1/6	480	9.2	15.3	11.5	18.4	11.5
16,000	36	8,000	1/4	410	13.3	22.2	16.6	26.6	16.6
24,000	42	12,000	1/3	300	20.0	33.3	25.0	40.0	25.0
30,000	48	15,000	1/2	290	25.0	41.7	31.2	50.0	31.2
34,000	54	17,000	1/2	260	28.3	47.2	35.4	56.6	35.4
46,000	60	23,000	3/4	245	38.3	63.8	48.0	76.6	48.0

Installation of Attic Fans. Install the fan near an opening through the ceiling of the attic (Fig. 18) or directly above a stairway. The opening should be centrally located so that good circulation of air can be secured in all parts of the house. The attic fan discharges the air to the outside through open windows or louvers in the attic walls or dormers. Place the fan over or near a partition wall or other solid support to prevent undue sagging of the ceiling and excessive vibration. An excellent position is over the closet or close to an attic door or ceiling opening. Do not install the fan less than three feet from the nearest side of the ceiling grille or opening. (If the fan is located too close to the grille, it may operate too noisily.) Place the fan the distance of its diameter from the nearest side of the ceiling opening, so that the air will be circulated as much as possible before it escapes through the exhaust openings. Construct the housing around the fan to create good suction. Mount the fan on rubber cushions or pads to keep noise and vibration to a minimum.

Due to the large volume of air that the attic fan will move per minute, there must be a sufficiently large opening into the attic for the air to enter; otherwise, back pressure will develop, causing poor circulation and excessive noise. The recommended sizes of ceiling inlet openings to the attic for efficient air circulation are shown in Table 2. To prevent back pressure, the outlet from the attic through the outside wall must be at least as large as recommended sizes shown in the table.

The attic inlet is usually a ceiling grille. It may be, however, just an open stairway or hole in the ceiling. The outlet for air to leave the house is called an exhaust opening. It is usually a louver constructed in the wall or windows, or just an opening in the wall (Fig. 18). The ceiling opening should be located so that air can be readily pulled through it from any or all of the rooms as desired. The opening or the open part of the grille should never be smaller than the fan. The opening into the attic must be at

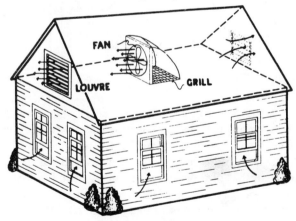

Fig. 18.

least as large as the fan. If a grille opening is used, the opening must be larger than the fan to provide for the lack of air flow through the partially obstructed area. For quiet operation, the net grille area should be of sufficient size to

assure air velocity through the grille of not more than 750
cubic feet per minute. For general practice, 500 to 750
cubic feet per minute is satisfactory. Ceiling fan openings
can be installed so that, when the fan is turned on, the grille
opens automatically, and, when the fan is turned off, the
grille closes. Certain types of grilles require metal shutter
equipment for best results.

Satisfactory operation of an attic fan also depends upon
adequate exhaust openings from the attic. Where possible,
these exhaust openings should be built into the side of the
house away from the prevailing winds. These openings can
be open windows or louvers. To secure adequate outlet

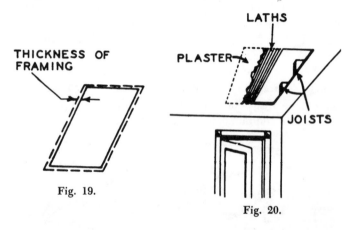

Fig. 19.

Fig. 20.

capacity, the construction of louvers is recommended.
Louver openings are usually constructed of wood, galvanized sheet iron, or copper. Metal louvers offer less obstruction to air flow and consequently their over-all size may be
smaller than that of other types. In every attic fan installation, it is necessary to provide a definite amount of free air
space in the attic openings so that the air may be expelled
easily. The free air space required for the louver openings
is space between the louver slats. This area depends upon
the amount of air to be moved and, therefore, upon the size
of the fan. Because of the obstruction to air flow caused by
the slats or louvers, their gross area will need to be more
than the actual free area required. Sometimes the net area

of the exhaust opening is too large for one louver or window, and, therefore, two or more such openings will be needed. (The required size of the exhaust area is given in Table 2.) Wire screen placed inside the exhaust louver will keep out birds, insects, and leaves but will retard the air flow. If wire screens are used, at least one-fourth additional louver area should be provided.

Installation of Fans with Plenum Chambers. The 24″ and 30″ fans may be constructed with plenum or suction chambers. A space of 5′ in front of the discharge opening of the fan is necessary so that the volume of air will not be obstructed. Place the fan unit over a solid wall partition. Locate the ceiling opening in the central part of the house. Install the fan discharge in the opposite direction from the largest or majority of openings in the attic so that it will discharge toward a blank wall.

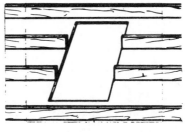

Fig. 21.

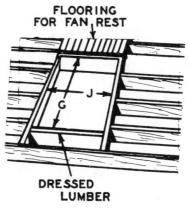

FLOORING FOR FAN REST

J

G

DRESSED LUMBER

Fig. 22.

For a 24″ fan, mark off the dimensions 44″ x 28¼″ square with the walls. For a 30″ fan, the dimensions should be 46″ x 33¾″. On the outside of the first rectangle mark off another rectangle (Fig. 19) larger than the first by a distance equal to the thickness of the framing used. Cut the ceiling opening on the dotted line.

Cut the plaster clean and completely through on the outside rectangle marked off in the ceiling (Fig. 20). Remove plaster in small portions with a sharp chisel. Cut the first two laths with a keyhole saw. A larger saw can be used for the other laths and for the joists.

Lay a false flooring around the opening (Fig. 21) on top of the ceiling joists to reinforce the joists and to protect the plaster during work in the attic.

The opening should be framed as shown in Fig. 22. Use 2″ dressed lumber measuring the same in width as the ceiling joists for the headers. The top of the headers should be flush with the top of the ceiling joists. Check the inside dimensions of the framing to correspond with the dimensions (Fig. 22, G, for a 20″ fan, and J, for a 30″ fan) required for the size of unit to be installed. Nail plaster stops on the bottom side of the framing, flush with the framing on the inside.

If the attic is not floored, provide a flooring on which to place the fan and plenum chamber. Use 1″ lumber to extend the flooring over 3 or 4 joists. When installing the unit, remove the protective metal grille on the back of the fan. To release the grille, remove the molding that is held in place with eight sheet-metal screws. Then replace the molding of the fan unit over the canvas and connect it to hold it in place. Turn the fan several times by hand to make

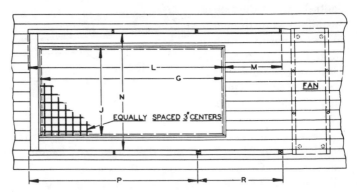

Fig. 23.   Floor plan of plenum chamber.

sure that all moving parts are clear of obstructions. The necessary wiring connection to the fan units and to the hinged door switch should be made by an electrician.

Construction of Plenum Chamber. Use Celotex, wallboard, or similar material to construct a plenum chamber (Figs. 23 and 24). Place the material over the ceiling open-

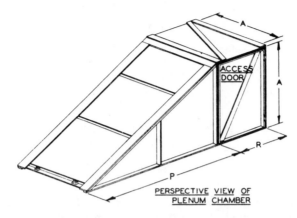

PERSPECTIVE VIEW OF
PLENUM CHAMBER

Fig. 24.

ing. Connect the plenum chamber to the fan frame with a canvas boot (Fig. 25), which should be securely fastened to the fan housing (with the molding on the housing to prevent air leakage).

A ceiling grille (either metal or wood) should be installed in the ceiling opening and held in place with a wide molding. Use a trap door (Fig. 26) for closing the ceiling opening when the fan is not being used. It is installed over the opening and serves as the top of the plenum chamber. A fusible link should be used to close the door automatically in case of fire.

Electric Wiring. Connect the fan motor to a separate circuit to carry the increased load, using No. 12 or heavier gauge wire as specified by the manufacturer. Locate the switch as near as possible to the most central and convenient location.

LUBRICATION

Various types of fans up to the 16″ size may have grease cups on both ends of the motor bell housings on the under-

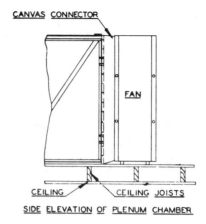

CANVAS CONNECTOR

FAN

CEILING — CEILING JOISTS

SIDE ELEVATION OF PLENUM CHAMBER

Fig. 25.

side. Use a special fan motor cup grease (No. F2) which is made exclusively for fan motor bearings and should be used only in the grease cups. Each grease cup has a wick and spring. The wick extends up through the bearing housing and into a small hole in the underside of the bearing. Any excess grease which is supplied to the bearings works its way out to the end of the bearing, drops down into the bearing housing, and returns through a channel to the grease cup. The grease supply in each cup is sufficient to last from 1500 to 2000 hours of operation before it needs replacing. Grease that appears to be dried up or hardened should be replaced.

Some types of fans are equipped with a grease cup on the front end of the motor housing and a small oil hole at the back end of the motor housing, on the upper side. A small amount of SAE 20 oil should be supplied to the back bearing occasionally, through this oil hole; then the bearings should be lubricated as indicated above.

There are small fans with motor bearings that are lubricated from oil holes at each end of the motor bearings. Use SAE 20 oil for lubricating. These fans have small felt washers inside the bearing housing which absorb the oil and keep the bearings lubricated over a long period of time.

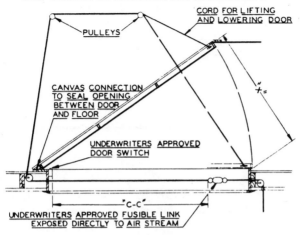

Fig. 26.

Large air circulator fans (16″ or over) should be oiled with a small amount of SAE 20 oil at least once a year. On the upper side of the motor, at each end, there is a blind cap which has a small knock-out plug at the center. This plug should be pushed in to permit oiling. Motors for these fans have a generous supply of wicking in the bearing housing to absorb the oil. Care must be taken not to over-oil the motor, for the oil may leak from the bearing and shaft and cause the fan blade to throw oil.

The oscillating gear case in all models uses a special non-fluid grease. A sufficient quantity of this grease should be used to cover all working parts. When repairing any oscillating type of fan, always check the lubricant in the oscillating mechanism. If the grease looks dry or appears to be hard, clean it out and replace it with new grease.

Never use the gear case grease in grease cups; it is not the proper lubrication. Never use the grease cup lubricant in the gear case, for it will liquefy and leak out.

CHAPTER 23

Air Conditioners

A room air conditioner of the type shown in Fig 1. is a completely self-contained refrigeration unit. In hot, humid summer weather it cools, dehumidifies, cleans, and circulates the air within a room; supplies fresh filtered outside air to the room; and exhausts stale smoky air from the room. A unit of this type is designed to lower the temperature of a room from eight to ten degrees below that of the outside. When cooling is not required, a well-designed unit filters, ventilates, exhausts stale air, and keeps the air in the room clean and fresh.

Fig. 1. Air conditioner.

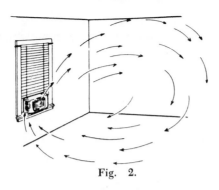

Fig. 2.

COMPONENTS

An air conditioner circulates the air so that all parts of the room are cooled and dehumidified (Fig. 2). Outside air enters the unit by means of a manually controlled *damper door* (Fig. 3). The filters in the type of unit shown in Fig. 4 insure clean fresh air in the room. The *evaporators* or *cooling coils* reduce the temperature of the air as it passes through the coil (Fig. 5). The *cold evaporator coil* dehumidifies or dries the room air by condensing the water vapor (Fig. 6).

SELECTION OF AIR CONDITIONERS

Rooms of the same size cannot be conditioned adequately by identical air conditioners. In determining the air conditioning requirements of a room, the following factors should be considered:

1. Total window area exposed to the sun.
2. Type of shading for window.
3. Amount of outside wall exposed to the sun.
4. General type of wall construction.
5. Type of roofing or ceiling over the room.
6. Underfloor construction.
7. Amount of heat generated in the room by other equipment.
8. Amount of door or arched space open to unconditioned rooms.
9. Number of occupants normally using the room.
10. Whether cooling is desired for day and/or night.

Because of the many factors involved, the homeowner should have the manufacturer's service representative make a preliminary survey to determine the type and size of unit to be installed.

OPERATION

Single-room air conditioners require the use of motive power, a refrigerating system to cool the air and reduce its

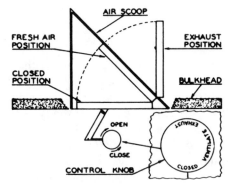

Fig. 3.

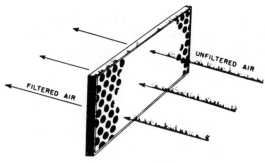

Fig. 4.

moisture content, and a duct system to guide the air through required channels.

Electrical System. The motive power in most single-room air conditioners is derived from two electric motors—the compressor motor (an a.c. motor of the capacitor-start, induction-run type) and the ventilating fan motor. Either of two a.c. motors—a shaded pole motor or a capacitor-start, capacitor-run motor—may be used for the ventilating fan drive.

Refrigeration System. The refrigerating system is divided into condensing and cooling units. The condensing unit consists of the compressor, the condenser coil, and the mounting assembly. The cooling unit has an evaporator

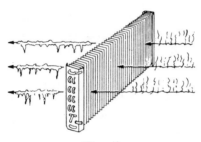

Fig. 5.

coil and an expansion valve. Freon-12 (a liquid refrigerant) is used in most standard air conditioners.

The *compressor* increases the temperature and pressure of the refrigerant vapor.

The *condenser* (condensing coil) consists of a series of finned tubes. It cools the vaporized, high-pressure, high-temperature refrigerant necessary for condensation and has finned tubing arranged for easy passage of air. The condenser air fan circulates air over the surfaces of the finned tubes.

Fig. 6.

The *evaporator* (cooling coil) also has a series of finned tubes. Liquid refrigerant flows into the evaporator from the expansion valves under greatly reduced pressure. The reduction of this pressure causes the refrigerant to vaporize rapidly and to absorb heat from the tubes and fins, thus chilling the evaporator. Warm air passing over the evaporator becomes cooled as it comes in contact with the tubes and fins of the evaporator.

The *thermostatic expansion valve* is a nonadjustable automatic valve actuated by a thermostatic element to control the flow of refrigerant. The thermostatic element of the valve functions to maintain the required flow of liquid refrigerant to the evaporator. The expansion valve is preset and hermetically sealed by the manufacturer.

Refrigerating Cycle. An air conditioner refrigerating cycle has both a high-pressure and a low-pressure side. The refrigerant is under high pressure from the compressor discharge valve through the condenser, the liquid receiver tank, and the liquid line to the expansion valve. The rest of the system, from the expansion valve through the evaporator and suction line to the compressor suction valve, is under low pressure. The refrigerant vapor passes from the evaporator to the suction valve of the compressor to complete the cycle.

Capacity of Refrigerating System. The capacity of an air conditioner refrigerating system depends upon the amount of Freon-12 circulated per hour of operation. For every pound of liquid converted to gas in the evaporator, 60 B.T.U.'s (British thermal units) are removed in the form of latent heat. This heat represents the amount of energy required to change one pound of liquid (Freon-12) to a vapor. The heat required for this change does not raise the temperature but enters the substance in the form of added energy. The heat energy added to the vapor is carried from the evaporator through the compressor to the condenser, where it is released by condensation. When the refrigerant is automatically returned to its liquid state, it is ready for re-use in the system.

Condensate Disposal. The air that passes over the tubes and fins of the evaporator is sufficiently cooled to condense moisture from the atmosphere. This moisture is then drawn automatically through a slot and blown as a vapor through the condenser to the outside.

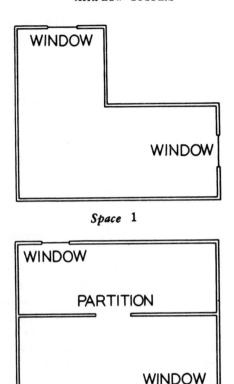

Space 1

Space 2

Fig. 7.

Airflow System. The airflow system of an air conditioner is composed of ducts that provide a flow of outside air through the condenser. The ducts also circulate the room air through the filter and evaporator. To cool the condenser, outside air in the unit is circulated over the motors and then returned through the condenser to the outside. This air stream, which is created by a blower fan of the compressor motor, picks up condensed moisture from the evaporator. It drives the moisture in the form of vapor to the outside together with the hot air from the condenser and motor-compressor compartment.

A fan draws the room air to be cooled through the room-air inlet grille, filter, and evaporator. The filtered, cooled, and dehumidified air is then forced into the room through the conditioned-air outlet grille or grilles.

INSTALLATION OF THE AIR CONDITIONER

A single-room air conditioner should be installed, if possible, in a window on the shady side of the room. If the unit must be exposed to the sun, use a ventilated awning to shield the outer casing. Venetian blinds or shades will cut down a certain amount of sun heat transmitted to the room through the windows.

The unit must be securely fastened to a good, sound window sill so that the weight will be safely supported and vibration reduced to a minimum. The window sill must be leveled to insure proper disposal of the condensate.

Multiple installations are sometimes required for rooms that have construction or layout features that would hinder the cooling efficiency of a single-room air conditioner. For example, spaces 1 and 2 shown in Fig. 7 were found after a survey to require a one-ton unit each. Space 1, with its irregular shape, and space 2, with its partitioned area, would have ideal cooling if two one-half-ton units were installed. Installation of a single one-ton unit in either of these spaces would, because of the restricted airflow, result in hot and cold spots.

Installation Tools. Tools required for uncrating and installing the equipment are as follows:

Claw hammer or nail puller.
Carpenter's level.
Tinsnips.
Hacksaw.
Scriber.
6' rule.
Screwdrivers:
   10" electrician type.
   heavy duty.
Hand drills and drill bits.

Wrenches:
   Ratchet wrench with following:
      $\frac{9}{16}$" socket for skid bolt.
      $\frac{5}{8}$" Universal socket with 9" extension for compressor hold-down nuts.
      $\frac{1}{2}$" open end for adjusting angle bolt heads.
      $\frac{9}{16}$" open end for adjusting angle nuts.
      $\frac{1}{8}$" Allen wrench.

Uncrating Procedures. Air conditioners are delivered from the manufacturer in crates with all the necessary installation fixtures (Fig. 8).

To uncrate the equipment, pull all staples in the bottom

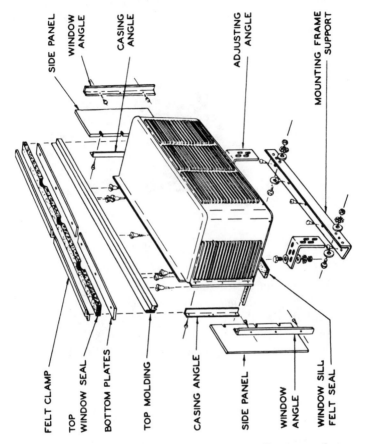

Fig. 8. Exploded view showing parts for installation in relative positions.

edge of the crate and remove the case. Be careful not to damage the cabinet. Unwind the service cord. Raise one side of the skid off the floor with blocks. Use a ¾″ socket wrench to remove the skid bolts in the counterboxes. Tip up the other side of the skid and remove the remaining bolts. It is not necessary to remove the room cabinet, but be careful in handling the chassis with the cabinet attached (Fig. 9). Place a block under the unit to protect the lower edges of the cabinet.

Installation in Double-Hung Windows. Make sure that the window sill is level (Fig. 10). The unit can be mounted on sills that are not level, but shims may have to be added

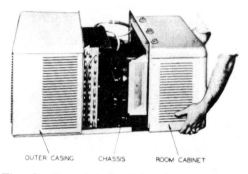

OUTER CASING   CHASSIS   ROOM CABINET

Fig. 9. Removing chassis from outer casing.

to level the unit itself, and additional sealing may be required. Scrape all paint runs and other foreign matter from the sill step so that the surfaces will be smooth for flush alignment. Check the depth of the inside sill. If it is less than 1⅜″, use wood stripping as shown in Fig. 11.

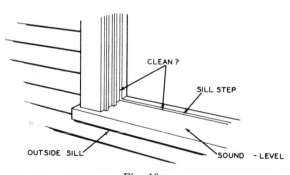

CLEAN ?

SILL STEP

OUTSIDE SILL

SOUND - LEVEL

Fig. 10.

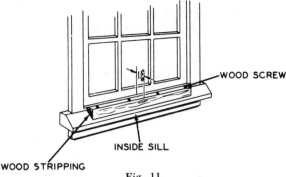

WOOD SCREW

INSIDE SILL

WOOD STRIPPING

Fig. 11.

Insert a side panel in the sash channel flush with the sill step (Fig. 12). Outside of the panel, place a window seal angle; adjust it for height and outline and drill pilot holes for the wood screws. Screw the angle firmly in place. The side panel should fit snug. Repeat the operation on the opposite side of the window frame. Remove the side panels. Do not alter the dimensions of these panels at this time.

Measure and mark the center of the inside sill with a line (Fig. 13). Extend the line to the outside sill as an aid in installing the angle mounting frame. Place the angle mounting frame on the outside sill with its inside angle toward you. Align the center hole over the center line on the sill. Outline and drill pilot holes for the wood screws. Insert

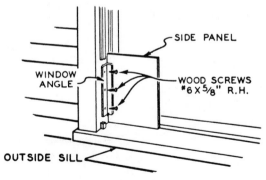

SIDE PANEL

WINDOW ANGLE

WOOD SCREWS #6 X ⅝" R.H.

OUTSIDE SILL

Fig. 12.

screws and tighten them. For stone or brick, use expansion plugs and bolts.

On some installations it may be necessary to shorten the vertical dimensions of the adjusting angles. To determine the required height, extend a level or straightedge from the top of the inside sill out over the angle mounting frame (Fig. 14). Measure from the bottom of the straightedge at A to the bottom of the angle mounting frame at B. This

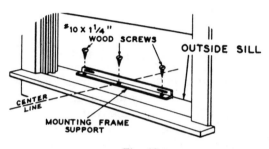

Fig. 13.

measurement minus ⅛″ equals the vertical dimension of the adjusting angle. Cut it to the required size with a hacksaw. Do not change the horizontal dimension. Estimate the position of the adjusting angles on the bottom of the

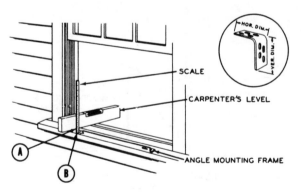

Fig. 14.

mounting frame track. Bolt the angles in place with the carriage bolts, keeping the heads of the bolts up. Place the

outer casing assembly of the air conditioner on the sill, with
the adjusting angles resting on the angle mounting frame
(Fig. 15).

Center the assembly by locating the center line on the
inside sill beneath the center hole in the mounting frame
cross member. Pull the casing toward you until the locating

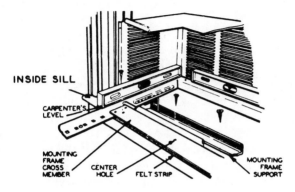

Fig. 15.

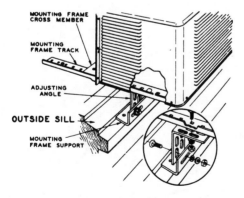

Fig. 16.

lip which extends down from the front cross member is
tight against the sill edge. Holding the casing firmly, outline the three screw holes and drill them with a pilot drill.
Insert the felt seal between the front cross member and the

inside sill. Insert wood screws and tighten them securely. Bolt the adjusting angles to the angle mounting frame. Level the assembly by moving the adjusting angles up or down (Fig. 16).

The sill clamp secures the mounting track to the edge of

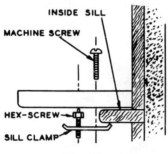

INSIDE SILL

MACHINE SCREW

HEX-SCREW

SILL CLAMP

Fig. 17.

the inside sill with screws (Fig. 17). After the clamp is in position, cut off the excess lengths of screws where they come through the bottom of the clamp.

Measure the side panels to fit the space remaining between the outer casing and window channel (Fig. 18). To cut a side panel, scribe a deep line; place the panel so that the scribed line rests on the edge of a flat surface. Hold the panel securely with the palm of one hand and strike the overlapping section a sharp blow with the heel of the other hand. With the

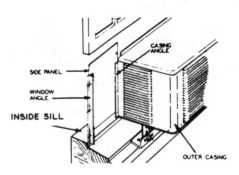

CASING ANGLE

SIDE PANEL

WINDOW ANGLE

INSIDE SILL

OUTER CASING

Fig. 18.

smooth surface facing the room side, set the panels in place. Put the casing angles in place, align the holes and slots, then insert screws and tighten them securely. Cut the top molding to the proper length and install it over the casing top

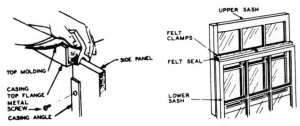

Fig. 19.

flange and the top of the side panels (Fig. 19). The rounded section of the top molding fits only on the casing top flange. The square portion fits only on the side panels. Lower the window, pressing it down behind the molding until the bottom edge of the sash rests on the ledge of the molding.

Remove the window lock. Seal the opening between the upper and lower sashes with felt and clamps furnished for this purpose. Cut or notch the wood as required. Nail the materials in position.

On three-fourths-ton and one-ton units, the hold-down nuts are tightened at the factory to hold the compressor firmly in place during shipment. Failure to release the nuts and remove the slide-out washers (Fig. 20) will result in noisy operation. Back off the nuts until they are flush with the tops of the studs. This position allows full expansion of the springs and provides ample space for movement of the com-

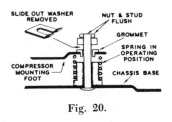

Fig. 20.

pressor. The chassis is now ready to be installed in the outer casing. Check the fans by hand spinning. Plug the unit in an electric outlet for a trial test before installing it in the outer casing. Inspect the bottom pan for foreign matter. Lift the chassis and slide it into the casing, holding the front of the unit up slightly to allow clearance for the felt sealing strip attached to the bottom pan. Push the chassis until the bulkhead flange is tight against the casing flanges.

Installation in Casement Windows: First Method. Because there are many different types of casement windows, it is possible to outline only the general procedures for installing air conditioners in them. Remove sufficient glass

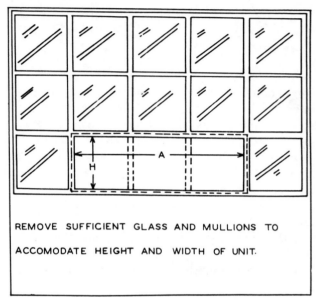

REMOVE SUFFICIENT GLASS AND MULLIONS TO
ACCOMODATE HEIGHT AND WIDTH OF UNIT.

Fig. 21.

and mullions to allow for passage of the unit (Fig. 21). Build up the room side of the window sill until its top is above the horizontal cross member forming the bottom of the frame on the metal window (Fig. 22). Measure the height and width of the opening left by the removed glass and cut a filler panel of ¼-inch masonite or equivalent material to fit (Fig. 23). Cut out the center of the filler panel to the exact outside dimensions of the outer casing. In cutting the height of the filler panel, allow for the height of the bottom cross member on the window frame.

Install the cross angle outer casing and adjusting angles in the same manner as that described for the double-hung windows. Install the filler panel in the opening and seal the edges to the window frame with putty or caulking com-

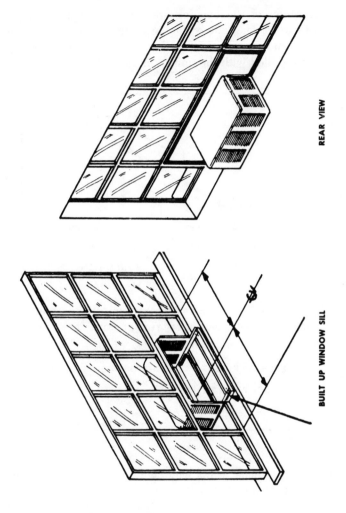

REAR VIEW

BUILT UP WINDOW SILL

Fig. 22.

pound. Install casing angles to hold the filler panel against the outer casing flanges. Insert the chassis, install the cabinet, and plug in the unit.

Installation in Casement Windows: Alternate Method. Remove the glass and mullions, as indicated above, to fill in the resulting opening. Install the outer casing in the exact center of the opening and secure it rigidly. Install filler board of masonite or similar material across the top of the unit and cement the board in place. Cut regular filler panels supplied with the unit to fill in at the sides. Cement

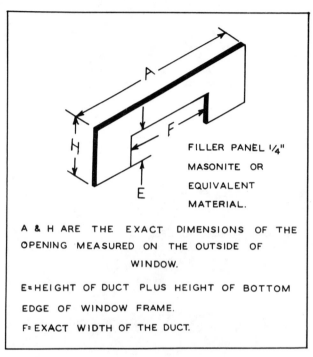

FILLER PANEL ¼"

MASONITE OR

EQUIVALENT

MATERIAL.

A & H ARE THE EXACT DIMENSIONS OF THE OPENING MEASURED ON THE OUTSIDE OF WINDOW.

E = HEIGHT OF DUCT PLUS HEIGHT OF BOTTOM EDGE OF WINDOW FRAME.

F = EXACT WIDTH OF THE DUCT.

Fig. 23.

the panels in place at the window frame. Use casing angles against the outer casing.

Some casement windows (where the glass panels are small) will require the cutting out of a horizontal cross member as well as vertical mullions. In such instances, reinstall the cross member at a height equal to the height of the outer casing. Cut and reinstall the glass above the cross members. The regular metal filler panels supplied with the unit may be used to fill in the sides.

Wall Installations. To install an air conditioner through a wall, cut a hole through the wall (Fig. 24) large enough so that when it is boxed with a wooden frame it will be 4″ wider than the width of the outer casing. The resulting two-inch clearance on each side of the outer casing is necessary for sufficient cooling air to flow to the condenser. The bottom board of the box must be pitched slightly to shed rain water.

In cutting through a stud or a frame construction wall,

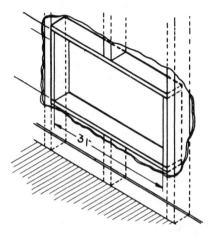

Fig. 24. Wall boxing.

make the height of the opening conform with the height of the unit, allowing for the thickness of the framing. In cutting through a masonry wall, make the opening slightly higher and wider. After the wood boxing has been installed (Fig. 25), the surrounding surface should be refinished, weatherproofed, and allowed to dry before the installation is completed. Make an inside frame that will serve to finish the wall. At the same time seal the opening between the casing edges and the wall (Fig. 26). Use ¾″ × 4″ facing wood, and miter and round the corner so that the fit at the corners of the casing is airtight. Fasten the inside frame on the boxing. It is recommended that the frame be assembled around the casing, as the outer casing louvers and the track depressions prevent passage of the casing. Center the casing in the wall boxing. Leave it loose so it can be ad-

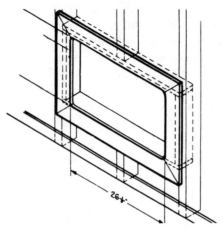

Fig. 25.   Wall boxing and inside frame.

justed slightly while installing the frame. Fit the bottom side between the locating lip and the depression in the tracks. Then fit the sides between the casing side flanges and the wall. Fit the top piece between the casing top flange and the wall. Make sure the frame is airtight against the casing and at the mitered joints since it must be a barrier against the weather.

Finally, level and secure the casing. Push the casing so that the flanges are tight against the inside framing. Put a screw through each side of the casing. Level the casing with the mounting frame angles. Put additional screws through the sides of the casing, if necessary, and plug in the unit.

MAINTENANCE

Elimination of dust, lint, and pollen from the air is the function of the *filters,* and they should be replaced periodically. Under normal conditions this is usually done once every season, but in areas where there is a great concentration of dust or pollen, it may be necessary to change the filters more often. Clogged filters will greatly reduce cooling capacity and may cause formation of ice on the cooling coils.

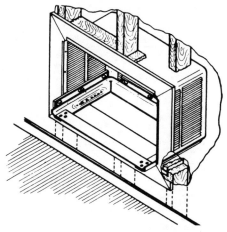

Fig. 26.   Completed installation.

The compressor motor requires no oiling. The condenser and evaporator fan motors should be oiled every season with a good grade of SAE 20 motor oil as directed by the manufacturer.

Air conditioners are designed to provide year-round service. The unit may be left in the window since all parts are thoroughly protected against exposure. Make sure the damper door is closed to prevent drafts.

As an added precaution, consult the following check list periodically:

1. Proper voltage and current should be checked against data plate.

2. Check the voltage under load.

3. Check the tightness of set screws on both the evaporator and the condenser fans.

4. Clear any foreign material from the bottom pan.

5. Check nuts, screws, and bolts throughout the unit for tightness.

6. All window sealing strips should be tight.

7. All curtains, drapes, and other window decorations should be adjusted away from the side louvers on the room cabinet of the air conditioner.

8. Make sure that each unit is connected to its own outlet.

TROUBLE AND REMEDY CHART

Trouble	Cause	Remedy
Compressor and fan motor do not start.	No power.	Check supply line fuses, circuit switches, and wall outlet.
	Faulty attachment cord.	Replace. Be sure the trouble is not due to the fact that the prongs of the plug are too close together or too far apart.
	Faulty switch.	Replace.
	Loose connections at switch terminals.	Tighten.
	Loose connections at wire nut connection.	Tighten.
Fan runs, but compressor does not start.	Low voltage.	Check as above.
	Faulty switch.	Replace.
	Faulty wire nut connection.	Tighten.
	Faulty connection at switch terminals.	Tighten.
	Defective ripcord to thermostat.	Replace.
	Thermostat set too low.	Adjust.
	Thermostat defective.	*Replace.
	Ripcord to compressor terminal box defective.	Replace.
	Loose connections on overload.	Tighten.
	Defective operating coil on relay.	*Replace relay.
	Defective contacts on relay.	*Replace relay.
	Defective starting capacitor (open circuited, short circuited, loss of capacity).	*Replace.
	Loose connections at compressor terminals.	Tighten.
	Defective compressor motor (short circuited, open circuited, grounded).	*Replace chassis.
	Compressor stuck.	Momentarily reverse run and start leads at compressor.
Fans run, compressor starts but stops after an interval.	Operation of overload due to high voltage.	Check voltage supply. Clean condenser. Check operation of relay. *If defective, replace. *Check compressor for short circuit. Replace, if defective. Check outside (condensing) air temperature. *Check condenser fan setting on shaft of motor. Check fan motor speed. Remove obstructions to air flow around outer casing. Unblock moisture trough from evaporator to condenser. Check running capacitor. *Replace if short or open circuited.
Fans run, compressor starts and runs, but compressor occasionally stops (on overload).	Low voltage due to overloaded circuits within the house or through local power demands. This condition might exist only at certain times during the day).	Run separate line to unit (12 AWG or larger). Consult local power suppliers.

* Consult service department of manufacturer.

TROUBLE AND REMEDY CHART—*Continued*

Trouble	Cause	Remedy
	High voltage due to fluctuations in local power system (usually occurs at various low load periods of the day.	Consult local power suppliers.
	Extreme heat on outer casing of unit during hottest part of the day.	Shade outer casing with a ventilated awning.
	Partial short circuit in compressor motor.	*Check compressor wattage.
	Open circuited running capacitor.	*Replace.
Compressor starts and runs, but fan does not run.	Faulty switch.	Replace.
	Open circuited fan motor.	*Replace.
	Fan binding on shroud or ring.	*Adjust fan setting.
	Bearings on fan motor frozen.	Where oil holes are provided, lubricate with SAE 20 oil. *Replace fan motor if necessary.
	Loose connection at switch terminals.	Tighten.
	Loose connections at wire nut.	Tighten.
Compressor starts and runs, but fan motor starts— then stops.	Operation of the internally connected fan overload device due to short circuit in fan motor windings, binding of fan on shroud or venturi, or lack of lubrication in fan motor bearings.	Adjust fan on shaft. Where oil holes are provided, lubricate with SAE 20 oil. *Replace fan if necessary.
Unit gives electrical shock.	Grounded electrical circuit.	Eliminate ground.
Noisy unit.	Fan hitting condenser or evaporator coils, shrouds, or venturi ring.	Adjust fan position on motor shaft; reposition fan motor-bracket assembly.
	Fan motor bearings defective.	*Replace fan motor.
	Loose hold-down nuts on fan motor-bracket assembly.	Align fan and tighten nuts.
	Broken or bent fan.	*Replace.
	Refrigerant absorbed in compressor oil after extended shutdown.	Operate compressor. (Noise will disappear after unit runs a while.)
	Internal parts of compressor defective.	*Replace chassis.
	Hold-down nuts on externally spring-mounted compressor not released, or slide-in shipping washers not removed.	Loosen nuts and remove slide-in washers.
	Loose room cabinet or outer casing.	Tighten.
	Unit improperly installed (loose side panels, adjusting angles; mounting track not securely fastened to sill).	Make necessary adjustment to components of installation.
	Screen-in damper opening vibrating.	Fasten securely.
	Damper control mechanism loose.	Tighten.
	Loose terminal box cover on side of compressor.	Tighten.
	Copper tubing vibrating.	Adjust or tape.
	Loose electrical components.	Fasten securely.
	Copper tubing vibrating.	Adjust or tape.
	Loose shrouds or venturi ring.	Tighten.

* Consult service department of manufacturer.

Trouble and Remedy Chart—*Continued*

Trouble	Cause	Remedy
Insufficient cooling.	Unit standing too long without being run. (It is possible for all the freon to become absorbed in the oil.)	Working pressures have to be established by getting the freon out of the oil. This may take several hours of continuous running and must be allowed for.
	Insufficient airflow through condenser, due to:	
	1—Dirty condenser.	Clean.
	2—Obstructed louvers on outer casing.	Remove obstructions.
	3—Fan motor not running.	Check electrical system.
	4—Fan motor not up to speed.	Check electrical system.
	5—Condenser fan slipping on motor shaft.	Adjust fan position; tighten set screw.
	Moisture in base pan not being picked up by slinger ring on condenser fan, or lack of moisture in trough below slinger ring.	Unblock moisture trough between evaporater and condenser; adjust position of condenser fan on motor shaft.
	Insufficient airflow through evaporator, due to:	
	1—Dirty evaporator.	Clean.
	2—Ice on evaporator coils (condition, with insufficient cooling, indicates restricted airflow through the evaporator).	Defrost.
	3—Dirty filter.	Clean. Replace if necessary.
	4—Obstructed louvers on room cabinet.	Remove obstructions.
	5—Fan motor not running.	Check electrical system.
	6—Fan motor not up to speed.	Check electrical system.
	7—Evaporator fan slipping on motor shaft.	Adjust fan position; tighten set screw.
	Damper opened, or felt damper seals defective.	Check and adjust.
	Room cabinet not attached securely to bulkhead.	Tighten.
	Heat load in room exceeds capacity of unit.	Refer to original load calculations; recalculate heat load.
	Windows and doors in room open.	Shut.
	Installation parts incorrectly used.	See installation data.
	Restricted capillary tube or strainer, indicated by:	*Check unit wattage, and take wet and dry bulb thermometer reading.
	1—Frost on capillary or strainer.	
	2—Low wattage.	
	3—Condenser not warm, evaporator only partially cool, or not at all.	
	Compressor not pumping, or insufficient charge of refrigerant, indicated by:	*Replace chassis if necessary. Check as above.
	1—Low wattage.	
	2—Condenser not warm, evaporator only partially cool, or not at all.	

* Consult service department of manufacturer.

TROUBLE AND REMEDY CHART—*Continued*

Trouble	Cause	Remedy
Too much cooling (not to be confused with insufficient cooling—ice on evaporator coils).	Defective thermostat. Outside (condensing) air temperature very low. Overcharge of refrigerant, indicated by high wattage and sweating of the return line all the way to the compressor.	*Replace. Operate unit for cooling with outside air (damper on ventilate). *Check wattage. Replace chassis.
Sweating.	See above. Moisture trough from evaporator to condenser clogged. Insulating seals on unit damaged. Fan motor not up to speed. Fans incorrectly positioned.	Remove obstructions to moisture flow. Inspect. Adjust or replace. Check electrical system. *Adjust.

* Consult service department of manufacturer.

CHAPTER *24*

Electric Water Heaters

Automatic electric water heaters are as dependable and safe as electric lights. The outer surface (the only part that can be touched) is thoroughly insulated and never gets warm. The required temperature of 150° can be safely maintained for dishwashing and clothes washing. No special installations are necessary other than wiring and normal piping.

CONSTRUCTION

The construction of a modern automatic electric water heater is simple and efficient. It has a rugged tank designed

Fig. 1. Electric round water heater.

Fig. 2. Electric table top water heater.

to withstand the high test pressure of 300 lbs. per square inch and to operate under all conditions. Water is heated by either one or two heating elements that are thermostatically controlled. The tank is completely surrounded by heavy insulation that keeps the heat in and the cold out. (Two standard types of heaters are shown in Figs. 1 and 2. The various parts of a typical twin unit storage-type heater are shown in Figs. 3 and 4.)

CAPACITY REQUIREMENTS

Individual families vary greatly in their consumption of hot water. In an average household, the hot water consumed amounts to approximately one-fourth to one-third of the total water usage. Water bills may thus be checked to estimate possible requirements. The following data indicate the average amounts of hot water required in relation to the size of the family.

PERSONS IN FAMILY	AVERAGE GALLONS USED PER MONTH
2	720
3	960
4	1200
5	1440
6	1680
7	1920
8	2160
10	2640

The following data on the average amount of hot water used for various home activities can also indicate probable daily or monthly hot-water requirements.

HOT WATER USED (Purpose)	AMOUNT USED EACH TIME (In Gallons)
Tub bath	7.0
Washing hands	.9
Shaving	2.0
Washing dishes	2.0-7.0
Laundry	30.0

It has been established that one kilowatt-hour (1000 watts for one hour) will raise approximately 3.5 gallons of

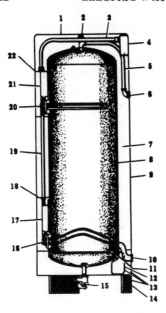

Fig. 3. Cross section of twin round shell storage type of water heater. (1) provision for installation of temperature-pressure relief valve; (2) removable cover; (3) rigid conduit (when used); (4) concealed outlet box; (5) "built-in" heat trap; (6) hot water outlet; (7) rock wool insulation; (8) tank; (9) steel outer shell; (10) cold water inlet with ¾" standard pipe thread. "S" trap design or straight nipple; (11) baffle. Cold water deflector; (12) tank supports; (13) tank supports, bottom cover, and base legs; (14) legs; (15) drain valve (or straight nipple, capped); (16) lower calrod unit; (17) lower thermostat; (18) steel housing around thermostat and unit, with insulation pad; (19) rigid conduit (when used); (20) upper calrod unit; (21) upper thermostat; (22) steel housing for upper thermostat and unit section.

water 100°F. Therefore, in one hour a 2000-watt unit will raise seven gallons of water from 50° to 150°F. Most tank sizes and unit wattages are designed to allow for complete recovery or reheating of the entire tank contents in an eight-hour period, which is usually during the night. This provision permits the homeowner to start the day with a full tank of hot water and have sufficient heating capacity during the day to prevent complete depletion.

INSTALLATION

The method of installing any water heater has a great deal to do with its efficiency and economy. Specific plumbing specifications furnished by the manufacturer should be followed. Basically, the water heater should be placed close to the point where most water is used. The proper place for the water heater, therefore, is in the kitchen, adjoining the sink. If it is placed in the basement, it should be located as near as possible to a point directly under the sink.

As a general rule, hot-water connections on water heaters are ¾″ (or ½″ for some heaters with very small capacity). For economy, the entire exposed hot-water pipe line should be insulated with air-cell asbestos or any other suitable insulating material. The shorter the hot-water supply lines, the more economical the installation and operation will be for the homeowner.

Relief Valves. Local codes require various types of pressure or temperature-relief valves, or combination temperature and relief valves. Generally these valves are not furnished with water heaters and must be secured from a local supplier.

Temperature Limit Switch. A temperature limit switch is mounted onto the tank by means of a soldered bracket and furnishes protection in case of an excessively high

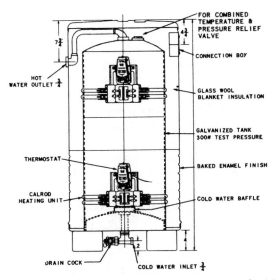

Fig. 4. Cross-section, schematic view of a typical 15-30-40-52-66-82-gallon water heater.

temperature. An adequate method of pressure relief must also be used with this switch. The temperature limit switch will shut off the units at a tank water temperature of 195°, plus or minus 5°F. After the limit switch has been tripped, one must reset it manually by pressing down the reset tab.

REPAIR OF DRAIN LEAKS

In case of drain leakage, both the sill cock and the straight nipple (shown at 15 in Fig. 3) that are used as drain fittings can be removed and replaced without dismantling the entire water heater. The water heater should be drained before a fitting is removed. Be sure to use fresh graphite or other sealing cement when replacing a fitting. Excessive water-system pressure can result in a leak at the drain. Modern water heaters are constructed for a 300-lb. test pressure and a 127-lb. working pressure.

Do not drain the tank completely before pulling the switch or removing the fuse to disconnect the electric service to the water heater. Do not reconnect the electric service until the tank is again full of water.

If there is a drain leak from within the heater, and it cannot be stopped by replacement of the fitting shown at 15, in Fig. 3, remove the shell of the water heater and the insulation and make the necessary repairs as follows:

1. Disconnect the heater from the power supply.
2. Remove the shell thermostat covers.
3. Disconnect the wiring at the electrical connection box (at back or bottom of heater).
4. Stand in front of the water heater and pull out the wiring (leaving only the ends still attached to the thermostat and unit).
5. Loosen the bushing on the conduit connecting the top steel thermostat housing with the lower thermostat housing.
6. Pull out the back steel thermostat housings.
7. Let all available water flow into the bathtub or other containers to conserve it.

8. Disconnect plumbing connections at the water-heater shell. Allow any remaining water to run out through the cold-water inlet.

9. Remove the heater shell.

10. Take out the screws at the bottom of the heater shell.

11. Place newspapers or large section of cardboard or board around the water heater to catch rock wool.

12. Lift the shell upward and off the heater. This may require a slight amount of prying at the cold water inlet. Such prying should be done on the underside of the fitting in order to prevent visible scratches or marks.

13. If the shell is difficult to remove, take out some of the rock wool through the thermostat housing openings. This permits additional movement and loosening of the shell. Also take some wool out through the top opening, as this may permit easier removal of the shell.

14. Move rock wool on the floor to one side of the water heater; work in a dry area that is covered with a newspaper.

15. Any wet rock-wool insulation should be thoroughly dried out before being reused for repairing the heater.

16. Check the cold water inlet to determine the point of leakage by connecting the heater again to the water lines. In this way you can test the pressure.

17. Should the cold water fitting be cracked, or the leak come from a threaded section, replace the inlet and retest.

18. If there is a leak in the tank, the necessary repairs should be made by a qualified mechanic.

19. Replace the shell of the heater and screw it in place.

20. Sift in some of the rock wool up to the bottom unit opening.

21. Replace the thermostat housings and conduit.

22. Complete the insertion of the rock wool. Add more rock wool if it does not cover the top of the heater.

23. Pull the wiring through the outlet box and connect it to the service.

24. Replace the top of the heater shell.

25. Reconnect the plumbing.

26. Replace the thermostat housing insulation pads and covers.

For heaters that have conduction-type units (Fig. 5), the procedure outlined above may be used, with the exception that the insulation is in blanket form and must be handled differently. In replacing the insulation blankets, hold them

Fig. 5. Conduction type of unit.

in place by shims while replacing the outer shell. If shims are not available, hold the blankets in their proper position by tying them with string until the shell is replaced. The string should be placed in such a position that it may be cut and removed from the heater after the shell is replaced.

CALROD HEATING UNITS

The Calrod water-heating unit shown in Fig. 4 has a resistance wire that is enclosed in and electrically insulated from a noncorrosive sheath. A complete immersion-type unit may consist of one or two hairpin-shaped coils. Before the construction of modern water heaters with conduction-type units, the Calrod heating units were of the direct-immersion type. They extended directly into the water and gave the maximum possible heat transfer. The modern Calrod heating units are similar in size and type and are adaptable to both round and square units. In case these heating units fail, they must be tested and, if necessary, repaired or replaced by the manufacturer's service department.

WIRING

An electric water heater is completely wired to the terminal box at the bottom or in the back of the heater. The branch circuit to the heater should be either 115-120 volts or 230-240 volts a.c., to correspond with the data on the heater nameplate. Single-phase operation is standard for most domestic water heaters. A water heater should be operated on a voltage not over 5 per cent above or below the nameplate rating. It should be on an individual branch circuit and should be controlled by an indicating switch which is not part of the appliance.

WATER HEATER RATING (In Watts)	115-120 VOLTS 2 WIRE	230-240 VOLTS 2 WIRE
SINGLE-UNIT HEATERS		
2000 or less	No. 12	No. 12
2500 or less	No. 10	No. 12
3000 or less	No. 10	No. 12
3500 or less	No. 8	No. 12
4000 or less	No. 8	No. 12
4500 or less	No. 8	No. 12
TWO-UNIT HEATERS *Top Unit*		
2000 or less	No. 12	No. 12
2500 or less	No. 10	No. 12
3000 or less	No. 10	No. 12
3500 or less	No. 8	No. 12
4000 or less	No. 8	No. 12
4500 or less	No. 8	No. 12
Bottom Unit		
2000 or less	No. 12	No. 12
2500 or less	No. 10	No. 12
3000 or less	No. 10	No. 12
3500 or less	No. 8	No. 12
4000 or less	No. 8	No. 12
4500 or less	No. 8	No. 12

Note: Heaters with units of more than 3000 watts each at 115-120 volts require magnetic switches and must be wired according to diagram furnished with the heater.

The preceding table shows the size of the rubber-covered wire recommended for the branch circuits to the heater. The ratings are based on the current-carrying capacity of rubber-covered conductors as specified by the National Electrical Code.

Limited Demand Type of Wiring. All standard two-unit water heaters are equipped with a double-throw thermostat in the upper position. Figure 6 shows a wiring diagram of the heater with the thermostat wired to give double-throw or "limited demand" operation. When a tank is filled with cold water, the double-throw thermostat closes the circuit to the upper unit while the circuit to the lower unit is open. As soon as the top section of the tank is heated to the required temperature, the double-throw thermostat

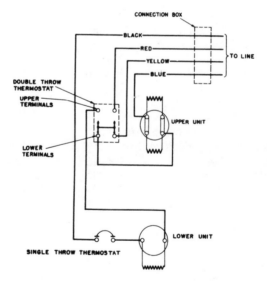

Fig. 6. Wiring diagram for a "limited demand" type of two-unit water heater.

opens, thus closing the circuit to the lower unit and cutting out the upper unit. Since only one unit is permitted to operate at a time, there is a reduction in the maximum demand to the connected load of the larger unit. Unless specified otherwise, the upper thermostat is wired at the factory to provide double-throw operation.

Non-limited Demand Type of Wiring. If single-throw operation is desired, the two leads to the upper terminals of the double-throw thermostat should be connected to

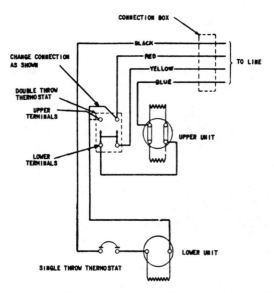

Fig. 7. Wire diagram for a "nonlimited demand" type of two-unit water heater.

the same terminal (Fig. 7). Whether the double-throw thermostat is opened or closed to the top unit, the circuit to the lower unit is closed. The maximum demand of the heater is the sum of the wattages for the top and bottom units.

THERMOSTAT ADJUSTMENT

The temperature setting may be adjusted externally by removing the thermostat cover and rotating the setting shaft. The cover plate on the shell thermostat indicates the direction and degrees of change. The normal factory

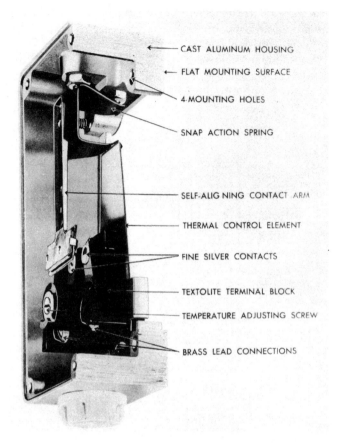

CAST ALUMINUM HOUSING

FLAT MOUNTING SURFACE

4 MOUNTING HOLES

SNAP ACTION SPRING

SELF-ALIGNING CONTACT ARM

THERMAL CONTROL ELEMENT

FINE SILVER CONTACTS

TEXTOLITE TERMINAL BLOCK

TEMPERATURE ADJUSTING SCREW

BRASS LEAD CONNECTIONS

Fig. 8. Cutaway view of a single-throw thermostat.

setting is 150° F. This is satisfactory and efficient for all domestic requirements.

The only recommended internal adjustment of the thermostat is cleaning or aligning the contacts as follows: Disconnect the heater from the power supply. Remove the shell thermostat cover. Remove the thermostat cover plate and inspect the contacts. If the contacts are burned or pitted, they can be cleaned and smoothed provided that all filings are carefully removed and the original contour is maintained. If operation of the thermostat is erratic, inspect the **U** spring for the correct placement of the pivot point at the inside top of the thermostat and knife edge at the top of the bimetallic blade.

Do not remove the thermostat arm or **U** spring from any thermostat. Specialized equipment is necessary for proper calibration.

In case a thermostat fails, it must be replaced. Disconnect the heater from the power supply and remove the shell thermostat cover. Remove the thermostat cover plate and disconnect the wiring at the terminals. (Mark the position of the wiring before disconnecting. Reconnection at the same point saves time and possible trouble.)

All bimetallic thermostats are single-pole and break only one side of the line. Single-throw and double-throw types are available. Single-throw thermostats of the type shown in Fig. 8 are normally used on single-unit heaters and in the lower position on twin-unit heaters. Double-throw types (Fig. 6) are used in the upper position on twin-unit heaters to permit operation of the booster unit on a limited-demand installation.

(*Note:* Consult Trouble and Remedy Chart for electric water heaters on page 532.)

TROUBLE AND REMEDY CHART

Trouble	Cause	Remedy
Heater does not operate or insufficient hot water available.	Fuse blown or circuit breaker tripped.	Replace or reset.
	One or more units inoperative.	*Check and replace if necessary.
	Thermostat inoperative or set too low.	*Check and replace if necessary.
	Insufficient capacity or slow recovery.	Check excessive water wastage.
	Leaks at hot water tap.	Repair.
	Careless use of hot water.	Eliminate.
	Long uninsulated pipe runs.	Insulate pipes with asbestos.
Water too hot.	Improper thermostat setting or thermostat out of adjustment.	*Check. Repair or replace.
	Connection to hot-water furnace coil.	Replace as directed by manufacturer.
	Heating unit grounded.	Remove ground.
High bills.	High power rate.	Consult local current supplier.
	Insufficient capacity for off peak or special rate service.	Consult service department of manufacturer.
	Leaks at hot-water outlets or careless use of hot water.	Repair leaks and check usage.
	Long uninsulated pipe runs.	Insulate.
	Heating unit grounded.	Remove ground.
	Thermostat set too high.	Reset thermostat.
Rusty water.	Rusty mains or household plumbing lines.	Check with local plumber.
	Galvanized tank installed in soft water area.	Replace with noncorrosive tank.
Water heater leaks.	Leaks at fittings or piping.	Repair or replace.
	Leak at tank.	Repair.
	High-pressure rupture.	Repair and, if possible, reduce pressure.

* Consult service department of manufacturer.

CHAPTER *25*

Electric Space Conditioning Systems in the Home

The great variety of compatible electric systems and equipment that can be freely intermixed provides the designer, builder, and the home owner with unusual flexibility. The choice of systems and equipment depends upon so many variables that each housing design must be considered separately.

Electric systems employ one or more of three basic heat transfer methods, as follows.

1. *Forced air*. Heated or conditioned air is circulated by built-in fans.

2. *Convection*. The air temperature is increased by contact with a heated surface, becomes lighter and rises, causing a circulation of air in the space.

3. *Radiation*. The transmission of heat energy by means of electromagnetic waves. This heat energy is transferred directly from a heated element to the occupants, furnishings, and other objects and surfaces without heating the air through which it passes, in much the same way the sun heats the earth and its inhabitants. The air in spaces with radiant systems picks up some heat from people and objects, then circulates by natural convection.

DECENTRALIZED OR IN-SPACE SYSTEMS

With in-space electric systems the heat is produced in the space to be heated. In-space systems permit wide latitude in structural design, and they blend inconspicuously with surfaces, furnishings, and decor. No space is required for remote heating units, pipes, or ducts. Independent room-by-room or zone control makes possible comfort conditions to suit individual preferences.

Baseboard Units. *Electric baseboard units* (Fig. 1) have heating elements encased in metal housings. Room air is heated by contact with the heating elements and most of

Fig. 1. Baseboard unit.

the heat is distributed by convection. There is some radiation, the proportion depending upon construction of the unit. Normal installation is along the bottom of outside walls, generally beneath windows where heat loss is greatest.

Baseboard units offer advantages in addition to those they share with other in-space systems. *For example,* they are flush-mounted and can be installed after walls are finished and in place, and they offer minimum interference with window placement arrangement.

Optional features and accessories include built-in raceways for wiring, end caps, corner and blank sections for joining lengths of baseboard units, and convenience outlets.

The control of room temperature with electric baseboard heating systems is accomplished by means of wall-mounted thermostats operating at either line or low voltage, or by integral line voltage thermostats. To prevent overheating the units, should air circulation through them be blocked for any reason, built-in thermal overload protection is provided.

Cooling can be provided by an independent central system or individual through-the-wall units. A built-in heat/cool selector switch and three-wire outlet are provided as integral parts of some baseboard units to make separate wiring unnecessary for through-the-wall cooling units. (*See* Table 30.)

TABLE 30

rating range	dimension range (inches)		
(120, 208, 240 or 277 volts)	height	depth	length
100-400 watts/ linear ft (341-1365 Btuh/ linear ft)	3⅜-10	2-4	24-144

Wall Units. *Electric wall heating units* (Fig. 2) are concentrated heat sources characterized by high output-to-size ratios. They contain resistance elements and provide heat by radiation, natural convection, or forced air. They are applicable both for general room heating and for smaller areas such as entryways, bathrooms, and kitchens. Some are designed for fully or partially recessed installation between

Fig. 2. Wall unit.

wall studs. Larger cabinet-type convectors are floor mounted flush against the wall. (*See* Table 31.)

TABLE 31

type	rating range (120, 208, or 240 volts)	style	dimension range (inches)		
			height	width	depth
natural convection	500-2000 watts (1707-6826 Btuh)	horizontal	9-13	13-30	2-3
		vertical	12-30	8-19	2-3
forced air	750-2400 watts (2560-8191 Btuh)	horizontal	6-13	12-17	3-5½
		vertical	12-25	7-17	3-5½

Control of wall heaters is accomplished by means of *thermostats* which can either be built in or wall mounted, line or low voltage. *Cooling* can be provided by a separate system.

Ceiling Units. *Ceiling heating units* for homes are uniquely electric (Fig. 3). They make available to the designer, builder, and home owner a versatile type of concentrated heat source that is not practical with flame fuel systems. Ceiling units are compact and designed for recessed or surface installation. They are generally centrally located within the space and are capable of providing a very even pattern of heat distribution even in irregular areas. They are well suited for smaller spaces such as bathrooms. (*See* Table 32.)

Ceiling units are offered in a great variety of standard models in rectangular or round shapes. They may be

Fig. 3. Ceiling unit.

radiant or forced-air types and may incorporate fluorescent or incandescent lights and exhaust fans.

Some ceiling units are controlled by wall-mounted or integral thermostats, either line or low voltage, and others by input regulators, timers, or manual switches. Cooling requires a separate system.

TABLE 32

rating range	dimension range (inches)			
(120, 208, 240 volts)	round diameter	rectangular width/length	pro- jection	in-ceiling depth
200-1650 watts (683-5631 Btuh)	10-15	7-16	½-4	3-8

Floor Insert Units. *Floor insert heaters* are designed for spaces with large glass areas, to offset the normal down-draft of cold air, and for other locations where wall space for heating equipment is limited. (*See* Table 33.)

Fully recessed into the floor (Fig. 4), they occupy no room space and distribute heat by natural convection or forced air.

Control is accomplished by integral line-voltage or wall-mounted line or low-voltage thermostat. A built-in high-

Fig. 4. Floor insert unit.

limit switch turns the unit off if the air circulation is inadvertently prevented. *Cooling* is by a separate system.

TABLE 33

rating range	dimension range (inches)		
(120, 208, 240, and 277 volts)	width	depth	length
350-2000 watts (1195-6830 Btuh)	5-6¼	8-10⅛	14-107

Cove or Valance Units. *Cove or valance heaters* are surface mounted on walls near the ceiling to provide the maximum direct radiant effect. By convection they also warm the ceiling, which then radiates heat to the rest of the room. (*See* Table 34.)

These units are mounted separately, rather than continuously, in numbers and ratings required to obtain the desired thermal output. Like baseboard units, they are practically inaudible. *Control* is accomplished by wall-mounted thermostat, either line or low-voltage. *Cooling* is by a separate system.

TABLE 34

rating range	dimension range (inches)		
(120, 208, 240 or 277 volts)	height	depth	length
450-1500 watts (1540-5100 Btuh)	2¼-6	1-6⅞	42-126

Radiant Ceiling Systems. *Electric radiant ceiling* systems make use of the phenomenon of radiation to transfer a major portion of the heat generated by resistance elements. Two types of the system are in common use for homes— ceiling cable and panels. In these systems the ceiling or the panels provide low intensity radiation, which is absorbed directly by the occupants, furnishings, and other surfaces of the room.

Among the features of such systems is a very uniform floor-to-ceiling temperature. They are inaudible and usually invisible and provide complete flexibility in the location of windows, doorways, closets, and furniture. *Control* of ceiling systems is by wall-mounted line-or-low-voltage thermostats. *Cooling* can be accomplished by a separate system.

Radiant ceiling systems are available in several forms.

Heating cable. The heating cable is one very effective form of radiant ceiling system. Electric heating cable buried in the ceiling becomes a permanent, integral part of the structure. It is attached to the ceiling in serpentine rows and connecting leads are carried down through the wall to the thermostat in plaster ceilings. The cable is covered with heat-conducting plaster before the white finish coat is applied. In drywall ceilings, heat-conducting fire-resistant sections are fastened to the ceiling, after a layer of masonry-base filler has been applied to the top surface of each section.

Heating cable can be embedded in concrete ceilings. The type of cable and embedding procedure for specific applications depend upon the concrete molding or casting process and thickness of ceiling.

Ceiling cable comes in factory-engineered lengths up to 1,800 feet to suit various room sizes and heat losses. It is available in 120, 208, and 240 volt ratings from 125 to 5000 watts.

Prewired panels. Integral drywall panels with resistance cable embedded in a gypsum core are compatible with conventional ⅜-inch drywall construction techniques. Panels are complete with connecting leads and are marked to indicate nailable areas. (*See* Table 35.)

TABLE 35

rating range (240 volts)	dimension range (feet) (all panels are ⅝-inch thick)	
	length	width
245-720 watts (836-2460 Btuh)	6-12	4

Panels with a conductive metallic layer laminated between two sheets of gypsum wallboard are also compatible with conventional ⅜-inch drywall construction techniques. Current is supplied from two copper electrodes which run the full length of the panel. Cutouts for lighting fixtures may be made, but are limited as to size, number, and location. (*See* Table 36.)

TABLE 36

rating range (240 volts)	dimension range (feet) (all panels are ⅝-inch thick)	
	length	width
240-720 watts (820-2460 Btuh)	6-12	4

Self-controlled enameled steel panels with heating cable embedded in a mineral core have higher heat intensity than other ceiling systems. They can be built into a conventional drywall ceiling (Fig. 5) or be surface or T-bar

Fig. 5.

mounted. These panels can be located near high heat loss areas. (*See* Table 37.)

TABLE 37

rating range	dimension range (inches)		
(120, 208, 240 and 277 volts)	length	width	thickness
500-1000 watts (1710-3413 Btuh)	48-96	24	⅞-1¼

PACKAGED HEATING/COOLING UNITS

Packaged through-the-wall units provide year-round conditioning of individual spaces—heating, cooling, dehumidification, filtration, and ventilation. All components are contained in a housing which consists of a floor or wall mounted interior cabinet and a wall sleeve. (*See* Figs. 6 and 7.)

Through-the-wall sleeves are built into the structure during the manufacturing process. Temporary weather plates and removable steel reinforcement prevent distortion during construction. The wall sleeve is of heavy-gauge galvanized steel, coated with mastic or thermosetting plastic for corrosion resistance and sound absorption. The depth of the sleeve varies to match wall thickness.

Fig. 6. Packaged heating/cooling unit.

Fig. 7. Packaged heating/cooling unit.

Outside louvers mount flush with the exterior wall and can be attached to the wall sleeves from inside. They are usually made of anodized extruded aluminum, but they are also obtainable in cast aluminum or wrought iron to provide the designers with an aesthetic choice.

The heating/cooling section is a factory-assembled unit mounted on a chassis. An insulated bulkhead isolates the compressor sounds from the occupied space and all air passages within the unit are acoustically deadened. Room-side blowers are large, low speed centrifugals. Large heating and cooling coils and low air velocities also help minimize sound. Packaged units are easy to install and service from the interior space.

The room cabinet is made of metal panels in baked enamel or other decorative finish with top, angle, or front discharge grilles. Built-in selector switches and other controls permit operation at two or more fan speeds over a range of temperatures. (*See* Table 38.)

TABLE 38

rating range (120, 208, 240 and 277 volts)		dimension range (inches)				
heating	cooling	wall sleeve		cabinet		
		height	width	height	width	depth
2-10 kw (6800-34,130 Btuh)	6500-27,000 Btuh	14-16	29-37	20-25	47-48	6-14

CENTRAL SYSTEMS

Central systems involve heating or cooling air or water in a remote unit and then conveying it by ducts or pipes to the spaces to be conditioned. There is a much wider choice of electric central equipment as compared to the choice of equipment using other fuel, including the uniquely electric heat pump which has no fossil fuel counterpart. And electric central equipment is much more compact and easy to install or service. *Electric central systems* fall into two broad categories—ducted air and hydronic.

Ducted Air Systems. The heat sources for electric ducted air systems for individual dwelling units are either furnaces, duct heaters, or air-to-air heat pumps. Designed to be installed with zero clearance, electric equipment for ducted systems comes in a wide variety of types and sizes, for horizontal or vertical installation. Up-flow or down-flow configurations are available for some types of equipment. (*See* Figs. 8, 9, 10, 11, and 12.)

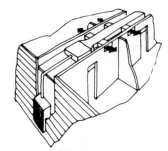

Fig. 8.

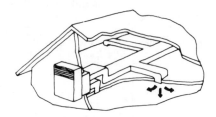

Fig. 9.

Fig. 10.

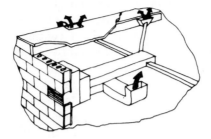

Fig. 11.

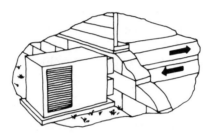

Fig. 12.

These features give the designer and builder of housing, and the home owner, great freedom in locating indoor equipment. Closets, furred-down ceilings, crawl spaces, attics, and basements are all suitable places for installing electric ducted air space conditioning equipment. With ducted air systems, it is a relatively easy task to provide for ventilation, filtration. and humidity control. Where heat pumps are used, cooling is automatically provided, and cooling is easily incorporated in other electric ducted air systems.

Electric Furnaces. The *electric furnace* is an unusually compact unit consisting of coiled wire resistance heating elements, a blower, filter, and controls, all housed in an insulated cabinet. Because of their low thermal inertia and low resistance to air flow, electric coils provide heat very quickly.

Furnaces are equipped with one or more coils, which are energized at present intervals by a sequencer to prevent high inrush currents. A high-limit control prevents overheating. The sequencer is activated by a wall-mounted low voltage thermostat, which also controls the fan unless its operation is continuous. Also obtainable as an option are controls that provide a very fine degree of room temperature stability by continuously regulating heat output in response to signals from a proportional thermostat. (*See* Table 39.)

The cabinets of electric furnaces are designed to accommodate the installations of direct expansion evaporator coils for cooling. Some furnaces come with all of the controls necessary for cooling as well as heating as standard equipment. In others, the additional controls required for cooling are supplied as an option. (*See* Figs. 13 and 14.)

TABLE 39

rating range (120, 208, and 240 volts)	dimension range (inches)			
	style	height	width	depth
3.7-40 kw (12, 628- 136, 520 Btuh)	vertical	16-48	15-26	12-28
	horizontal	12-42	20-32	22-42

Air Handling Units with Duct Heaters. The *air handling unit* consists of a blower housed in an insulated cabinet with openings for connections to the ducts (Fig. 15). *Electric resistance duct heaters* are installed either in the primary or branch ducts leading from the air handling unit to the rooms in the dwelling unit.

Duct heaters consist of parallel rows of open or sheathed spirals of resistance wire supported in a rectangular steel frame. Heaters are completely factory assembled units available in two configurations. The slip-in type is inserted into a hole cut in the side of the duct. The heater terminal box covers the opening and the heater is secured in place

Fig. 13. Electric furnaces.

Fig. 14.

Fig. 15. Air handling unit with duct heaters.

with sheet metal screws. The flanged type consists of a similar heater, factory mounted in a section of ducting which is flanged for easy connection into the duct run.

Where the heaters are installed in the primary duct, operation and control are similar to those of an electric furnace. Where they are installed in two or more branch ducts, much greater control flexibility is possible, since each branch supplies a separate zone with independent temperature control. Heat loss from the ducts is minimized when the heating coils are located near the interior spaces, because hot duct runs are shorter. As with the electric furnace, heat is available immediately when the coils are energized. Protection against excessive temperature and interrupted air flow is provided by limit switches.

Cooling may be accomplished by the addition of a cooling coil in or adjacent to the air handling unit. Exceptionally effective moisture control can be achieved by supercooling the air for dehumidification and reheating it by means of the duct heaters to produce air at the desired temperature and very low in vapor content. (*See* Tables 40 and 41.)

TABLE 40

AIR HANDLERS

style	dimension range (inches)		
	height	width	depth
horizontal	10-25	24-54	15-41
vertical	22-58	23-51	14-33

TABLE 41

DUCT HEATERS

rating range (120, 208, 240, 277 and 240 volts)	dimension range
100 watts and up (341 Btuh and up)	to fit duct sizes from 3 x 5 inches and up.

CENTRAL DUCTED AIR SYSTEMS

Air-to-Air Heat Pumps. Essentially, the *heat pump* is a reversible-cycle refrigeration machine which supplies useful heat to interior spaces during cold weather and removes unwanted heat from the spaces during warm weather. Its basic mode of operation—except for the reversibility feature—is identical to that of any conventional refrigeration machine and it has the same three basic components: compressor, condenser, and evaporator.

In a heat pump, the refrigeration cycle can be reversed by the operation of values, so the roles played by the heat exchangers are reversed when the interior spaces switch from heating to cooling or vice versa.

The heat pump has the unique ability to furnish more energy in the form of heat than is put into it in the form of electric power. This is so because it removes heat from other sources. In the case of the air-to-air heat pump, this source is outside air, which has some heat even at relatively low temperatures. In a sense, the heat absorbed by the evaporator coil is *free* and electric energy is required only to move it. This free heat is what makes it possible for the heat pump to supply more energy than it consumes. With easily attained coefficient of performance of 2 to 1 (ratio of heat output to energy input) or even higher, the heat pump has the lowest operating cost of any electric heating/cooling equipment. Another feature of a heat pump system is that it occupies less space than a system with separate heating and cooling components.

Air-to-air heat pumps (Fig. 16) are installed in ducted air systems in a manner similar to furnaces or air handling

Fig. 16. Air-to-air heat pump.

units. Like standard air conditioning units, they are available in both single package or split system configurations. The single package units may be installed indoors with one heat exchanger ducted to outside air, or through the wall. Or they may be located outdoors (Fig. 17), on a pad or on the roof adjacent to the space to be conditioned to minimize duct losses.

Fig. 17.

A split unit consists of an indoor and an outdoor section connected by refrigerant tubing and electric wiring. Such units provide greater flexibility in location of equipment, since they are sectionalized.

A separately mounted low-voltage heating/cooling thermostat controls operation of the heat pump. Available as an option is a selector switch that permits automatic operation but also provides a means of limiting operation to heating or cooling, or shutting the unit off.

A heat pump is normally sized on the basis of the summer cooling load, which means that for some climates, integral or separate supplemental resistance heating elements are required to handle peak winter conditions. (*See* Table 42.)

HYDRONIC SYSTEMS

Central hydronic systems utilize water (or steam) as the heat transfer medium. In the case of electric hydronic systems, the heat source is an electric boiler with electric immersion heaters. Interconnections between dwelling units are easily made with quick-disconnect fittings.

TABLE 42

rating range (208 to 480 volts)	dimension range (inches)			
heating or cooling	types	height	width	depth
17,000-60,000 Btuh	single package	22-23	28-55	32-50
	split system outdoor section	21-38	17-70	21-50
	indoor section	10-58	22-51	18-28

Electric Boilers. *Electric hydronic systems* using boilers as the heat source can be designed to serve an individual dwelling unit or a multi-family structure (Fig. 18).

For individual dwelling units. An electric boiler for a single dwelling unit consists of an insulated steel or cast iron generator, replaceable immersion heating elements, expansion tank, circulating pump, prewired controls, and other accessories, all factory assembled into a packaged unit. Immersion elements are energized by a sequence control to maintain preset water temperature and to limit power inrush (Fig. 19). Limit controls hold temperatures and pressures within preset ranges.

A circulating pump, activated by a low-voltage thermostat, controls the flow of water from the boiler to room heating units, which are available in many types and sizes for baseboard, wall, or valance installation.

These packaged units are so compact that a unit as small as a portable TV set will heat a six-room house. Because of their size, and because they require no combustion air or flues and are approved for zero-clearance installation, electric boiler assemblies can be located practically anywhere, such as under kitchen counters or in closets. *Cooling* can be provided by a separate system. (*See* Table 43.)

For multi-family structures. With a *central electric hydronic system* serving a multi-family structure, piping delivers hot water or steam from a remotely located boiler to

Fig. 18. Electric boiler.

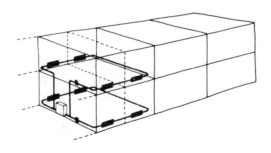

Fig. 19. Individual dwelling unit.

TABLE 43

rating range	dimension range (inches)		
(208 and 240 volts)	height	width	depth
2-40 kw (6826-136, 520 Btuh)	14-36	6-25	7-24

heating units in each dwelling unit (Fig. 20). A low-voltage thermostat in each dwelling unit actuates a zone control valve to regulate the flow of hot water or steam to the interior spaces.

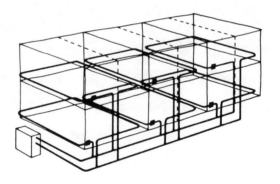

Fig. 20. Central electric hydronic system for multi-family structures.

Like the small electric boilers for single family dwelling units, the larger boilers for multi-family structures are more compact than corresponding fossil-fuel boilers and are not restricted as to location. Built-in sequence and limit controls for operation of the immersion of electrode heaters are standard. *Cooling* can be provided by chilled water from central refrigeration equipment, or by individual through-the-wall units. (*See* Table 44.)

TABLE 44

rating range (240-480 volts)	dimension range (inches)		
	height	width	depth
40-1800 kw (136, 522- 6,151,600 Btuh)	33-76	85-75	8-82

Water-to-air heat pumps. Water-to-air heat pumps are the key elements in a space conditioning system that is well suited to multi-family structures. This system is actually a hybrid configuration. While it has some of the characteristics of a decentralized system, it requires some remote heating and cooling equipment that is common to all of the dwelling units.

With this system there is a heat pump unit in each room or in each dwelling unit. The water-to-refrigerant heat exchangers in all of these units are connected together by a closed loop of circulating water. Heat is rejected into the water by heat pumps on the cooling cycle and absorbed from the water by units on the heating cycle. (*See* Fig. 21.)

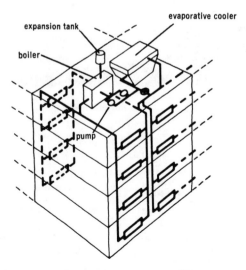

Fig. 21. Water-to-air heat pumps in space conditioning system.

Also connected into the water loop at a central location are an electric boiler and an evaporative cooling tower, which operate as necessary to maintain loop water temperature between 60°F and 95°F year 'round. Water at temperatures within this range make possible very efficient and reliable performance of the heat pumps.

A major advantage of this system is that it conserves energy by transferring excess heat from one dwelling unit to another that requires it. Since no exposure to outside air is necessary, there is great latitude in locating the heat pumps. Console types can be installed in any wall in the dwelling unit (Fig. 22), and others can be concealed in a closet or furred-down ceiling. (*See* Table 45.)

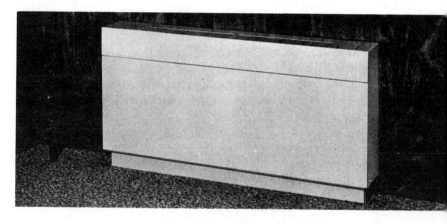

Fig. 22. Console-type heat pump.

TABLE 45

type	rating range, Btuh (120, 208 and 240 volts)		dimension range (inches)			
	heating	cooling	style	height	width	depth
console	6000-30,000	6000-20,000		20-25	40-58	6-12
concealed	7000-80,000	6000-60,000	horizontal	12-23	19-46	26-44
			vertical	16-55	20-27	19-25

This system provides all the operating flexibility of three- or four-pipe systems with but two pipes which require no insulation because of the relatively constant water temperature in the loop. Consequently, piping costs and space requirements are less. The heat pump units are rugged and simple, with heat exchangers, valves, pumps, and other components enclosed in a thermally and acoustically insulated cabinet. Control for both heating and cooling is by wall-mounted, low-voltage thermostats, and the control system may include an off-heat-cool-automatic selector switch. The choice of temperature in each dwelling unit is completely independent regardless of the season or mode of operation in other spaces.

HOW TO USE ELECTRIC HEATING

Single-Family Detached. The typical floor plans shown in Figs. 23, 24, 25, and 26 illustrate the flexibility in the application of electric comfort conditioning for the single-family home. Figs. 23, 24, and 25 are variations of central ducted air systems which utilize electric furnaces, air handling units with duct heaters—both with cooling options—or heat pumps which provide both heating and cooling.

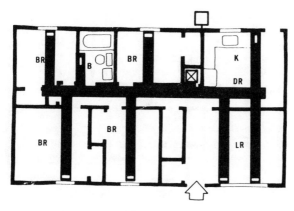

Fig. 23.

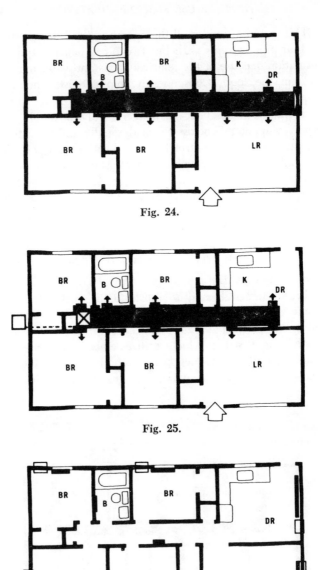

Fig. 24.

Fig. 25.

Fig. 26.

The plan shown in Fig. 26 illustrates the application of baseboard units with the option of through-the-wall cooling or central cooling by means of a furred-down hallway. This floor plan is representative of the in-space or decentralized systems utilizing in-space units, including radiant ceilings for heating.

Central ducted air system. In the central ducted air system a counterflow unit is installed in a closet with the ducts between or below the floor joists. An upflow unit can be used when a basement is provided (Fig. 23). *Cooling* may be provided by a split system with the condenser unit mounted externally, or a heat pump.

Central simplified ducted air system. A packaged heating/cooling unit is installed in the ceiling or above the rafters adjacent to an external wall or in the external wall itself (Fig. 24). This self-contained package unit may be a furnace or air handler with duct heaters (for zone control) —both with cooling options—or a heat pump to provide heating, cooling, and ventilation.

A *horizontal unit* is installed in the ceiling of the hall, a closet, or utility area (Fig. 25). The ducting in the hall ceiling serves all rooms either directly or through duct extensions. Cooling may be provided by a split system with the condenser unit mounted externally, or a heat pump to provide heating and cooling.

Decentralized or In-Space System. *Decentralized units* such as baseboard, wall, or radiant ceiling are installed as required in each room. Cooling can be provided by through-the-wall units (Fig. 26) or by a compact central unit located in the furred-down hallway, as shown in Fig. 25.

TOWNHOUSES

The linear grouping of a series of two-story dwelling units into *townhouses* with common walls achieves high land use and provides compact dwellings.

The *central ducted systems* shown in Figs. 27 and 28 illustrate two different methods of comfort conditioning interior dwelling units with the major heat loss/heat gain

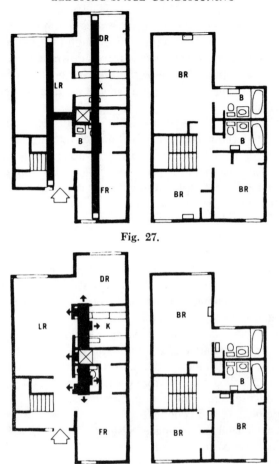

Fig. 27.

Fig. 28.

areas at the front and rear. Dwelling units with three exposed sides may require additional ducts to exposed wall areas.

The *decentralized or in-space systems* (Figs. 29 and 30) illustrate two methods of simplifying the structure by locating individual units in each space to be conditioned and, in addition, providing individual temperature control in each zone. Dwelling units with three exposed sides may require additional baseboard or wall units on exposed wall areas.

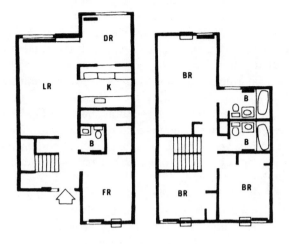

Fig. 29.

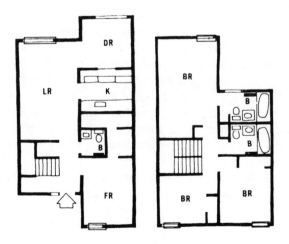

Fig. 30.

Central Ducted Air System. An *upflow unit* is installed in a closet. Ducts in the ceiling of the first floor, or in the space provided between rafters and joists by some designs, distribute conditioned air as shown in Fig. 27. *Cooling* may be provided by a split system with the condenser unit mounted externally.

Central Simplified Air System. A *horizontal unit* installed in the ceiling of the utility core distributes air through a plenum and/or ducts to all rooms as shown in Fig. 28.

Decentralized or In-Space System. *Decentralized units* such as baseboard, wall, or radiant ceiling systems provide heating in each room (Fig. 29). *Cooling* is provided by through-the-wall units.

Packaged heating/cooling units installed in front and rear exposed walls. In some rooms they may be installed below the windows. Supplementary heat in the bathrooms is provided by decentralized baseboard, wall, or ceiling units. (*See* Fig. 30.)

GARDEN APARTMENTS

The floor plans shown in Figs. 31, 32, 33, 34, and 35 are typical of the many ways in which electric comfort conditioning can be utilized in garden apartments. The floor plans shown in Figs. 31, 32, and 33 illustrate ducted air systems. Figure 31 requires extensive duct work to distribute air to the peripheral exposed walls. The plan shown in Fig. 32 has the advantage of being contained essentially within the utility core. The third plan shown in Fig. 33 utilizes a packaged heating/cooling unit installed at the top of a closet on an exposed wall for easy access to outside air.

The plans shown in Figs. 34 and 35 illustrate the use of decentralized or in-space units—baseboard, wall units or radiant ceilings to provide heating and through-the-wall or packaged heating/cooling units for cooling.

Central Ducted Air System. A vertical unit is installed in the closet. Circulation may be upflow or counterflow,

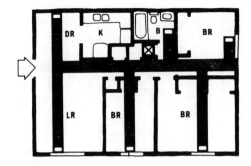

Fig. 31.

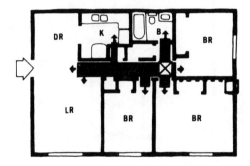

Fig. 32.

depending upon the design of the modules (Fig. 31). *Cooling* is available with a split system.

Central Simplified Ducted Air System. A *horizontal unit* is installed in the ceiling of the hall (Fig. 32). Properly located registers near the ceiling in each room provide air distribution. This type of ceiling installation also provides additional storage or closet space. *Cooling* is available with a split system. External (condenser) unit may be mounted flush to exterior wall in the top of the closet, as shown in Fig. 32.

A *packaged heating/cooling unit* is mounted in the ceiling of a closet adjacent to or in an exposed wall (Fig. 33).

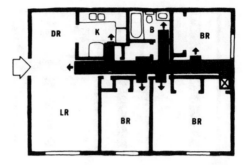

Fig. 33. Central simplified ducted air system with a packaged heating/cooling unit.

Fig. 34. Decentralized or in-space system.

This self-contained packaged unit may be a furnace or air handler with duct heaters—both with cooling options— or a heat pump to provide heating, cooling, and ventilation.

Decentralized or In-Space System. *Baseboard units* or other decentralized units are installed in each room (Fig. 34). *Cooling* may be provided by through-the-wall units (Fig. 35).

Individual packaged heating/cooling units, either heat pumps or air conditioning units with strip heaters, are installed through exposed walls, usually below the windows.

EFFICIENCY APARTMENTS

Decentralized or In-Space Systems. A *packaged* heating/ cooling unit is installed under the window in the living

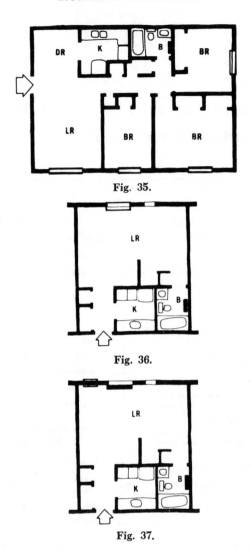

Fig. 35.

Fig. 36.

Fig. 37.

room. A baseboard, wall, or ceiling unit is installed in the bathroom. (*See* Fig. 36.)

Baseboard units are installed under the window in the living room and in the bathroom. A through-the-wall unit is installed in the living room in the exposed wall. (*See* Fig. 37.)

APPENDIX **I**

Graphical Electrical Symbols for Residential Wiring Plans

(These symbols have been extracted or adopted from American Standards Association Standard ASA Z32.9-1943, wherever possible. Adaptations and new symbols included in this list have been proposed for inclusion in the next revision of that standard.)

General Outlets

Lighting Outlet

Ceiling Lighting Outlet for recessed fixture (Outline shows shape of fixture.)

Continuous Wireway for Fluorescent Lighting on ceiling, in coves, cornices, etc. (Extend rectangle to show length of installation.)

Lighting Outlet with Lamp Holder

Lighting Outlet with Lamp Holder and Pull Switch

Fan Outlet

Junction Box

Drop-Cord Equipped Outlet

Clock Outlet

To indicate wall installation of above outlets, place circle near wall and connect with line as shown for clock outlet.

Convenience Outlets

Duplex Convenience Outlet

S_D Automatic Door Switch

S_P Switch and Pilot Light

S_{WP} Weatherproof Switch

S_2 Double-Pole Switch

Low-Voltage and Remote-Control Switching Systems

$\underline{S}$ Switch for Low-Voltage Relay Systems

$\underline{MS}$ Master Switch for Low-Voltage Relay Systems

O_R Relay—Equipped Lighting Outlet

— · — · · — Low-Voltage Relay System Wiring

Auxiliary Systems

Push Button

Buzzer

Bell

Combination Bell-Buzzer

Chime

566

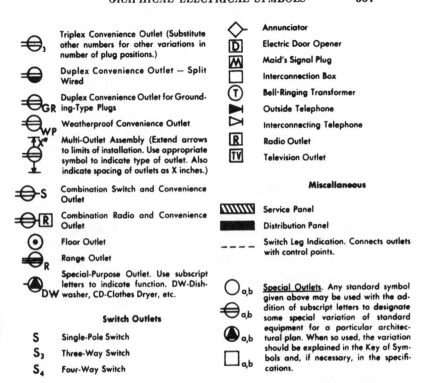

Triplex Convenience Outlet (Substitute other numbers for other variations in number of plug positions.)

Duplex Convenience Outlet — Split Wired

Duplex Convenience Outlet for Grounding-Type Plugs

Weatherproof Convenience Outlet

Multi-Outlet Assembly (Extend arrows to limits of installation. Use appropriate symbol to indicate type of outlet. Also indicate spacing of outlets as X inches.)

Combination Switch and Convenience Outlet

Combination Radio and Convenience Outlet

Floor Outlet

Range Outlet

Special-Purpose Outlet. Use subscript letters to indicate function. DW-Dishwasher, CD-Clothes Dryer, etc.

Switch Outlets

S   Single-Pole Switch

S_3 Three-Way Switch

S_4 Four-Way Switch

Annunciator

Electric Door Opener

Maid's Signal Plug

Interconnection Box

Bell-Ringing Transformer

Outside Telephone

Interconnecting Telephone

Radio Outlet

Television Outlet

Miscellaneous

Service Panel

Distribution Panel

Switch Leg Indication. Connects outlets with control points.

Special Outlets. Any standard symbol given above may be used with the addition of subscript letters to designate some special variation of standard equipment for a particular architectural plan. When so used, the variation should be explained in the Key of Symbols and, if necessary, in the specifications.

Glossary of Terms

Absorption. When light strikes an object, some of the light is reflected and some is absorbed. The darker the object, the more light it absorbs. Dark colors absorb large amounts of light, much as a blotter soaks up ink.

Alternating current. Current that continually changes in magnitude from zero to a certain value and periodically reverses in direction.

Alternation. One-half cycle of alternating current.

Alternator. An alternating current generator.

Ammeter. The instrument for the measurement of current.

Ampere. The unit of electrical current.

Ampere-hour. The quantity of electricity equivalent to a current of one ampere flowing past a point in a conductor in one hour.

Ampere-turn. The magnetizing force produced by a current of one ampere flowing through a coil of one turn.

Anode. The electrode in a cell (voltaic or electrolytic) that attracts the negative ions and repels the positive; the positive pole.

Apparent power. Product of volts and amperes in a-c circuits where the circuit and voltage are out of phase.

Arc. The luminous glow between incandescent electrodes.

Armature. The movable part of a motor or the removable part of a magnetic circuit, such as the iron placed across the poles of a horseshoe magnet.

Auto-transformer. A transformer in which the primary and secondary are connected together in one winding.

Ballast. A device used with a fluorescent lamp to obtain the necessary circuit conditions for starting and operating.

Bare lamp. Incandescent-filament or fluorescent lamp with no shielding.

Battery. A device for converting chemical energy into electrical energy; two or more cells connected together as a unit.

Beam spread. In any plane, the angle between the two directions in which the candlepower is equal to a stated percentage (usually 10 per cent) of the maximum candlepower in the beam.

Bobêche. A saucer-shaped element at the base of or below the candle socket in a candle-type fixture. In some modern fixtures of traditional styling the bobêche is often used to conceal light sources.

Bowl. Diffusing glass or plastic used to shield light sources from view.

Branch circuit. One of the conductors in a parallel circuit.

Brightness. The degree of apparent lightness of any surface emitting or reflecting light. Everything that is visible has some brightness.

Brush. The conducting material, usually a block of carbon, bearing against the commutator or sliprings through which the current flows in or out.

Bulb. Glass enclosure of incandescent-filament lamp.

Candela. A unit of luminous intensity, equal to 1/60 of the luminous intensity of a square centimeter of a black body heated to the temperature of the solidification of platinum (1773.5° C).

Candlelight. Fluorescent lamp color similar to deluxe warm white.

Candlepower. Luminous intensity.

Candlepower distribution curve. A curve showing the variation of luminous intensity of a lamp or luminaire.

Canopy. Shield (usually metal) covering joint of lighting fixture to ceiling.

Capacitance. Property of a circuit that opposes any change of voltage.

Capacitive reactance. The effect of capacitance in opposing the flow of alternating or pulsating current.

Cathode. A fluorescent lamp part which serves to conduct the electricity from the wires into the gas. Fluorescent lamps are of two types, depending on whether they have *hot* or *cold* cathodes. The hot cathode is a coiled tungsten filament, operating at a higher temperature than the cold cathode. The latter is an iron cylinder resembling a large thimble. The electrical characteristics of the two are somewhat different, and each has certain advantages. The electrode in a cell (voltaic or primary) that attracts the positive ions and repels the negative ions; the negative pole.

Certified. Term applied to portable lamps and fixtures made to meet certain specifications for quantity of light, quality of lighting, sturdy construction, and safety. Fluorescent ballasts and starters meeting certain performance requirements are also *certified*. Such products usually carry a certification tag or label.

Channel. A common term for the metal enclosure containing the ballast, starter, lampholders, and wiring for a fluorescent lamp.

Choke coil. A coil of low ohmic resistance and comparatively high impedance to alternating current.

Chroma. The attribute of perceived color used to describe its departure from gray of the same lightness.

Circuit. The complete path of an electric current, including, usually, the generating device.

Circuit breaker. A device that opens a circuit while it is carrying current; often used in abnormal conditions, such as overloads.

Circular mil. An area equal to that of a circle with a diameter of 0.001 inch. Used for measuring the cross section of wires.

Coffer. Recessed panel in ceiling or dome.

Color rendering. General expression for the effect of a light source on the color appearance of objects in conscious or subconscious comparison with their color appearance under a reference light source.

Commutator. That part of the armature of a dynamo which converts an alternating into a direct current.

Condenser. A device consisting of two or more conductors separated by non-conductor material which holds or stores an electric charge.

Conductance. The reciprocal of electrical resistance. Conducting power.

Conductivity. The ease with which a substance transmits electricity.

Conductor. A material capable of transmitting electric current.

Contactor. A device for closing and opening electrical circuits remotely; a magnetically operated switch.

Control panel. An upright panel, open or closed, where switches, rheostats, and meters are installed for controlling and protecting electrical devices.

Convenience outlet. Electric receptacle (often along baseboard) used for connecting portable lamps and appliances.

Converter, rotary. An electrical machine having a commutator at one end and sliprings at the other end of the armature, used for the conversion of alternating to direct current.

Cool white. Fluorescent lamp *whiter* than filament lamp. Blends well with daylight; red colors may not appear as red under it.

Core. A mass of iron placed inside a coil to increase its magnetism.

Cornice. The crown molding used both to cover and to embellish the intersection between side walls and ceiling. In modern lighting practice, it is a horizontal member of wood or plaster attached to the ceiling approximately 6 to 8 inches from the wall, with lighting incorporated between it and the wall.

Cornice lighting. Comprises light sources shielded by a panel parallel to the wall and attached to the ceiling, distributing light downward over the vertical surface.

Coulomb. The unit of static electricity; the quantity of electricity transferred by one ampere in one second.

Counter electromotive force (emf). An emf induced in a coil or armature that opposes the applied voltage.

Coupling. Term used to represent the means by which energy is transferred from one circuit to another.

Cove. In modern lighting practice, an element of design

mounted on an upper wall to conceal light sources and provide indirect lighting.

Cove lighting. Comprises light sources shielded by a ledge and distributing light upward over the ceiling.

Current. *Alternating current* (a.c.) reverses its direction of flow periodically. *Direct current* (d.c.) flows continuously in one direction. Gradual drift of free electrons, under pressure (voltage), along a conductor.

Cut-off. The angle up from the vertical at which the reflector or shielding medium cuts off the view of the light source.

D'Arsonval galvanometer. A galvanometer in which a moving coil swings between the poles of a permanent horseshoe magnet.

Deluxe. Trade term applied to fluorescent lamps with good color rendition of all colors. There are cool white and warm white deluxe lamps.

Dielectric. Material that will not conduct an electric current.

Diffuse reflection (diffusion). That process by which incident flux is redirected over a range of angles.

Diffuser. A device to scatter the light from a source primarily by the process of diffuse transmission.

Dimmer. A device for providing variable light output from lamps.

Diode. A vacuum tube containing the filament and the plate. It serves as a rectifier of alternating current.

Direct current. An electric current that flows in one direction only.

Direct glare. Glare resulting from areas of excessive luminance and insufficiently shielded light sources in the field of view.

Direct lighting. A lighting system in which all or nearly all of the light is distributed downward by the luminaire. Fixtures that send somewhat more light upward but are still predominantly of the direct type are called *semidirect*.

Downlight. Luminaire which directs all the luminous flux down. Usually recessed, though may be surface mounted.

Dynamo. A machine for converting mechanical energy into electrical energy or vice versa.

Eddy current. A current induced in the core of an armature of a motor, dynamo, or transformer caused by changes in the magnetic field.

Efficacy. The ratio of luminous flux from lamp to the electrical power (watts) consumed.

Electricity. One of the fundamental things in nature, which is manifested as a force of attraction or repulsion, and also in work, that can be performed when electrons are caused to move; a material agency which when in motion exhibits magnetic, chemical, and thermal effects, and when at rest is accompanied by an interplay of forces between associated localities in which it is present.

Electrode. The solid conductors of a cell or battery which are placed in contact with the liquid; conductor that makes electrical contact with a liquid or gas.

Electrolyte. A substance that conducts a current by the movement of ions. The liquid in a battery or other electrochemical device.

Electromagnet. A magnet made by passing current through a coil of wire wound on a soft iron core.

Electromotive force (emf). The electrical force that moves or tends to move electrons.

Electron. The smallest particle of negative electricity.

Energy. The ability or capacity for doing work.

Farad. Unit of capacitance equal to the amount of capacitance present when 1 volt can store 1 coulomb of electricity.

Field. The region where a magnet or electrical charge is capable of exerting its force.

Field magnet. The magnet used to produce a magnetic field (usually in motors or generators).

Filter. A device which changes, by transmission, the magnitude and/or the spectral composition of the flux incident upon it.

Fixture. *See* Luminaire.

Flood lamp (R or PAR). Incandescent filament lamp providing a relatively wide beam pattern.

Fluorescent lamp (tube). A low-pressure mercury electric-discharge lamp in which a fluorescing coating (phosphor) transforms ultraviolet energy into visible light.

Flux. Magnetic lines of force, assumed to flow from the north pole to the south pole of a magnet.

Footcandle (fc). A quantitative unit for measuring illumination. The illumination on a surface one foot square on which there is a uniformly distributed flux of one lumen.

Footlambert (fL). A quantitative unit for measuring luminance. The footcandles striking a diffuse reflecting surface, times the reflectance of that surface, equals the luminance in footlamberts.

Frequency. The number of cycles of an alternating current per second.

Fuse. A part of a circuit made of a material that will melt and break the circuit when current is increased beyond a specific value.

Galvanometer. An instrument used to measure small currents.

General lighting. The lighting designed to provide a substantially uniform level of illumination throughout an area, exclusive of any provision for special local requirements.

Generator. A machine that converts mechanical energy into electrical energy.

Glare. Any brightness or brightness relationship that annoys, distracts, or reduces visibility.

Glitter. Small areas of high brightness sometimes desirable to provide sensory stimulation.

Grid. A metal wire mesh placed between the cathode and plate.

Grid battery. The battery used to supply the desired potential to the grid.

Grid leak. A very high resistance placed in parallel with the grid condenser.

Ground. A connection made directly to the earth or to a frame or structure which serves as one line of a circuit.

Harp. A rigid wire device which fits around the socket and bulb and supports the shade on some styles of floor and table lamps.

Henry. Unit of inductance; the inductance that will cause 1 volt to be induced if the current changes at the rate of 1 ampere per second.

Horsepower. The English unit of power, equal to work done at the rate of 550 foot-pounds per second. Equal to 746 watts of electrical power.

Hue. The attribute of perceived color which determines whether it is red, yellow, green, blue, or the like.

Humidifier. A device designed to increase the humidity within a room or a house by means of the discharge of water vapor. It may consist of individual room-size units or larger units attached to the heating plant to condition the entire house.

Hydrometer. Device for measuring the specific gravity of liquids.

Illumination. The density of luminous flux on a surface.

Impedance. The total opposition to the flow of alternating or pulsating current.

Incandescent filament lamp (bulb). A lamp in which light is produced by a filament heated to incandescence by an electric current.

Indirect lighting. Lighting provided by luminaires which distribute all light upward. Luminaires sending some luminous flux downward but still predominantly of the indirect type are called semi-direct.

Inductance. Property of a circuit that opposes a change in current.

Induction. The act or process of producing voltage by the relative motion of a magnetic field and a conductor.

Induction coil. Two coils so arranged that an interrupted current in the first produces a voltage in the second.

Inductive reactance. The opposition to the flow of alternating or pulsating current due to the inductance of the circuit.

Instantaneous value. The value at any particular instant of a quantity that is continually varying.

Insulation board, rigid. A structural building board made of coarse wood or cane fiber in 1/2- and 25/32-inch thicknesses.

Insulation, reflective. Sheet material with one or both surfaces of comparatively low heat emissivity, such as aluminum foil.

Insulation, thermal. Any material high in resistance to heat transmission that, when placed in the walls, ceiling, or floors of a structure, will reduce the rate of heat flow.

Insulator. A medium that will not conduct electricity.

Interreflectance. The portion of the lumens reaching the work plane that has been reflected one or more times in the space.

Interrupter. A device for the automatic making and breaking of an electrical circuit.

Ion. An electrically charged atom.

Joule. A unit of energy or work. A joule of energy is liberated by one ampere flowing for one second through a resistance of one ohm.

Kilo. A prefix meaning 1,000.

Kilowatt. Unit of power equal to 1,000 watts.

Kilowatt-hour. Unit of electrical energy; equal to 1 kilowatt multiplied by 1 hour.

Lag. The number of degrees an alternating current lags behind voltage.

Laminations. The thin sheets or discs making up an iron core.

Lamps. A generic term for a man-made source of light. By extension, the term is also used to denote sources that radiate in regions of the spectrum adjacent to the visible. (A lighting unit consisting of a lamp with shade, reflector, enclosing globe housing, or other accessories is also called a *lamp*. To distinguish between the assembled unit and the light source within it, the latter is often called a *bulb* or a *tube*, if it is electrically powered.)

Leakage. Current lost through imperfect insulators.

Lens. Device for optical control of luminous flux by the process of refraction.

Lighting distribution. Luminaires are classified according to the manner in which they control or distribute the luminous flux.

Lighting outlet. Means by which branch circuits are made available for connection to lampholders, to surface—mounted fixtures—to flush or recessed fixtures, or for extension to mounting devices for light sources in valances, cornices, or coves.

Line of force. A line in a field of force that shows the direction of the force.

Load. The power that is being delivered by any power

producing devices. The equipment that uses the power from the power producing device.

Local lighting. The lighting used to provide illumination over a relatively small area or confined space.

Louver. A series of baffles used to shield a source from view at certain angles or to absorb unwanted light.

Lumen. The unit of luminous flux.

Lumiline. A tubular incandescent lamp with a filament extending the length of the tube and connected at each end to a disc base.

Luminaire. A complete light unit consisting of a lamp, or lamps, together with parts designed to distribute the light, to position and protect the lamps and to connect the lamps to the power supply.

Luminaire efficiency. The ratio of the luminous flux leaving a luminaire to that emitted by the lamp, or lamps, used therein.

Luminance (photometric brightness). The luminous intensity of any surface in a given direction per unit area of that surface as viewed from that direction.

Luminous ceiling. A ceiling area lighting system comprising a continuous surface of transmitting material of a diffusing or light controlling character with light sources mounted above it.

Luminous flux. The descriptive term for the time rate of flow of light.

Lux (lx). A quantitative unit for measuring illumination; the illumination on a surface of one meter square on which there is a uniformly distributed flux of one lumen.

Magnetic circuit. The complete path followed by magnetic lines of force.

Magnetic flux. The total number of lines of force issuing from a pole.

Magnetic pole. Region where the majority of magnetic lines of force leave or enter the magnet.

Magnetism. The property of the molecules of certain substances, as iron, by virtue of which they may store energy in the form of a field of force and which is due to the motion of the electrons in the atoms of the substance; a manifestation of energy due to the motion of a dielectric field of force.

Magnetomotive force. The force necessary to establish flux in a magnetic circuit or to magnetize an unmagnetized object.

Matte surface. A surface from which the reflection is predominantly diffuse, with or without a negligible specular component.

Maximum value. The greatest instantaneous value of an alternating voltage or current.

Megohm. A large unit of resistance; equal to one million ohms.

Microfarad. Practical unit of capacitance; one-millionth of a farad.

Mil. One thousandth of an inch.

Milliammeter. An ammeter reading thousandths of an ampere.

Milliampere. Small unit of electric current; equal to one-thousandth of an ampere.

Millivoltmeter. A voltmeter reading thousandths of a volt.

Motor. A device for converting electrical energy into mechanical energy.

Motor-generator (M-G). A generator driven by an electric motor.

Motor starter. Device for protecting electric motors from excessive current while they are reaching full speed.

Munsell color system. A system of object color specification based on perceptually uniform color scaled for the three variables—hue, value, and chroma.

Mutual inductance. Inductance associated with more than one circuit.

Mutual induction. The inducting of an electromotive force (emf) in a circuit by the field of a nearby circuit.

Negative charge. The electrical charge carried by a body which has an excess of electrons.

Neutron. A particle having the weight of a proton but carrying no electric charge.

Nucleus. The heavy or central part of an atom.

Ohm. Fundamental unit of resistance.

Ohmmeter. Device for measuring resistance by merely placing test prods across the resistor and reading the indication on a calibrated scale.

Parabolic lamp (PAR). Parabolic aluminized reflector lamp, spot or flood distribution, made of hard glass for indoor or outdoor use.

Peak value. *See* Maximum value.

Perceived object color. The color perceived to belong to an object, resulting from characteristics of the object, of the incident light, and of the surround, the viewing direction and observer adaptation.

Period. The time required for the completion of one cycle.

Permanent magnet. Piece of steel or alloy that has its molecules lined up so that a magnetic field exists without the application of a magnetizing force.

Pole. One of the ends of a magnet where most of its magnetism is concentrated.

Portable lighting. Table or floor lamp, or wall unit, which is not permanently affixed to the electrical power supply.

Positive charge. The electrical charge carried by a body which has become deficient in electrons.

Potential. A characteristic of a point in an electric field or circuit indicated by the work necessary to bring a unit positive charge to it from infinity; the degree of electrification as referred to some standard, as that of the earth.

Potential difference. The arithmetical difference between two electrical potentials; same as electromotive force, electrical pressure, or voltage.

Power. The rate of doing work or the rate of expending energy. The unit of electrical power is the watt.

Power factor. Ratio of true power to apparent power; equal to the cosine of the phase angle between the voltage and current.

Primary line of sight. The line connecting the point of observation and the fixation point.

Proton. A positively charged particle whose charge is equal, but opposite, to that of the electron.

Pulsating direct current. Current that varies in magnitude but not in direction.

Quality of lighting. The distribution of brightness and color rendition in a visual environment. The term is used in a positive sense and implies that these attributes contribute favorably to visual performance, visual comfort, ease of seeing, safety, and aesthetics for the specific visual tasks involved.

Radiant heating. A method of heating, usually consisting of a forced hot water system with pipes placed in the floor, wall, or ceiling, or with electrically heated panels.

Recessed or flush unit. A luminaire mounted above the ceiling (or behind a wall or other surface) with the opening of the luminaire level with the surface.

Rectifier. Device for changing alternating current to pulsating direct current.

Reflectance. The ratio of the flux reflected by a surface or medium to the incident flux. This general term may be restricted by the use of one or more of the following adjectives: specular (regular), diffuse, and spectral.

Reflected glare. A glare resulting from specular reflections of high luminance in polished or glossy surfaces in the field of view.

Reflection. The process by which flux leaves a surface or medium from the incident side.

Reflector. A device used to redirect the luminous flux from a source by the process of reflection.

Reflector lamp (R). Incandescent filament lamp with reflector of silver or aluminum on inner surface.

Refraction. The process by which the direction of a ray of light changes as it passes obliquely from one medium to another.

Regressed unit. A luminaire designed with the control

medium above the ceiling line.

Relay. Device for controlling electrical circuits from a remote position; a magnetic switch.

Reluctance. A measure of the opposition that a material offers to magnetic lines of force.

Resistance. The opposition of a conductor to an electric current.

Rheostat. A variable resistance for limiting the current in a circuit.

Rotor. The rotating part of an a-c induction motor.

Self-inductance. Inductance associated with but one circuit.

Self-induction. The process by which a circuit induces an electromotive force (emf) into itself by its own magnetic field.

Sensitivity. The degree of responsiveness; in connection with current meters it is the current required for full-scale deflection; in connection with voltmeters it is the ohms per volt of scale on the meter.

Series connection. An arrangement of cells, generators, condensers, or conductors so that each carries the entire current of a circuit.

Series-wound. A motor or generator in which the armature is wired in series with the field winding.

Shielding angle of a luminaire. The angle between a horizontal line through the light center and the line of sight at which the bare source first becomes visible.

Silver bowl lamp. Incandescent filament lamp with silver reflector on the lower half of bowl. Provides indirect distribution.

Solenoid. A tubular coil for the production of a magnetic field; electromagnet field; electromagnet with a core that is free to move in and out.

Special-purpose outlet. A point of connection to the wiring system for a particular equipment.

Specific gravity. The ratio of the mass of a body to the mass of an equal volume of water at 4°C.

Specular angle. That angle between the perpendicular to a surface and the reflected ray. It is numerically equal to the angle of incidence.

Specular reflection. That process by which incident flux is redirected at the specular angle.

Specular surface. Shiny or glossy surfaces (including mirror and polished metals) that reflect incident flux at a specular angle.

Speed. Time rate of motion measured by the distance moved in unit time; in rotating equipment it is the revolution per minute, or r.p.m.

Spot lamp (R or PAR). Incandescent filament lamp providing a relatively narrow beam pattern.

Starter. A device used in conjunction with a ballast for the purpose of starting an electric discharge (fluorescent) lamp.

Stator. The part of an a-c generator or motor that has the stationary winding on it.

Step-down transformer. A transformer with fewer turns in the secondary than in the primary.

Step-up transformer. A transformer with more turns in the secondary than in the primary.

Subjective brightness. The subjective attribute of any light sensation giving rise to the percept of luminous intensity including the whole scale of qualities of being bright, light, brilliant, dim, or dark. (The term brightness occasionally is used when referring to measurable *photometric brightness.*)

Surface mounted unit. A luminaire mounted directly on the ceiling.

Suspended unit. A luminaire hung from the ceiling by supports.

Switch. A device for opening or closing an electrical circuit.

Synchronous. Having the same period and phase; happening at the same time.

Task plane. *See* Work plane.

Temperature. The condition of a body that determines the transfer of heat to or from other bodies; condition as to heat or cold; degree of heat or cold.

Thermocouple. A device for directly converting heat energy into electrical energy.

Torchere. An indirect floor lamp which directs all, or nearly all, of the luminous flux upward.

Torque. The effect of a force to produce rotation about a center.

Transformer. Device for raising or lowering a-c voltage.

Transmission. The characteristic of many materials such as glass, plastics, and textiles. The process by which incident flux leaves a surface or medium on a side other than the incident side.

Transmittance. The ratio of the flux transmitted by a medium to the incident flux.

Troffer. A long recessed lighting unit, usually installed with the opening flush with the ceiling.

True power. The actual power consumed by an a-c circuit; equal to I^2R; expression used to distinguish from apparent power.

Tube. *See* Fluorescent lamp.

Tungsten Halogen lamp. Compact, incandescent filament lamp with initial efficacy essentially maintained over life of the lamp.

Unidirectional. As applied to a current of electricity, a current that flows in one direction only.

Vacuum tube. A tube from which the air has been pumped out. The tube contains an element that emits electrons when properly excited and an electrode to attract the electrons and set up a current in an external circuit.

Valance. A longitudinal shielding member mounted across the top of a window. Usually parallel to the window, it conceals the light sources and usually gives both upward and downward distributions. The same device applied to a wall is a wall bracket.

Value. The attribute of perceived color by which it seems to transmit, or reflect, a greater, or lesser, fraction of the incident light.

Veiling reflection. Regular reflections which are superimposed on diffuse reflections from an object and which partially, or totally, obscure the details to be seen by reducing the contrast.

Visual acuity. The ability to distinguish fine details.

Visual angle. The angle which an object or detail subtends at the point of observation.

Visual field. The focus of objects or points in space which can be perceived when the head and eyes are kept fixed.

Visual surround. All portions of the visual field except the visual task.

Visual task. Those details and objects which must be seen for the performance of a given activity.

Volt. Unit of potential, potential difference, emf, or electrical pressure.

Voltage regulator. Device used in connection with generators to keep voltage constant as load or speed is changed.

Voltages. When mentioned as 115 or 230 volts, voltages should be understood to be nominal and to include, respectively, voltages of 110 to 125 and 220 to 250.

Voltmeter. An instrument for measuring potential difference or electrical pressure.

Wall switch. A switch on the wall, not a part of any fixture, for the control of one or more outlets.

Watt. The unit of electrical power, equal to a joule per second.

Wattmeter. An instrument for measuring electrical power in watts.

Weight. The force with which a body is attracted toward the center of the earth by the gravitational field of force.

Work plane. The plane at which work is done and at which illumination is specified and measured. Unless otherwise specified, this is assumed to be a horizontal plane at the level of the task.

Full Load Currents of Motors

(The following data are approximate full-load currents for motors of various types, frequencies, and speeds. They have been compiled from average values for representative motors of their respective classes. Variations of 10 per cent above or below the values given may be expected.)

Amperes—Full-load current

Hp. of motor	Direct-current motors			Single-phase motors		Squirrel-cage induction motors										Slip-ring induction motors									
						Two-phase					Three-phase					Two-phase					Three-phase				
	115-v	230-v	550-v	110-v	220-v	110-v	220-v	440-v	550-v	2,200-v	110-v	220-v	440-v	550-v	2,200-v	110-v	220-v	440-v	550-v	2,200-v	110-v	220-v	440-v	550-v	2,200-v
¼				4.8	2.4																				
⅓	4.5	2.3		7	3.5																				
½	6.5	3.3	1.4	9.4	4.7	4.3	2.2	1.1	0.9		5.0	2.5	1.3	1.0											
1	8.4	4.2	1.7	11	5.5	4.7	2.4	1.2	1.0		5.4	2.8	1.4	1.1											
1½	12.5	6.3	2.6	15.2	7.6	5.7	2.9	1.4	1.2		6.6	3.3	1.7	1.3		6.2	3.1	1.6	1.3		7.2	3.6	1.8	1.5	
2	16.1	8.3	3.4	20	10	7.7	4.0	2	1.6		9.4	4.7	2.4	2.0		6.7	3.4	1.7	1.4		7.8	3.9	2.0	1.6	
3	23	12.3	5.0	28	14	10.4	5	3	2.0		12.0	6	3.0	2.4		11.7	5.9	3.0	2.3		14.4	7.2	3.6	2.9	
5	40	19.8	8.2	46	23		8	4	3.0			9	4.5	4.0		12.5	6.3	3.1	2.5		20.2	10	5.0	4	
7½	58	28.7	12	68	34		13	7	6			15	7.5	6.0			8.7	4.3	3.5			15	7.5	6	
10	75	38	16	86	43		19	9	8			22	11	9.0			13.0	6.5	5.2			25	13	10	
15	112	56	23				24	12	10			27	14	11			20.0	10.0	7.6			28	14	11	
20	140	74	30				33	16	13			38	19	15			24.3	12.1	10.0			45	23	18	
25	185	92	38				45	23	19			52	26	21	5.7		39	19.5	15.6			56	28	22	
30	220	110	45				55	28	22	6		64	32	26	7		49	24.7	19.8			67	34	27	7.5
40	294	146	61				67	34	27	7		77	39	31	8		60	30.0	24.0	6.4		82	41	33	10
50	364	180	75				88	44	35	9		101	51	40	10		72	36.0	28.8	7.8		106	53	42	11
60	436	215	90				108	54	43	11		125	63	50	13		93	46.5	37.3	9.5		128	64	51	14
75	540	268	111				129	65	52	13		149	75	60	15		113	57	45	12.1		150	75	60	16
100		357	146				156	78	62	16		180	90	72	19		135	68	54	14.0		188	94	75	19
125		443	184				212	106	85	22		246	123	98	25		164	82	65	17.3		246	123	99	25
150			220				268	134	108	27		310	155	124	32		214	108	87	21.7		310	155	124	31
175							311	155	124	31		360	180	144	36		267	134	108	27		364	182	145	37
200			295				415	208	166	43		480	240	195	49		315	158	127	32		490	245	196	82

Common Abbreviations and Letter Symbols

Term	Abbreviation or Symbol
alternating current (noun)	a.c.
alternating-current (adj.)	a-c
ampere	a
audiofrequency (noun)	AF
capacitance	C
capacitive reactance	X_C
centimeter	cm
conductance	G
coulomb	Q
counterelectromotive force	cemf
current (d-c or rms value)	I
current (instantaneous value)	i
dielectric constant	K, k
difference in potential (d-c or rms value)	E
difference in potential (instantaneous value)	e
direct current (noun)	d.c.
direct-current (adj.)	d-c
electromotive force	emf
frequency	f
henry	h
hertz	Hz
horsepower	hp
impedance	Z
inductance	L
inductive reactance	X_L
kilovolt	kv
kilovolt-ampere	kva
kilowatt	kw
kilowatt-hour	kwhr
magnetic field intensity	H
magnetomotive force	m.m.f.
microampere	ma
microfarad	uf

```
microhenry ............................................ uh
microvolt ............................................. uv
microwatt ............................................ uw
milliampere .......................................... ma
millihenry ........................................... mh
millivolt ............................................. mv
milliwatt ............................................ mw
mutual inductance .................................... M
picofarad ............................................ pf
power ................................................ P
resistance ........................................... R
revolutions per minute ............................... rpm
root mean square ..................................... rms
time ................................................. t
torque ............................................... T
volt ................................................. v
volt-ampere .......................................... va
watt ................................................ w
```

For computing the current or the horsepower of a d-c motor, use the formula given for single phase, but omit the power factor.

To find the value of an alternative current when voltage, power (in watts) and power factor are known, use the formulas:

$$\text{amp} \ = \ \frac{\text{watts}}{\text{volts times P.F.}} \ , \text{ or}$$

$$\text{amp} \ = \ \frac{\text{watts}}{\text{volts times 1.73 times P.F.}} \ ,$$

according to whether the equipment is single phase or three phase.

Index